Display Interfaces

Wiley-SID Series in Display Technology

Editor:
Anthony C. Lowe
The Lambent Consultancy, Braishfield, UK

Display Systems:
Design and Applications
Lindsay W. MacDonald and Anthony C. Lowe (Eds)

Electronic Display Measurement:
Concepts, Techniques and Instrumentation
Peter A. Keller

Projection Displays
Edward H. Stupp and **Matthew S. Brennesholz**

Liquid Crystal Displays:
Addressing Schemes and Electro-Optical Effects
Ernst Lueder

Reflective Liquid Crystal Displays
Shin-Tson Wu and **Deng-Ke Yang**

Colour Engineering:
Achieving Device Independent Colour
Phil Green and Lindsay MacDonald (Eds)

Display Interfaces:
Fundamentals and Standards
Robert L. Myers

Published in Association with the
Society for Information Display

Display Interfaces

Fundamentals and Standards

Robert L. Myers
Hewlett-Packard Company, USA

JOHN WILEY & SONS, LTD

Copyright © 2002   John Wiley & Sons Ltd, The Atrium, Southern Gate, Chichester,
West Sussex PO19 8SQ, England

Telephone (+44) 1243 779777

Email (for orders and customer service enquiries): cs-books@wiley.co.uk
Visit our Home Page on www.wileyeurope.com or www.wiley.com

All Rights Reserved. No part of this publication may be reproduced, stored in a retrieval system or transmitted in any form or by any means, electronic, mechanical, photocopying, recording, scanning or otherwise, except under the terms of the Copyright, Designs and Patents Act 1988 or under the terms of a licence issued by the Copyright Licensing Agency Ltd, 90 Tottenham Court Road, London W1T 4LP, UK, without the permission in writing of the Publisher. Requests to the Publisher should be addressed to the Permissions Department, John Wiley & Sons Ltd, The Atrium, Southern Gate, Chichester, West Sussex PO19 8SQ, England, or emailed to permreq@wiley.co.uk, or faxed to (+44) 1243 770571.

This publication is designed to provide accurate and authoritative information in regard to the subject matter covered. It is sold on the understanding that the Publisher is not engaged in rendering professional services. If professional advice or other expert assistance is required, the services of a competent professional should be sought.

Other Wiley Editorial Offices

John Wiley & Sons Inc., 111 River Street, Hoboken, NJ 07030, USA

Jossey-Bass, 989 Market Street, San Francisco, CA 94103-1741, USA

Wiley-VCH Verlag GmbH, Boschstr. 12, D-69469 Weinheim, Germany

John Wiley & Sons Australia Ltd, 33 Park Road, Milton, Queensland 4064, Australia

John Wiley & Sons (Asia) Pte Ltd, 2 Clementi Loop #02-01, Jin Xing Distripark, Singapore 129809

John Wiley & Sons Canada Ltd, 22 Worcester Road, Etobicoke, Ontario, Canada M9W 1L1

ISBN 0-471-49946-3

Typeset in Times from files supplied by the author
Printed and bound in Great Britain by Antony Rowe Ltd, Chippenham, Wiltshire
This book is printed on acid-free paper responsibly manufactured from sustainable forestry
in which at least two trees are planted for each one used for paper production.

Contents

Series Editor's Foreword		xi
Preface		xiii

1 Basic Concepts in Display Systems — 1
- 1.1 Introduction — 1
  - 1.1.1 Basic components of a display system — 1
- 1.2 Imaging Concepts — 3
  - 1.2.1 Vector-scan and raster-scan systems; pixels and frames — 4
  - 1.2.2 Spatial formats vs. resolution; fields — 7
  - 1.2.3 Moving images; frame rates — 8
  - 1.2.4 Three-dimensional imaging — 10
- 1.3 Transmitting the Image Information — 11

2 The Human Visual System — 13
- 2.1 Introduction — 13
- 2.2 The Anatomy of the Eye — 14
- 2.3 Visual Acuity — 19
- 2.4 Dynamic Range and Visual Response — 22
- 2.5 Chromatic Aberrations — 23
- 2.6 Stereopsis — 24
- 2.7 Temporal Response and Seeing Motion — 25
- 2.8 Display Ergonomics — 30
- References — 31

3 Fundamentals of Color — 33
- 3.1 Introduction — 33
- 3.2 Color Basics — 34
- 3.3 Color Spaces and Color Coordinate Systems — 37
- 3.4 Color Temperature — 42

VI CONTENTS

3.5	Standard Illuminants	44
3.6	Color Gamut	45
3.7	Perceptual Uniformity in Color Spaces; the CIE L*u*v* Space	46
3.8	MacAdam Ellipses and MPCDs	48
3.9	The Kelly Chart	49
3.10	Encoding Color	49

4 Display Technologies and Applications 53
4.1	Introduction	53
4.2	The CRT Display	55
4.3	Color CRTs	57
4.4	Advantages and Limitations of the CRT	60
4.5	The "Flat Panel" Display Technologies	61
4.6	Liquid-Crystal Displays	64
4.7	Plasma Displays	69
4.8	Electroluminescent (EL) Displays	71
4.9	Organic Light-Emitting Devices (OLEDs)	72
4.10	Field-Emission Displays (FEDs)	73
4.11	Microdisplays	75
4.12	Projection Displays	78
	4.12.1 CRT projection	79
4.13	Display Applications	80

5 Practical and Performance Requirements of the Display Interface 83
5.1	Introduction	83
5.2	Practical Channel Capacity Requirements	84
5.3	Compression	86
5.4	Error Correction and Encryption	88
5.5	Physical Channel Bandwidth	89
5.6	Performance Concerns for Analog Connections	92
	5.6.1 Cable impedance	92
	5.6.2 Shielding and filtering	95
	5.6.3 Cable losses	96
	5.6.4 Cable termination	98
	5.6.5 Connectors	100
5.7	Performance Concerns for Digital Connections	102

6 Basics of Analog and Digital Display Interfaces 105
6.1	Introduction	105
6.2	"Bandwidth" vs. Channel Capacity	106
6.3	Digital and Analog Interfaces with Noisy Channels	107
6.4	Practical Aspects of Digital and Analog Interfaces	109
6.5	Digital vs. Analog Interfacing for Fixed-Format Displays	111
6.6	Digital Interfaces for CRT Displays	112
6.7	The True Advantage of Digital	113
6.8	Performance Measurement of Digital and Analog Interfaces	113
	6.8.1 Analog signal parameters and measurement	114

| | | 6.8.2 | Transmission-line effects and measurements | 119 |
| | | 6.8.3 | Digital systems | 121 |

7 Format and Timing Standards — 123

7.1 Introduction — 123
7.2 The Need for Image Format Standards — 123
7.3 The Need for Timing Standards — 125
7.4 Practical Requirements of Format and Timing Standards — 126
7.5 Format and Timing Standard Development — 130
7.6 An Overview of Display Format and Timing Standards — 131
7.7 Algorithms for Timings – The VESA GTF Standard — 135

8 Standards for Analog Video – Part I: Television — 139

8.1 Introduction — 139
8.2 Early Television Standards — 139
8.3 Broadcast Transmission Standards — 141
8.4 Closed-Circuit Video; The RS-170 and RS-343 Standards — 144
8.5 Color Television — 146
8.6 NTSC Color Encoding — 147
8.7 PAL Color Encoding — 154
8.8 SECAM — 155
8.9 Relative Performance of the Three Color Systems — 156
8.10 Worldwide Channel Standards — 157
8.11 Physical Interface Standards for "Television" Video — 157
  8.11.1 Component vs. composite video interfaces — 157
  8.11.2 The "RCA Phono" connector — 158
  8.11.3 The "F" connector — 159
  8.11.4 The BNC connector — 159
  8.11.5 The N connector — 160
  8.11.6 The SMA and SMC connector families — 160
  8.11.7 The "S-Video"/mini-DIN connector — 160
  8.11.8 The SCART or "Peritel" connector — 161

9 Standards for Analog Video – Part II: The Personal Computer — 163

9.1 Introduction — 163
9.2 Character-Generator Display Systems — 164
9.3 Graphics — 165
9.4 Early Personal Computer Displays — 166
9.5 The IBM PC — 167
9.6 MDA/Hercules — 167
9.7 CGA and EGA — 168
9.8 VGA – The Video Graphics Array — 168
9.9 Signal Standards for PC Video — 170
9.10 Workstation Display Standards — 173
9.11 The "13W3" Connector — 176
9.12 EVC – The VESA Enhanced Video Connector — 177
9.13 The Transition to Digital Interfaces — 179
9.14 The Future of Analog Display Interfaces — 181

VIII CONTENTS

10 Digital Display Interface Standards — 183
- 10.1 Introduction — 183
- 10.2 Panel Interface Standards — 184
- 10.3 LVDS/EIA-644 — 185
- 10.4 PanelLink™ and TMDS™ — 188
- 10.5 GVIF™ — 191
- 10.6 Digital Monitor Interface Standards — 191
- 10.7 The VESA Plug & Display™ Standard — 191
- 10.8 The Compaq/VESA Digital Flat Panel Connector – DFP — 193
- 10.9 The Digital Visual Interface™ — 194
- 10.10 The Apple Display Connector — 196
- 10.11 Digital Television — 197
- 10.12 General-Purpose Digital Interfaces and Video — 197
- 10.13 Future Directions for Digital Display Interfaces — 199

11 Additional Interfaces to the Display — 203
- 11.1 Introduction — 203
- 11.2 Display Identification — 203
- 11.3 The VESA Display Information File (VDIF) Standard — 205
- 11.4 The VESA EDID and DDC Standards — 207
- 11.5 ICC Profiles and the sRGB Standard — 210
- 11.6 Display Control — 212
- 11.7 Power Management — 213
- 11.8 The VESA DDC-CI and MCCS Standards — 214
- 11.9 Supplemental General-Purpose Interfaces — 216
- 11.10 The Universal Serial Bus — 217
- 11.11 IEEE-1394/"FireWire™" — 219

12 The Impact of Digital Television and HDTV — 223
- 12.1 Introduction — 223
- 12.2 A Brief History of HDTV Development — 224
- 12.3 HDTV Formats and Rates — 227
- 12.4 Digital Video Sampling Standards — 229
  - 12.4.1 Sampling structure — 230
  - 12.4.2 Selection of sampling rate — 230
  - 12.4.3 The CCIR-601 standard — 231
  - 12.4.4 4:2:0 Sampling — 232
- 12.5 Video Compression Basics — 233
  - 12.5.1 The discrete cosine transform (DCT) — 235
- 12.6 Compression of Motion Video — 237
- 12.7 Digital Television Encoding and Transmission — 241
- 12.8 Digital Content Protection — 242
- 12.9 Physical Connection Standards for Digital Television — 244
- 12.10 Digital Cinema — 245
- 12.11 The Future of Digital Video — 247

13	**New Displays, New Applications, and New Interfaces**	**249**
	13.1 Introduction	249
	13.2 Color, Resolution, and Bandwidth	251
	13.3 Technological Limitations for Displays and Interfaces	253
	13.4 Wireless Interfaces	255
	13.5 The Virtual Display – Interfaces for HMDs	257
	13.6 The Intelligent Display – DPVL and Beyond	259
	13.7 Into The Third Dimension	261
	13.8 Conclusions	264

Glossary	**267**
Bibliography, References, and Recommended Further Reading	**279**
Printed Resources	279
Fundamentals, Human Vision, and Color Science	279
Display Technology	280
Television Broadcast Standards and Digital/High-Definition Television	280
Computer Display Interface Standards	281
Other Interfaces and Standards	281
On-Line Resources	281
Standards Organizations and Similar Groups	282
Other Recommended On-Line Resources	283
Index	**285**

Series Editor's Foreword

By their nature, display interfaces and the standards that govern their use are ephemeral. They are the more so because extremely rapid developments in the field have been driven by increasing pixel content of displays and by requirements for increased colour depth and update rates.

So, why write a book on this subject? There are several reasons, but foremost among them is the fact that the nature and the performance limitations of display interfaces are often ill understood by many professionals involved in display and display system development. That is why this latest addition to the Wiley-SID series in Display Technology pays particular attention to the principles that underlie display interfaces and their architecture.

In the first four chapters, the author includes information on basic concepts, the human visual system, the fundamentals of colour and different display technologies to enable an inexperienced reader to acquire sufficient background information to address the remaining nine chapters of the book. In these chapters, all aspects of display interfaces are addressed, starting with performance requirements and the basics of analogue and digital interfaces. Then follow discussions of standards for format and timing, analogue video (for TV and computers) and digital interfaces. Other interfaces than those used to convey image data to the display are also discussed; these are the interfaces, which, among other functions, enable a computer to identify and then correctly to address a newly connected display. The book concludes with a discussion of the impact of digital and HDTV and of the changes that will be necessary if future interface designs are to be able to deal with ever increasing display pixel content. Throughout the book, a great deal of practical information with examples of commonly used hardware is provided. This is backed up by a section containing references to source material available in print or from the web and a glossary in which all the commonly used terms are defined.

Interface architectures and the standards that govern them will certainly change. Even so, this volume will remain a valuable handbook for engineers and scientists who are working in the field and a lucid and easy to read introduction to the subject for those who are not.

Anthony C Lowe
Braishfield, UK 2002

Preface

Human beings are visual creatures. We rely on imagery and our sight for communication, entertainment, and practically every interaction with our environment and with other individuals. So it is not surprising that the single most important output device for electronic information and entertainment products is some form of display. For many years, this was almost always a cathode-ray tube (CRT) display, and the basics of those were more or less the same regardless of the particular application in question.

Today, the situation has changed. Many different display technologies have either opened new applications for electronic displays, or are challenging the CRT for supremacy in its traditional markets. With these new types of display, and with the new applications and usage models that they enable, a bewildering array of issues face the display designer and system integrator. Besides the obvious question of which display to use, how does one ensure that the displayed image will appear as expected – either in terms of being a recognizable facsimile of reality, or at the least appearing similar to some other display? What interface should be used, and how? What does the display really have to do – or not do – in order to provide a satisfactory image? While the display systems based on each of the various technologies must all perform the same basic functions, how the desired performance is actually achieved can vary greatly depending on the technology, application, or usage model in question.

This book is an attempt to address many of these issues, and is intended for anyone who needs to deal with electronic displays – both CRTs and the newer technologies – as a systems integrator, content provider, graphics hardware designer, or even as a serious amateur user or hobbyist. I will examine the basic operation of the more popular display technologies, but the inner workings of display will not be the main thrust of the discussion. Instead, I will approach the problem of using electronic displays primarily from a "functional description" perspective. Rather than being concerned with the details of the operation of each technology, this will look more at how each type of display behaves, and how to make the best use of them in various systems and environments.

From another perspective, this book is simply my attempt at producing both the tutorial and reference that I wished I had had when I started working with electronic displays. This does not mean that I will only cover the basics of display systems and interfaces, or limit the discussion to only the simplest aspects of the subject, but rather that I will at least try to present the material in a manner accessible to the person who does *not* spend all of their time working in this field, while at the same time trying to include as much of the commonly-needed reference material for display systems work as possible.

This begins with an overview of some of the basic concepts in display systems, the workings of human vision, and color science. Entire books can be (and have been!) written on these subjects, but at least the fundamentals in each area must be presented before an understanding of how best to make use of electronic displays can be developed. Ultimately, any display system must first and foremost be viewed as a human interface, and we should neither fail to address the needs and expectations of the viewer, nor provide capabilities in the display which greatly exceed those.

Next will be a review of the major display types and technologies, including a look at some of the typical applications of each and some of the more interesting new technologies now on the horizon. The non-CRT technologies now gaining market share, and especially some exciting new developments just now being introduced, permit electronic displays to be employed in applications never before thought possible. With each of these, come new challenges for the display designer and system integrator, not the least of which is meeting user expectations that remain primarily shaped by the CRT.

Having covered both those factors which define required display performance, and those technologies which will be used to deliver that performance, it will be time to get to the main subject of this work – the interface between the display device itself, and the systems which provide image information for presentation via that device. There is a surprisingly wide range of past and present interfaces which have been defined and used, with varying degrees of success and acceptance. These include some that have been defined as industry standards, and some that achieved nearly complete dominance in a given application even though they were never really intended as anything but a quick solution for a particular product.

Selected standards within the display industry are discussed, both in terms of their history and in the basic requirements and applications of each standard. The display industry has benefited greatly from these, particularly those standards that have established common interface, timing, and control definitions. As the numbers of different display types and technologies continue to grow, it is these standards which permit the easy use of practically any display within a single system design. Particular attention is given to the similarities and differences between standards in two of the largest display markets today – television and the computer industry - and what may be expected as these continue on the course of convergence. The question of "analog" vs. "digital" interfaces is one of the key questions in this area, and is dealt with at some length.

The field of display interfaces covers not only the means through which image information is conveyed to the display itself. In modern systems, there is also very often the need for supplemental interfaces, and specialized standards for the identification, configuration, and control of the display device by the host system. These are very often included in the same physical interface standards as the "video" connection itself,

Many of the difficulties and shortcomings common in practical applications and display system implementations are also reviewed, along with possible solutions for each where possible. Again, a large share of these, in the modern display market, arise from the need to integrate a wide range of display types and technologies within a single system, while obtaining optimum performance from each. Many of the concepts presented in earlier chapters come together at this point as we confront this problem of optimizing the complete display system.

Finally, although there is certainly a lot of risk in doing so, it is appropriate that we look at where the future of display and display interfaces might go, and try to make some predictions in that regard so that we can – hopefully – be better prepared to meet the future when it becomes the present (as it seems to continue to want to do).

My goal is that this book will prove a valuable reference, as well as an educational text, for a wide range of people who deal with electronic display technology and systems. In attempting to meet that goal, the knowledge and insight of a number of friends and colleagues in this industry has been invaluable. In truth, a book of this type can never be the original work of one author; it must be a compilation of knowledge and information produced and presented over years of development within the field in question, from a very large number of individuals and organizations. First, my thanks go to all the colleagues with whom I have worked on various standards committees and other such groups, past and present. My experiences from this work, and the knowledge that I gained by working with these people, were invaluable in the creation of this book. I would especially like to thank the following people who have served, and in many cases continue to serve, on the Video Electronics Standards Association's Display Committee, and who so often provided insights into the standards discussed here: Jack Hosek, of NEC-Mitsubishi; Ian Miller, of Samsung; Alain d'Hautecourt, of Viewsonic; Don Chambers, of Total Technologies; Gary Manchester, of Molex; Hans van der Ven, of Panasonic; Richard Cappels, formerly of Apple Computer; Shaun Kerigan, formerly of IBM, Andy Morrish, of National Semiconductor; Mary DuVal, of Texas Instruments; Jory Olson, of InFocus; Hugo Steemers and Joseph Lee, of Silicon Image; Joe Goodart, of Dell; and John Frederick of Compaq Computer (now HP). I would also like to thank Bill Lempesis and Joan Holewinski of the VESA staff for their support.

Many of the illustrations used here were graciously supplied by a number of companies, and my thanks must also go to them and to the people who worked to provide me with this material: Gary Manchester, Sharry Fisher, and Mike Finn of Molex Corp.; Jan Spence and Mary DuVal of Texas Instruments; Don Chambers of Total Technologies; and Mark Handschy of Displaytech, Inc.. Thanks also go to the VESA Board of Directors, for permission to publish excerpts from several VESA standards here.

Writing such books is not my full-time occupation (as will no doubt be evident when reading this one), and so my appreciation is also due to those who provided the additional support and understanding required in trying to fit the time needed for this task into and around other commitments. This group must include at least my management and colleagues within HP's Electronic Systems Technology Center, and especially Judy Glazer, Ken Knaus, and David Braun, and most certainly my family – my wife Jane and my daughter Meredith, who saw far too many evenings and weekends of my time given to work over a keyboard instead of being spent with them.

Last but not least, I would like to thank several people directly involved with the production of this book. I am very grateful for the help of Anthony Lowe, past president of the SID, who originally suggested this book and provided many helpful comments and reviews of the text as it developed. Peter Mitchell and Simone Taylor of John Wiley & Sons had, at different times through the course of this writing, editorial responsibilities for this line of books, and finally my deepest appreciation to Kathryn Sharples of John Wiley & Sons, who provided all manner of day-to-day support for this novice author, and whom I am sure doubted more than once that this would ever be finished!

1

Basic Concepts in Display Systems

1.1 Introduction

Regardless of the type of display used, its size, or the application in question, there are some concepts that are common to just about any display system. In the most general usage of the phrase, "display system" can be taken to mean any system through which information is conveyed to people through visual means. A book, a painting, or a sign could all be considered "display systems" in this sense, although admittedly they are not the sort of things with which this book will primarily be concerned.

But even these simpler examples of "displays" share some basic properties and concepts with the most advanced electronic display now being developed. If we use the above broad definition, and use the convenient term *image* to refer to any information which is being conveyed visually, then several of these basic concepts should become readily apparent.

1.1.1 Basic components of a display system

Remaining within the broad definition of a *display system,* again "a system through which information is conveyed to people through visual means," we could further divide this system into several basic elements. Any practical display system would include, in some form or another, all of the following (Figure 1-1):

1. An image source. This might be a real object, as viewed by an image transducer such as a camera. It might be a computer program (along with the hardware on which that program is running), as in the case of the completely synthetic images seen in many movies. It might even be information which is originally not in viewable form; an example might

2 BASIC CONCEPTS IN DISPLAY SYSTEMS

be spoken words transcribed to text (in which case other factors, such as the choice of font, etc., also could be considered a part of the "image source").

2. Image processing and rendering. Almost always, the basic information from the source must be processed in some manner before being delivered to the rest of the system, and then must be put into a form which is suitable for the intended display device. An example of basic image processing is the "gamma correction" performed as a basic part of the broadcast television system. The computer graphics (CG) field has given us the term "rendering" to refer to the process of finally changing the image information into the form required by the display; in the case of CG, "drawing" the image in nearly display-ready form into a *frame buffer*.

3. Image storage, compositing, and transmission. Within many, if not most, systems, there are provisions for the storage of the image information prior to display. This can range from the combination of digital memory and video tape storage common in the television industry, to the purely digital frame buffer of the computer graphics system. The image storage portion of a display system (in this broad sense of the term) may be used simply to delay the delivery of the image information to the viewer, but it may also be an important factor in the image processing and rendering step, and shared by that portion of the system. Between the two, the image storage subsystem is also quite often that point at which information from multiple sources is composited into the final single image which will be seen by the viewer. Lastly, the composited image must be delivered to the viewer via a transmission channel suited to the application in question.

4. The display itself. In the context of this particular discussion, also included within the display subsystem are those portions of the complete system which are used to translate the image information between the format in which it is stored (or in which it has been transmitted so far) and the format used by the display device proper, should such a step be required. For example, computer systems will most commonly store image information in digital form, which often must then be converted into an analog video signal for use by the display.

5. The viewer. It may seem odd to be including the viewer in a discussion of a display system, which might normally be assumed to be simply a collection of electronic equipment or some other inanimate construction. The point here is that the viewer is in reality the single most important factor in determining the performance requirements and other factors that define the rest of the system. It is pointless, for instance, to construct a system which could provide detail or differences in color far beyond the capability of human vision to resolve. Some display systems are even defined completely for the specific needs of a particular class of viewer, as in the case of those which aid the visually impaired. In any case, it is of utmost importance to keep in mind that any display system ultimately relies on, and is limited by, the needs and limitations of human vision. Display systems are a *human* interface.

The basic notion brought up in that last section is worth repeating for the overall system. It makes little sense to provide capabilities in any one part of the system that cannot be used by the whole. Conversely, a limit in any of these subsystems is a limit on the whole. Most often, unless it is anticipated that a given subsystem will be used with other "blocks" of varying performance, it is best to attempt to achieve a balance between all parts of the complete system. And again, in the case of display systems specifically, it is ultimately the capabilities of the viewer that will define the requirements for the system.

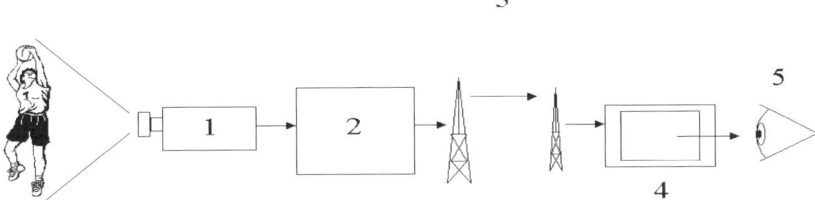

Figure 1-1 The elements of any imaging system include: (1) the image source, in this case a real object as seen by an image capture device (a camera); (2) image processing; (3) image storage and transmission; (4) image display; and (5) last but certainly not least, the viewer. As the entire purpose of an imaging system is to deliver visual information to the viewer, the viewer becomes the most important factor in the entire system, and dictates the required performance of the rest of the chain.

1.2 Imaging Concepts

The previous section defined a display system as one that is intended to convey image information to a viewer, but did not say anything about that image information itself. In what form should we expect it to be? How much information will there be, and at what rate must it be conveyed? What do we mean by "image" in the first place?

The most common-sense answer to the last question is that "image" simply means "what people see". We will look more deeply into just how human vision operates in a later chapter, but at this point this simple definition will serve as the basis for examining some concepts from imaging science. Humans see the world in color, and can discern motion, apparently at a fairly high level of resolution, and in three dimensions – how does this ability define that which is captured or produced through electronic or other means for human viewing?

First, while humans *do* possess visual systems that provide "three-dimensional" information, we should recognize that this comes from having two "image receptors" (eyes), each of which actually only captures a two-dimensional view of the world. We gain information about the third dimension through comparing these two views. So our basic working definition of *image* might be "a two-dimensional visual representation", using the word "visual" to hide a multitude of issues regarding the limitations of vision. This agrees with the common-sense definition as well: a picture on a wall is an image, that which appears on a television screen is an image, and the contents of a page of a book is an image.

More importantly, we draw a distinction between *image* and *reality*; the light reflected from an object, and captured by a lens, may form an image of that object on a sheet of paper – but the image is not the real object, and does not contain all of the information about that object. It contains only the information relating to the appearance of the object – "that which we see" – and often only a part of that.

4 BASIC CONCEPTS IN DISPLAY SYSTEMS

1.2.1 Vector-scan and raster-scan systems; pixels and frames

But how can we represent these images in a form which can be handled by mechanical or electronic equipment? We can answer that by looking at two means through which people have created images – or pictures – for years. The most obvious means of creating a picture is as a child does, through drawing lines which come together to form recognizable representations of shapes, objects, letters, and so forth. We can "teach" a machine to draw in this manner, simply by defining the image space (the "surface" onto which the image will be drawn) properly. We might, for example, create a two-dimensional coordinate system, and then "tell" the machine (through its programming, controls, or whatever) to create the image as follows:

1. Draw a red line from point (2,4) to point (4,4).
2. Draw a red line from point (2,4) to point (2,0).
3. Draw a red line from point (2,2) to point (3,2)...

and so forth. Rather than defining the lines' starting and ending points, we could also simply define the start point, the direction (relative to some agreed-upon reference), and the length of the line; the results would clearly be the same. This sort of drawing is called a *vector representation*, and display devices which create images in this manner are said to be *vector-scan* displays (Figure 1-2). The classic mechanical plotter or analog oscilloscope are excellent examples. For a more current example, we should note that computer graphics systems often use vector representations of lines and objects within an image, even if the final display device does not use this form.

But there's another way that people have used for centuries to make images. You can make a recognizable picture by placing small dots or tiles of pure color next to each other to

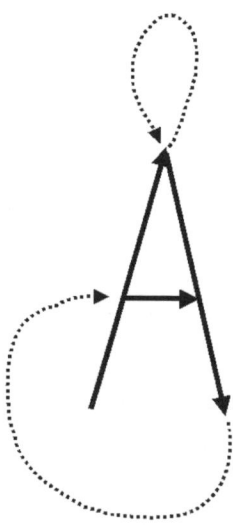

Figure 1-2 *Vector scanning.* The dotted lines represent portions of the scan in which no line is being drawn or illuminated, but rather the drawing device is being repositioned to create the next line.

form the desired shape or pattern; this is called a *mosaic*. To get a machine to draw a mosaic, it is again simplest if we define the image surface in terms of a two-dimensional coordinate system. If that is done, then we can instruct the machine as follows:

1. Place a blue tile at (4,3).
2. Place a blue tile at (3,2).
3. Place a green tile at (1,5)...

and so forth until the desired image has been created (Figure 1-3). (We assume in this exercise that all the tiles are the same size, and properly sized so that each one fills the "grid square" to which it is assigned.) But what if we are trying to create this mosaic from a "real" scene? One way to do this would be to project the image of the scene to be created, through a lens as was described above, onto a sheet of paper on which was also drawn a grid of lines using the chosen coordinate system. A person could then simply look at the projected image, determine the color, brightness, etc., at each point on the grid, and use that information to create the mosaic. (We assume we have tiles of any possible color, brightness, and so forth; this implies a very understanding tile supplier, not to mention a rather capable tile-making machine.)

This system of creating images in the manner of a mosaic is actually the basis behind practically all electronic imaging and display systems in use today. Information about the image is taken at a number of regularly spaced sampling points; this information may then be processed, translated, etc., until finally sent to a display which may itself create the desired

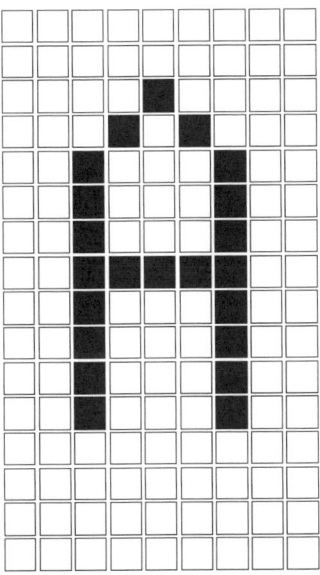

Figure 1-3 Creating an image as a mosaic. This forms the basis for the raster-scan display types; however, it is important to note that the picture elements ("pixels") we will be discussing are actually to be considered as dimensionless point samples, rather than something analogous to square tiles.

6 BASIC CONCEPTS IN DISPLAY SYSTEMS

output through controlling each of a regular array of points or cells making up the display's "screen". The term *picture element,* most often condensed into the convenient word *pixel,* has been used to refer both to the individual samples of the original image at each point, and to those individual elements in the display device which make up the final displayed image. It is important not to confuse these two meanings of the word – *pixel* as a point sample of an image, and *pixel* as a physical part of a display device – as they have significant differences. The most important of these is that, in the former definition, *pixel* truly refers to a *point sample* taken from the original image; it has neither size nor shape, but is only the information regarding the characteristics of the image at a theoretically dimensionless point. This is very different from the *pixel* of many display devices, which most definitely has a fixed size, shape, and other restrictions. As will be seen later, this distinction is at the heart of many differences in image appearance among different display technologies and imaging techniques.

At this point, we should also realize that there is a more efficient means of conveying the data in such a "mosaic" system than specifying the coordinates for each tile separately. Assuming that each location within the 2-D image space must receive a "tile", or pixel, it is much easier simply to specify a starting point – the first "tile to be placed", or the first pixel location to be sampled – and then to proceed through the array in a predefined, regular manner. If the space is defined by a rectangular coordinate system, an obvious method is to proceed by sampling each location in a row, and through each row in turn. This is the basic description of the *raster scan* technique (Figure 1-4). Almost every display system in use today employs raster scanning; the vast majority of these begin the scan at the upper left corner of the image, and proceed through all the pixels in a given horizontal row and thus through each row in the supposed array of pixels. One pass through the full array is generally referred to as a *frame*.

It should be noted that the above description, based on a uniform rectangular array of sample points or "pixels", is a close match to the reality of practical fixed-format display

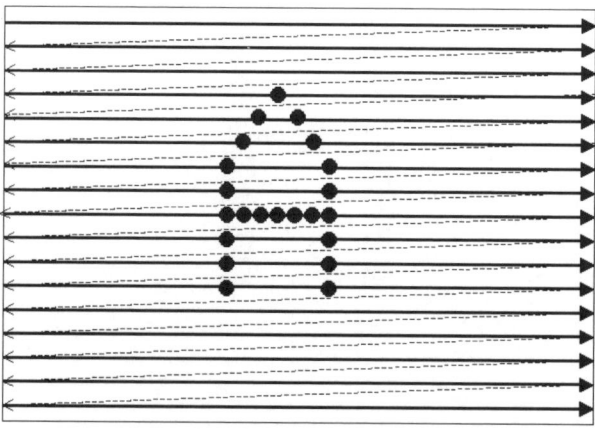

Figure 1-4 A *raster-scanned* image. In this form of display, the "drawing" mechanism follows a regular path across the imaging area, and the picture elements are placed or illuminated as required to build the desired image. Again, the dotted lines represent times during which no "drawing" is performed, but rather the drawing mechanism is being repositioned.

devices such as LCD panels, etc.. However, the notion of constructing an image through a regular, repeated scanning structure is not necessarily tied to the idea of fixed sampling points. In so-called "purely analog" systems such as the original television implementations, the image is scanned on a continuous basis, without distinct, discrete sampling points as have been assumed above. (There is much more to be said on the distinctions, both real and imagined, between "analog" and "digital" systems; this will be covered in a later chapter.) Even in systems which do employ discrete pixels, there is no requirement that these be in a simple rectangular array. Other sampling structures have been proposed, and in many cases may provide certain advantages over the common rectangular structure.

1.2.2 Spatial formats vs. resolution; fields

Regardless of whether or not the array of sampling points (pixels) is rectangular (i.e., the pixels are arrayed orthogonally, in linear rows and columns), a concept often encountered in imaging discussions is the notion of "square" pixels. This does not mean that the pixels are literally square in shape – again, a pixel in the strictest sense of the word is a dimensionless, and therefore shapeless, point sample of the original image. Instead, a sampling grid or *spatial format* is said to be "square" if the distance between samples is the same along both axes.

Images do not have to be sampled in a "square" manner, and in fact many sampling standards (especially those used in digital television) do not use square sampling, as we will see later. However, it is often easier to manipulate the image data if a given number of pixels can be assumed to represent the same physical distance in both directions, and so square sampling or square-pixel image format is almost always used in computer-generated graphics.

The term *format*, or *spatial format*, is used to refer to the overall "size" of the image, in terms of the number of pixels horizontally and vertically covered by the defined image space. For example, a common format used in the computer graphics industry is 1024 × 768, which means that the image data contains 1024 samples per horizontal row, or line, and 768 of these lines (or, in other words, there are 768 pixels in each vertical column of the array of pixels). Unfortunately, the convention used in the computer industry is exactly opposite to that used in the television industry; TV engineers more often refer to image formats by quoting the number of lines (the number of samples along the vertical axis) first, and then the number of pixels in each line. What the computer graphics industry calls a "1024 × 768" image, television would often call a "768 × 1024" image.

Readers who are familiar with the computer industry will also notice that what we are here referring to as a "spatial format" is often called a "resolution" by computer users. This is actually a misuse of the term *resolution*, which already has a very well-established meaning in the context of imaging. There is a very important distinction to be made between the *format* of a sampled image, and the *resolution* of that image or system, and so we will try to avoid giving in to the common, incorrect usage of the latter term.

Resolution properly refers simply to the amount of details which can be resolved in an image or by an imaging system; it is usually expressed in terms of the number of basic image elements (lines, pairs of lines, or pixels - also "dots" or "samples") per a definite physical distance. "Dots per inch", often seen in specifications for computer printers, is a legitimate measure of resolution – dots or pixels per line is not, since a "line" in a image does not have a definite, fixed physical length. Changing from a spatial format of, for instance, 800 × 600 pixels to one of 1600 × 1200 pixels does *not* necessarily double the resolution. It would do so

8 BASIC CONCEPTS IN DISPLAY SYSTEMS

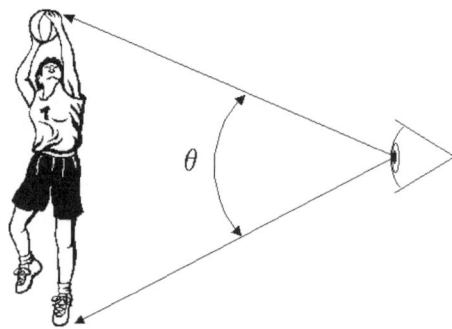

Figure 1-5 The portion of the visual field occupied by a given image may be expressed in terms of *visual angle,* or in *visual degrees,* which is simply a measure of the angle subtended by the image from the viewer's perspective.

only if the physical dimensions of the image and/or the objects within it remain constant, and so finer details within the image are now resolved.

Resolution may also be expressed in terms of the same sorts of image elements (dots, lines, etc.) per a given portion of the visual field, since this is directly analogous to distance within the image. The amount of the visual field covered by the image is often expressed in terms of *visual degrees,* defined as shown in Figure 1-5. A common means of indicating visual acuity – basically, the resolution capability of the viewer, himself or herself – is in terms of *cycles per visual degree*. Expressing resolution in cycles (or a related method, in terms of *line pairs*) recognizes that detail is perceived only in contrast to the background or to contrasting detail; one means of determining the limits on resolution, for example, is to determine whether or not a pattern of alternating white and black lines, or similarly a sinusoidal variation between white and black, is perceived as a distinct pattern or if it simply blurs into the appearance of a continuous gray shade. (It should be noted at this point that resolution limits are often different between cases of white/black, or *luminance,* variations, and differences between contrasting colors.)

1.2.3 Moving images; frame rates

To this point, we have considered only the case of a static image; a "still picture", which is often only a two-dimensional representation of a "real" scene. But we know that reality is not static – so we must also face the problem of representing motion in our images. This could be done by simply re-drawing the moving objects within the image (and also managing to restore the supposedly static background appropriately), but most often it is simply assumed that motion will be portrayed by replacing the entire image, at regular intervals, with a new image. This is how motion pictures operate, by showing a sequence of what are basically still photographs, and relying on the eye/brain system to interpret this as a convincing representation of smooth motion. Each individual image in the series is referred to as a *frame* (borrowing terminology originally used in the motion picture industry), and the rate at which they are displayed is the *frame rate*. Later, we will see that there may be two separate "frame rates" to

IMAGING CONCEPTS 9

consider in a display system – first, the *update rate*, which is how rapidly new images can be provided to or created in the *frame buffer*, which is the final image store before the display itself, and the *refresh rate*, which is how rapidly new images are actually produced on the "screen" of the display device. These are not necessarily identical, and certain artifacts may arise from the exact rates used and their relationship to one another.

Quite often in display systems, the entire frame will not be handled as a single object; it is often necessary to transmit or process component portions of the frame separately. For instance, the color information of the frame may be separated out into its primary components, or (for purposes of reducing the rate at which data must be transmitted), the frame may be broken apart spatially into smaller components. The term most often used to refer to these partial frames is *fields*, which may generically be defined as any well-defined subcomponent of the frame, but into which each of the frames can similarly be separated. (In other words, components which are unique only to a given frame or series of frames are not generally referred to as fields.)

We should note at this point that our discussion of images has now extended into a third dimension – that of time – and that the complete visual experience we typically receive from a display system actually comprises a regular three-dimensional array of samples in both time and space (Figure 1-6). Each frame represents a sample at a specific point in time; each pixel within that frame is a sample within a two-dimensional image space, representing certain visual qualities of the image at that point in space at that time. As in any sampling-based system, various difficulties and artifacts may be produced through the nature of the sampling methodology itself – the sampling rate, the characteristics of the actual implementation of the sampling devices, etc.. These will be covered in considerable detail in later chapters, but it is important to realize throughout that we are almost always in display work dealing with a series of samples rather than continuous data.

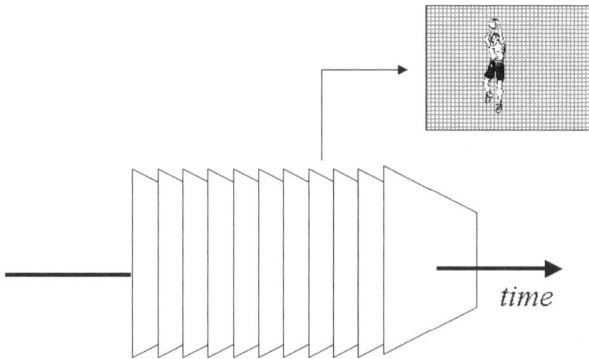

Figure 1-6   Many imaging applications involve the transmission of successive frames of information (as in the case of moving imagery), which introduces a third dimension (time) into the situation. In such cases, not only is each frame to be considered as a 2-D array of samples of the original image, but the frame itself may considered as one sample in time out of the sequence which represents a changing view.

10 BASIC CONCEPTS IN DISPLAY SYSTEMS

1.2.4 Three-dimensional imaging

Extending the functioning of display systems in a third spatial dimension, and thus providing images with the appearance of solid objects, has long been a goal of many display researchers, and indeed considerable progress has been made in this field. While true "3-D" displays are not yet commonplace, there has been more than enough work here for some additional terminology to have arisen. At the very least, rendering images in three dimensions (i.e., keeping track of 3-D spatial relationships when creating images via computer) has been used for many years. It is fairly common at this point, for instance, to treat the image space as extending not just in two dimensions but in three (Figure 1-7). In this case, the sample points might no longer be called "pixels", but rather *voxels* – a term derived from "volume pixel". (It should also be obvious at this point that in extending the image space into three dimensions, we have literally increased the amount of information to be handled geometrically.)

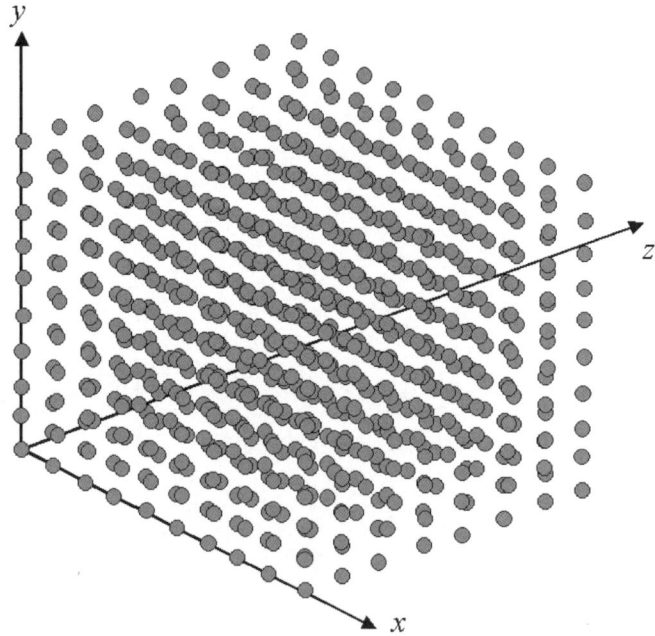

Figure 1-7 In a truly "three-dimensional" imaging system – in the sense of one having three spatial dimensions – the samples are viewed as occupying a regular array in space, and are generally referred to as "voxels" (for *volume pixel*). It is again important to keep in mind that these samples, as in the case of the pixels of a 2-D image, are dimensionless, and each represents the image information as taken from a single point within the space.

1.3 Transmitting the Image Information

Having reviewed some of the basic concepts regarding imaging, and especially how images are sampled for use within electronic display systems, it is time to consider some of the basics in communicating the information between parts of the complete system – and in particular, how this information is best transmitted to the display device itself.

As was noted earlier, almost all modern image sensors and display systems employ the "mosaic" model for sampling and constructing images. In other words, "real-world" images are sampled using a regular, two-dimensional array of sampling points, and the resulting information about each "pixel" within that array may be used to re-create the image on a display device employing a similar array of basic display elements. In practically all modern display systems, the image information is sent in a continuous stream, with all pixels in the array transmitted repeatedly in a fixed standard order. Again, this is generally referred to as a *raster scan* system.

For any form of interface to the display device, then, the data capacity required of the interface in a raster-scan system may be calculated by considering:

1. The amount of information contained in each sample, or pixel. This is normally stated in *bits* (binary digits), as is common in information theory, but no assumption should be made as to whether the interface in question uses analog or digital signalling based on this alone. Typically, each pixel will contain at least 8, and more often as much as 24–32 bits of information.
2. The number of samples in each image transmitted, i.e., the number of pixels per *frame* or *field*. For the typical two-dimensional rectangular pixel array, this is easily obtained by multiplying the number of pixels per line or row by the number of lines or rows per frame. For example, a single "1280 × 1024" frame contains 1,310,720 pixels.
3. The field or frame rate required. As has been noted, most display systems required repeated and regular updating of the complete image; the required rate is determined by a number of factors, including the characteristics of the display itself (e.g., the likelihood of "flicker" in that display type or application) and/or the need to portray motion realistically. These factors result in typical rates being in the upper tens of frames per second.
4. Any overhead or "dead time" required by the display; as used here, this term refers to any limitations on the amount of time which can be devoted to the transmission of valid image data. A good example is in the "blanking time" required in the case of CRT displays, which can reduce the time available for actual image data transmission by almost a third. (This is discussed in more detail in Chapter 5.)

For example, a system using an image format of 1024 × 768 pixels, and operating at a frame rate (or refresh rate) of 75 frames/s, with 24 bits/pixel, and with 25% of the available time expected to be "lost" to overhead requirements, would require an interface capable of supporting a peak data rate of

$$\frac{1024 \text{ pixels/line} \times 768 \text{ lines/frame} \times 24 \text{ bits/pixel} \times 75 \text{ frames/s}}{0.75} = \sim 1.89 \text{ Gbits/s}$$

(This actually represents a rather medium-performance display system, so one lesson that should be learned at this point is that display systems tend to require fairly capable interfaces!)

It should be noted at this point that these sorts of requirements are very often erroneously referred to as "bandwidth"; this is an incorrect usage of the term, and is avoided here. Properly speaking, "bandwidth" refers only to the "width" of an available transmission channel in the frequency domain, i.e., the range of the available frequency space over which information may be transmitted. The rate at which information is provided over that channel is the *data rate*, which may not exceed the *channel capacity*.

The information capacity of any channel may be calculated from the available bandwidth of that channel *and* the noise level, and is given by a formula developed by Claude Shannon in the 1930s:

$$\text{Channel capacity (bits/s)} = BW \log_2(1 + SNR)$$

where BW is the bandwidth of the channel in Hz, and SNR is the ratio of the signal power to the noise power in that channel. This distinction between the concepts of "bandwidth" and "channel capacity" will become very important when considering certain display applications, such as the broadcast transmission of television signals.

It is important to note that this formula gives the theoretical maximum capacity of a given channel; this maximum is never actually achieved in practice, and how closely it may be approached depends very strongly on the encoding system used. Note also that this formula says nothing about whether the data are transmitted in "analog" or "digital" form; while these two options for display interfaces are compared in greater detail in later chapters, it is important to realize that these terms fundamentally refer only to two possible means of encoding information. To assign certain attributes, advantages, or disadvantages to a system based solely on whether it is labelled "analog" or "digital" is often a serious mistake, as will also be discussed later.

In discussions of display interfaces and timings, the actual data rate is rarely encountered; it is assumed that the information on a per-pixel basis remains constant (or at least is limited to a fixed maximum number of bits), and so it becomes more convenient to discuss the pixel rate or *pixel clock* instead. (The term "pixel clock" usually refers specifically to an actual discrete clock signal which is used in the generation of the output image information or signal, but "pixel clock rate" is often heard even in applications in which no actual, separate clock signal is present on the interface.) In the example above, the pixel rate required is simply the data rate divided by the number of bits per pixel, or about 78.6 Mpixels/s. (If referred to as a pixel *clock* rate, this would commonly be given in Hz.)

Having covered the basics of imaging – at least as it relates to the display systems we are discussing here – and display interface issues, it is now time to consider two other areas which are of great importance in establishing an understanding of the fundamentals of display systems and their applications. These are the characteristics of human vision itself, and the role these play in establishing the requirements and constraints on our displays, and the surprisingly complex issue of color and how it is represented and reproduced in these systems. These are the topics of the next two chapters.

2

The Human Visual System

2.1 Introduction

It is very appropriate that a book dealing primarily with display interfaces devotes at least some time to a discussion of human vision, with at least a functional description of how it works and what its limitations are. As pointed out in the previous chapter, the viewer should be considered a part of the display system, and arguably the most important part. The ultimate objective of any display system, after all, is to communicate information in visual form to one or more persons; to be most effective in achieving that goal, we need to understand how these people will receive this information. In other words, the behavior and limitations of human vision determine to a very great extent the requirements placed on the remaining, artificial portions of the display system.

As we are primarily concerned with describing human vision at a practical, functional level, this chapter does not go into great detail of the anatomy and physiology of the eye (or more generically, the eye/brain system), except where relevant to that goal. It must also be noted that some simplification in these areas will unavoidably occur. Our aim is to provide a description of how well human vision performs, rather than to go in to the specifics of how it works.

From the perspective of one trying to design systems to produce images for human viewing, the visual system has the following general characteristics:

1. It provides reasonably high acuity – the ability to distinguish detail within an image – at least in the central area of the visual field. However, we will also see that even those blessed with the most acute vision do not have very good acuity at all outside of this relatively small area. In brief, we do not see as well as we sometimes think we do, over as large an area as we might believe.

2. Humans in general have excellent color vision; we can distinguish very subtle differences in color, and place great importance on this ability – which is relatively rare in mammals, at least to the extent that we and the other higher primates possess it.
3. Our vision is stereoscopic; in other words, we have two eyes which are positioned properly, and used by the full visual system (which must be considered as included the visual interpretation performed by the brain) to provide us with depth cues. In short, we "see in three dimensions".
4. We have the ability to "see motion" – in simple terms, our visual system operates quickly enough so as to be able to discern objects even if they are not stationary within the visual field, at least up to rates of motion commonly encountered in everyday life. Fast objects may not be seen in great detail, but are still perceived and quite often we can at least distinguish their general shape, color, and so forth. In fact, human vision is relatively good, especially in the periphery of the visual field, at detecting objects in motion even when they cannot be "seen" in detail. We will see, however, that this has a disadvantage in terms of the usability of certain display technologies – this ability to see even high-speed motion also results in seeing rapidly varying light sources as just that: sources of annoying "flicker" within the visual field.

While our eyes truly are remarkable, it is very tempting – and all-too common – to believe that human visual perception is better than it really is, and is always the standard by which the performance of such things should be judged. Our eyes are not cameras, and in many ways fall short of "cameralike" performance in many objective measures. They are nothing less, but also nothing more, than visual sense organs evolved to meet the needs of our distant ancestors. So while we should never consider the human eye as anything less than the amazing organ that it is, we must also become very aware of its shortcomings if we are to properly design visual information systems.

2.2 The Anatomy of the Eye

As this is not intended to be a true anatomy text, we simplify the task of studying the eye by concentrating only on those structures which directly have to do with determining the eye's performance as an image capture device. In cross-section, the human eye resembles a simple box camera, albeit one that is roughly spherical in shape rather than the cubes of early cameras (Figure 2-1). Many of the structures of the eye have, in fact, direct analogs in simple cameras (Figure 2-2).

Following the optical path from front to back, light enters the eye through a clear protective layer – the *cornea* – and passes through an opening in the front called the *pupil*. The size of the pupil is variable, through the expansion and contraction of a surrounding ring of muscular tissue, the *iris*. The iris is a primary means of permitting the eye to adapt to varying light levels, controlling the amount of light which enters the eye by varying the size of the pupil. Immediately behind the pupil is the *lens* of the eye, a sort of clear bag of jelly which is connected to the rest of the eye's structure through muscles which can alter its shape and thus change its optical characteristics. Light is focused by the lens on the inner rear surface of the eyeball, which is an optically receptive layer called the *retina*. The retina is analogous to film in a camera, or better to the sensing array in an electronic video camera. As we will see in greater detail, the retina is actually an array of many thousands of individual light-

THE ANATOMY OF THE EYE   15

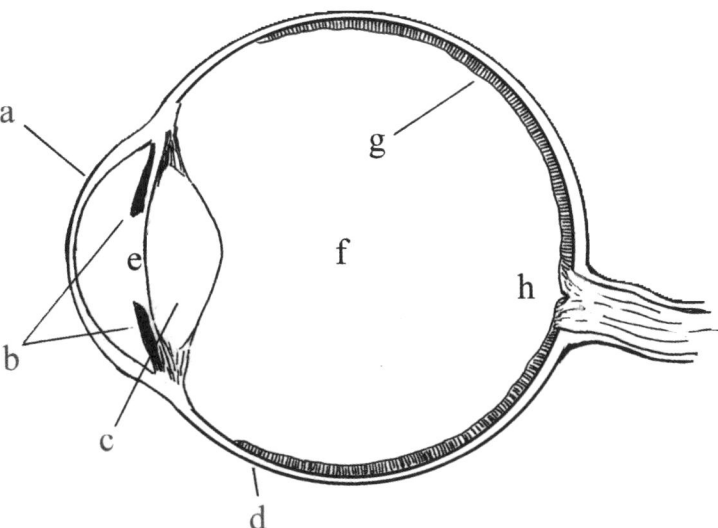

Figure 2-1   The basic anatomy of the human eye. This simplified cross-sectional view of the eye shows the major structures to be discussed in this chapter. (a) *Cornea;* this is the transparent front surface of the eyeball. (b) *Iris;* the colored portion of the eye, surrounding and defining the pupil as seen from the front. (c) *Lens;* the lens is a transparent, disc-shaped structure whose thickness is controlled by the supporting fibers and musculature, thereby altering its focal length. Along with the refraction provided by the cornea, the lens determines the focus of images on the retina. (d) The *sclera,* or the "white" of the eye; this is the outer covering, which gives the eyeball its shape. (e) *Pupil.* As noted, the size of the pupil, which is the port through which light enters the eye, is controlled by the iris. The space at the front of the eye, between the lens and the cornea, is filled with a fluid called the *aqueous humor.* (f) Most of the eyeball is filled with a thicker transparent jelly-like substance, the *vitreous humor.* (g) The *retina.* This is the active inner surface of the eye, containing the receptor cells which convert light to nerve impulses. (h) The point at which the optic nerve enters the eye and connects to the retina is, on the inner surface, the *optic disk,* which is devoid of receptor cells. This results in a *blind spot,* as discussed in the text.

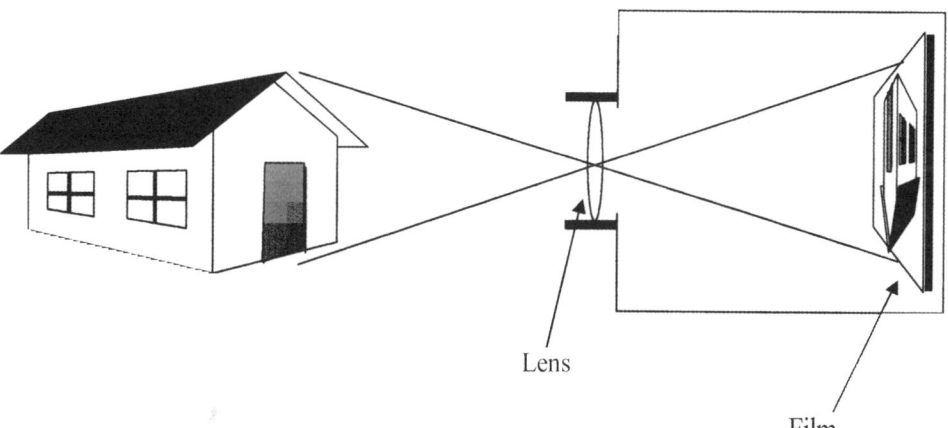

Figure 2-2   The eye is analogous to a simple box camera, with a single lens focusing images upside-down on the film (which corresponds to the retina).

sensing cells, each of which can be considered as taking a sample of the image delivered to this surface.

The actions of these three structures – the iris (and the pupil it defines), the lens, and the retina – primarily determine the performance of the eye from our perspective, at least in the typical, normal, healthy eye. To complete our review of the eye's basic anatomy, three additional features should be discussed. The eye is supported internally by two separate fluids which fill its internal spaces: the *aqueous humor*, which fills the space between the cornea and the lens, and the somewhat firmer *vitreous humor*, which fills the bulk of the eyeball behind the lens. Ideally, these are completely clear, and so do not factor into the performance of the eye in the context of this discussion. (And obviously, when they are not, it will be the concern of an ophthalmologist, not a display engineer!) Finally, the remaining visible portion of the eye, the white area surrounding the cornea and iris, is the *sclera*.

Of the three structures we are most concerned with, it is the retina that is the most complex, at least in terms of its impact on the performance of human vision in the normal eye. The retina may be the "film" in the camera, but its functioning is vastly different and far more complex than that of ordinary film. The retina incorporates four different types of light-sensing cells, which provide varying degrees of sensitivity to light across the visible spectrum, but which are not evenly distributed across the retinal surface. This is also the portion of the eye which directly connects to the brain and nervous system, via the *optic nerve*.

The cells of the retina fall into two main types, named for their shape: *rods* are the most sensitive to light in general, but respond across the entire range of visible wavelengths. In contrast, the *cones* are somewhat less sensitive overall, but exist in three different varieties, each of which is sensitive primarily to only a portion of the spectrum. Humans typically perceive light across slightly less than one octave of wavelengths, from about 770 nm (deep red) to perhaps 380–390 nm (deep blue or violet). The relative sensitivities of the rod cells and the three types of cones are shown in Figure 2-3. It should be clear from this diagram that the cones are what permit our eyes to discriminate colors (a subject which is covered in more detail in the next chapter), while the rods are essentially "luminance-only" sensors which handle the bulk of the task of providing vision in low-light situations. (This is why colors seem pale or even absent at night; as light levels drop below the point where the cones can function, it is only the rods which provide us with a "black and white" view of the world.) There are approximately 6–10 million cone cells across the retina, and about 120 million rods. (This imbalance has implications in terms of visual acuity, as will be seen shortly).

As noted above, the distribution of these cells is not even across the surface of the retina. The densest packing of cone cells occurs near the center of the visual field, in an area called the *fovea*, which is less than a millimeter in diameter. As we move out to the periphery of the retina, the number of cone cells per unit area drops off markedly. The distribution of rods also shows a similar pattern, but not as dramatically, and with the difference that there are no rods at all in the fovea itself (Figure 2-4). These distributions impact visual performance in several ways; first, visual acuity – the ability to distinguish detail within an image – drops off rapidly outside of the central portion of the visual field. Simply put, we are actually able to discern fine detail only in the central portion of our field of view. This is the natural result of having fewer light-sensing cells – which may be viewed as "sample points" for this purpose – outside of that area. The effective spatial sampling frequency resulting from the lower number of cells in the periphery can even lead to some interesting aliasing artifacts in this portion of the field of vision. The outside areas of our visual fields are basically best at detecting motion, and seeing it in low-light conditions – survival skills evolved for avoiding

THE ANATOMY OF THE EYE 17

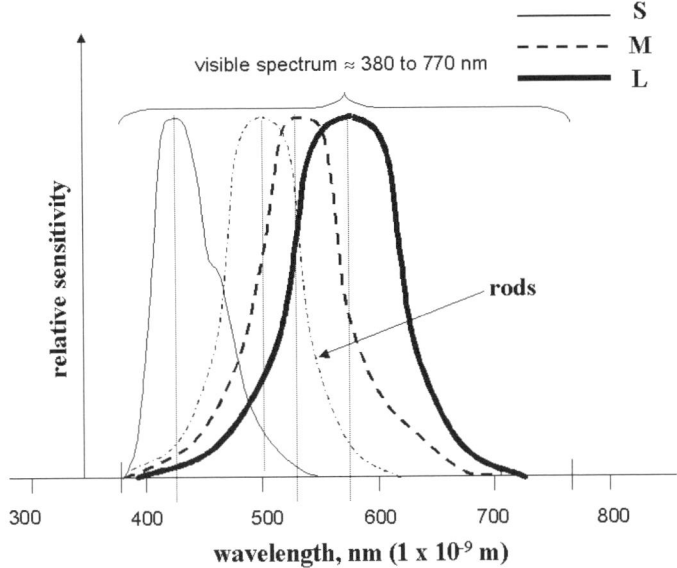

Figure 2-3 Approximate normalized sensitivities of the rods and the three types of cone cells in the human retina. Peak cone sensitivities are at approximately 420 nm for the "short-wavelength" (S) cells, 535 nm for the "medium" (M), and 565 nm for the "long" (L), and the peak for the rods is very close to 500 nm. The S, M, and L cones are sometimes thought of as the "blue," "green,", and "red" receptors, respectively, although clearly their responses are not strictly limited to those colors.

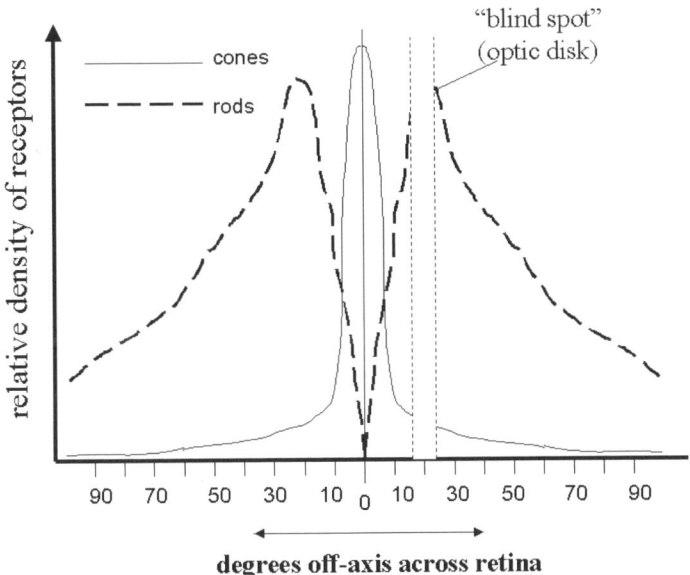

Figure 2-4 Typical distributions of the rod and cone cells across the retina. Note that there are essentially no rods in the very center of the visual field (the *fovea*), and no receptors at all at the point where the optic nerve enters the eye (the "blind spot").

predators. Primitive man had little need for seeing a fast-moving attacker in detail; the requirement was to quickly detect something big enough, and moving quickly enough, to be a threat.

The absence of rods in the very center of the visual field also means that we do not see low-luminance objects well if we look directly at them. This is the underlying factor behind an old stargazer's trick: when attempting to view dim stars in the night sky, you are advised to look slightly *away* from the target area. If you stare directly at such an object, you're placing its image on a portion of the retina which is not good at seeing in low light (there are no rods in the fovea); moving the direction of your gaze slightly away puts the star's image onto a relatively more "rod-rich" area, and it suddenly pops into view!

Even the central area of the retina, the part used for seeing detailed images in full color, is not without anomalies. Slightly off from the actual center of the retina (in terms of the visual field) is the point at which the optic nerve enters the eye. Surprisingly, there are no light-sensitive cells of any type in this location, resulting in a *blind spot* in this part of the field. The existence of the blind spot is easily demonstrated. Using the simple diagram below, focus the gaze of your right eye on the left-hand spot (or the left eye on the right spot), and then move the book in and out from your face. At a distance of perhaps three or four inches, the "other" spot will vanish – having entered the portion of the field of view occupied by the blind spot.

•   •

We are not normally aware of this apparently severe flaw in our visual system for the same reason we do not notice its other shortcomings; simply put, we have learned – or rather, evolved – to deal with them. In this case, the brain seems to "fill in" the area of the blind spot with what it *expects* to be there, based on the surrounding portion of the visual field. In the example above, when the spot "vanishes", we see what appears to be just more of the empty white expanse of the page, even though this is clearly not what is "really" there. Other examples of the compensations built into the eye/brain system will be discussed shortly.

One form of deficiency in the retina is common enough to warrant some attention here. As was noted above, the three types of cone cells are the means through which the eye/brain system sees in color, and the fact that these types have sensitivity curves which peak in different portions of the visible spectrum is what permits the discrimination of different colors. There are, speaking in approximate terms, cones which are primarily sensitive to red light, others sensitive to green light, and a third type for detecting blue light. The brain integrates the information given by these three types of cells to determine the color of a given object or light source. (These basics of color are examined in greater detail in the following chapter.) However, many people are either lacking in one or more types, have fewer than the normal number, or lack the usual sensitivity in a given type (through the partial or complete lack of the *visual pigments* responsible for the color discrimination capability of these cells). All of these result in a specific form of "color blindness", more properly known as color vision deficiency. Contrary to the popular name, this condition only rarely means a complete lack of all ability to perceive color; instead, "color blindness" more often refers to the inability to distinguish certain shades. The particular colors which are likely to be confused depend on the type of cell in which the person is deficient. By far the most common form of color vision deficiency is marked by an inability to distinguish certain shades of red and green. (This condition is linked to the absence of a gene on the X chromosome, and so is far more common in males than in females. For it to occur in a female, the gene must be missing in both X chromosomes.) Approximately 8% of the male population is to some degree "red/green color blind", or to give the condition its proper name are *dichromats*, having effectively only two types of color-distinguishing cells. (Red/green confusion is only the most common form of dichromatism; similar conditions exist which are characterized by confusion or lack of perception of other colors in the spectrum, resulting from deficiencies in different cone types.) In contrast, only about 0.003% of the total population are completely without the ability to perceive color at all – a condition called *achromatopsia*. Such people perceive the world in terms of luminance only; they see in shades of gray, although often with greater sensitivity to luminance variations than those with "normal" vision.

2.3 Visual Acuity

"Acuity" refers to the ability to discriminate detail; in measuring visual acuity, we generally are determining the smallest distance, in terms of the fraction of the visual field, over which a change in luminance or color can be detected by a given viewer, under a certain set of conditions. Acuity is to the eye what resolution is to the display hardware; the ability to discern detail, as opposed to the display's ability to properly present it. Measuring acuity generally involves presenting the viewer with images of increasingly fine detail, and determining at which point the detail in question can no longer be distinguished. One of the more useful types of images to use for this is produced by varying luminance in a sinusoidal fashion, along one dimension only; this results in a pattern of alternating dark and light lines, although without sharp transitions. Two such patterns, of different spatial frequencies, are shown in Figure 2-5. The limits to visual acuity under a given set of conditions are determined by noting the frequency, in terms of cycles per degree of the visual field, at which the variations are no longer detected and the "lines" merge into a perception of an even luminance level across the image.

Figure 2-5 Two patterns of different spatial frequencies. The pattern on the right has a frequency, along the horizontal direction, roughly twice that of the pattern on the left. (Both represent sinusoidal variations in luminance.) It is common to express both display resolution and visual acuity in terms of the number of cycles that can be resolved per unit distance or per unit of visual angle.

The acuity of the eye is ultimately limited by the number of receptors per unit area in the retina, since there can be no discrimination of spatial detail variations across a single receptor. For this reason, as noted above, visual acuity is always highest near the center of the visual field, within the fovea, where the receptor cells are the most densely packed. However, the actual acuity of the eye at any given moment will typically be limited to somewhat below this ultimate figure by other factors, notably the performance limits of the lens. Remembering also that a major part of the eye's ability to adapt comes from the action of the iris – which changes size of the pupil, and so the effective diameter of the lens – the ambient light level also affects the limit of visual acuity.

A graph of typical human visual acuity, under conditions typical of those experienced in the viewing of standard electronic displays, is shown as Figure 2-6. Note that there is a peak around 10–20 cycles per degree, and a rapid decline above this frequency. A commonly assumed limit on visual acuity is 60 cycles per degree, or one cycle per minute (limited by the size of the cone cells in the fovea, each of which occupies about half a minute of the field.). To better visualize this, one minute of angle within the visual field represent an object about the size of a golf ball, seen from a distance of about 150 m. One degree of the visual field is approximately the portion of the field occupied by the sun or the full moon, which is also roughly the portion of the field covered by the fovea. The ability to resolve details separated by one minute of visual arc is also the assumed acuity referred to when saying you have "20/20" vision – which simply means that you see the details on a standard eye chart (a test image with high contrast information) as well at a distance of 20 feet as a "normal" person is expected to. (In Europe, this same level of acuity is referred to as "6/6" vision, which is essentially the same thing using meters.)

Note, however, that this refers to acuity only in terms of luminance variations only – basi-

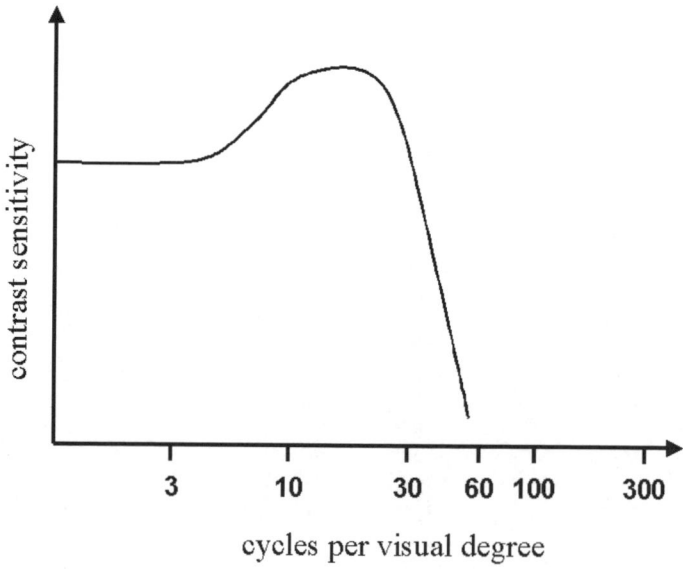

Figure 2-6 Relative human visual acuity, in terms of contrast sensitivity vs. cycles of luminance variation per visual degree, under viewing conditions typical of those experienced in normal office lighting, etc.. 60 cycles per visual degree is commonly taken as the practical limit on human visual acuity.

cally, how well one sees "black and white" detail – and then only for items at or very near the center of the visual field. Due to the lower numbers of the color-discriminating cone cells, and the larger area covered by each (plus the fact that three cells, one of each type, are required to distinguish between all possible colors), spatial acuity in terms of color discrimination is much poorer than that for luminance-only variations. Typically, spatial acuity measured with color-only variations (i.e., comparing details which differ only in color but not in luminance) is no better than half that of the luminance-only acuity. And, as the density of both types of receptors falls off dramatically toward the periphery of the visual field, so does spatial acuity – especially rapidly in terms of color discrimination. As the receptors can be viewed as sampling the image projected onto the retina, we can also consider the possibility of sampling artifacts appearing in the visual field. This can be readily demonstrated; grid or alternating-line patterns which appear quite coarse when seen directly can appear to change in spatial frequency or orientation when in the periphery of the field, due to aliasing effects caused by the lower spatial sampling frequency in this area.

Not all people, of course, possess this level of "normal" visual acuity. Distortions in the shape of the eyeball or lens result in the inability of the lens to properly focus the image on the retina. Should the eyeball be somewhat elongated, increasing the distance from lens to retina, the eye will be unable to focus on distant objects (Figure 2-7c), resulting in nearsightedness or *myopia*. This can also occur if the lens is unable to thicken (to become more curved), which is needed for a shorter focal length. If the eye is shorter, front to back,

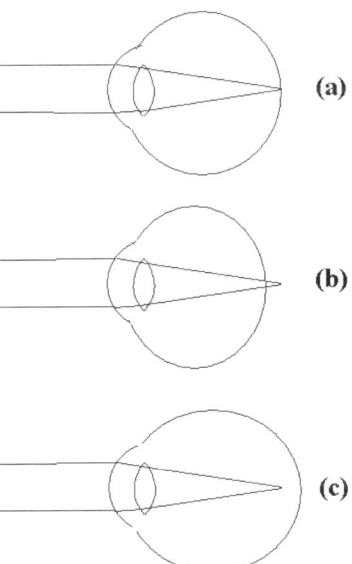

Figure 2-7 Loss of acuity in the human eye. (a) In the normal eye, images of objects at various distances from the viewer are focused onto the retina by the combined action of the cornea and the lens. (b) If the eyeball is effectively shorter than normal, or if the lens is unable to focus properly, nearby objects will not be properly focused, a condition known as *farsightedness* or *hyperopia*. (c) Similarly, if the eyeball is longer than normal or through an inability of the lens to focus on distant objects, the individual suffers from *nearsightedness* or *myopia*. (A general loss of visual acuity, independent of the distance to the object been seen, is called *astigmatism*, and commonly results from irregularities in the shape of the cornea.)

curved), which is needed for a shorter focal length. If the eye is shorter, front to back, than ideal (Figure 2-7b), or the lens is unable to become sufficiently flat, it will be unable to properly focus on nearby objects, and the person is said to be farsighted (the condition of *hyperopia*). A general distortion of focus (usually along a particular axis) results from deformations of the cornea, in the condition called astigmatism. In all but very severe cases of each condition, "normal" vision is restored through the use of corrective external lenses (eyeglasses or contact lenses), or, more recently, through surgically altering the shape of the cornea (typically through the use of lasers). The ability of the eye to properly focus images may also be affected by physical illness, exhaustion, eyestrain (such as through lengthy concentration on detailed, close-up work), and age.

2.4 Dynamic Range and Visual Response

At any given moment, the eye is capable of discriminating varying levels of luminance over a range of perhaps 100:1 or slightly higher. If the brightness of a given object in the visual field falls below the lower end of the range at any moment, it is simply seen as black. Similarly, those items outside the range at the high end are seen as "bright", with no possibility of being distinguished from each other in terms of perceived brightness. However, as we know from personal experience, the eye is capable of adapting with time over a much wider absolute range. The opening and closing of the iris varies the amount of light admitted to the interior of the eye, and (along with other adaptive processes) permits us to see well in conditions varying from bright sunlight to nearly the darkest of nights. In terms of the total range that can be covered by human vision with this adaptation, the figure is more like 10,000,000:1. The lowest luminance which is generally considered as visible under any conditions, to a fully dark-adapted eye, is about 0.0001–0.001 cd/m^2; the greatest, at least in terms of what can be viewed without permanent damage to the eye,[1] is on the order of 10,000 cd/m^2, a value achievable from a highly reflective white surface in direct sunlight. Adaptation of the eye to varying light levels within this range permits the 100:1 range of discrimination to be set anywhere within this total absolute range.

Within a given adapted range, however, the response of the eye is not linear. At any given instant, we are capable of better discrimination at the lower end of the eye's range than the higher – or in other words, it is easier to tell the difference between similar dimly lit areas of a given scene than to tell the difference between similar bright areas. (This is again as might be expected; it is more important, as a survival trait, to be able to detect objects – or threats - in a dimly lit area (such as a cave) than it is to be able to discriminate shadings on the same object in broad daylight.) The typical response curve is shown in Figure 2-8. This non-linear response has some significant implications for the realistic portrayal of images on electronic displays, as can be seen in Chapter 4. The non-linearity of the response also has an impact on the amount of information required to properly convey visual data. Given the ability to discriminate luminance over only a range of 100:1 or slight higher, we are tempted to assume that only about 7–8 bits per sample would be required to encode luminance. Tests with 7–8 bits per sample of luminance *with linear encoding*, however, will show clearly discernible bands (*contouring*), especially in the darker areas, due to the eye's ability to discern finer

---

[1] Of course, the level at which permanent damage may occur is not at all clear-cut, depending on the duration of the exposure and other factors, especially the wavelength of the light source in question.

CHROMATIC ABERRATIONS 23

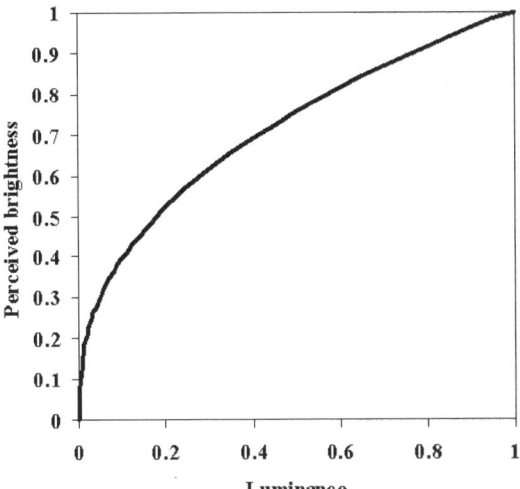

Figure 2-8 Typical normalized luminance response curve for the human eye, showing the non-linear relationship between absolute luminance (within the current adapted range) and perceived brightness. This shows that the eye is more sensitive to luminance changes in the "dark" end of the current range than to similar-sized changes in "bright" areas. The response curve shown here is a power function wherein the perceived brightness is given as $Y^{(1/2.5)}$, or $Y^{(0.4)}$. Different standard models have used different values for the exponent in this function, ranging from about 0.333 to 0.450.

differences at the low end of the luminance range. Ten to twelve bits of luminance information per sample, if linear encoding is to be used, is generally assumed to be required for the realistic portrayal of images. (Note, however, that this level of performance is very often well beyond the capability of many display and image-sampling devices; noise in these systems may limit the resolvable bits/sample to a lower value, especially for those operating at "video" (smooth-motion) sampling rates.) Encoding full-color images, as opposed to simply luminance information only, complicates this question considerably, as can be seen in Chapters 3 and 6.

2.5 Chromatic Aberrations

Color affects our vision in at least one other, somewhat unexpected, manner. The lens of the eye is a simple double-convex type, but made of a clear, jellylike material rather than glass. In most optical equipment, focusing is achieved by varying the spacing of the optical elements (lenses, mirrors, etc.); in the eyes of living creatures, images are focused by altering the shape of the lens itself, and so its optical characteristics. (The curved surface of the transparent cornea also acts to bend light, and is a major contributor in focusing the image – however, its action is not variable.) However, simple lenses of any type suffer from a significant flaw with respect to color. The refractive index of any optical material, and so the degree to which light is "bent" at the interface of that material and air, varies with the frequency of the light. Higher-frequency light, toward the blue end of the spectrum, is bent less than lower-frequency light. If not compensated for, this has the effect of changing the focal length of the

24 THE HUMAN VISUAL SYSTEM

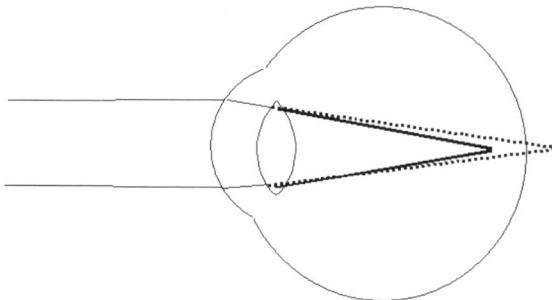

Figure 2-9 In a simple lens, higher-frequency light (i.e., blue) is refracted to a lesser degree than lower-frequency light (red). In the case of human vision, this results in the blue components of an image being focused effectively "behind" the red components, leading to a false sense of depth induced by color (*chromostereopsis*). This also makes it very tiring to look at images containing areas of bright red and blue in close proximity, as the eye have a very difficult time focusing!

lens for various colors of light. (In conventional lenses, this compensation comes in the form of an additional optical element with a slightly different refractive index, bonded to the original simple lens. Such a color-corrected lens is called an *achromat*.)

With the simple lens of the eye, this sort of *chromatic aberration* results in images of different color being focused slightly differently. Pure fields of any given color can be brought into proper focus, through the adaptive action of the lens, but if objects of very different colors are seen in close proximity, a problem arises. The problem is at its worst, of course, with colors at the extremes of the visual spectrum – blue and red. If bright examples of both colors are seen together, the eye cannot focus correctly on both; the blue image focuses "behind" the red, as seen in Figure 2-9. Besides being a source of visual strain (as the eye/brain system attempts to resolve the conflict in focus), this also creates a false sense of depth. The blue object(s) are seen as behind the red, through *chromostereopsis* (the perception of depth resulting solely from color differences rather than actual differences in the physical distance between objects). Due to these problems, the use of such colors in close proximity – bright red text on a blue background, for instance – is to be avoided.

2.6 Stereopsis

Besides the false sense of visual depth mentioned above, human beings are, of course, very capable of seeing true depth – we have "three-dimensional," or stereoscopic vision. By this we mean that human beings can get a sense of the distance to various objects, and their relative relationships in terms of distance to the viewer, simply by looking at them. This ability comes primarily (but not exclusively!) from the fact that we have two eyes which act together, seeing in very nearly the same direction at all times, and a visual system in the brain which is capable of synthesizing depth information from these two "flat", or two-dimensional, views of the world. In nature, stereo vision is most often found in creatures which are at least part-time hunters, and so need the ability to accurately judge the distance to prey (to leap the right distance, or to aim a spear, etc.). Most animal species which possess a sense of sight have two eyes (or at least two primary eyes), but relatively few have them properly located and working in concert so as to support stereo vision.

Perceiving depth visually (*stereopsis*, a general term covering such perception regardless of the basic process) is basically a matter of parallax. Both eyes focus on the same object, but due to the fact that they are spaced slightly apart in the head do not see it at quite the same angle. The eye/brain system notes this difference, and uses it to produce a sense of the distance to the object. This can also be used to impart a sense of depth to two-dimensional images; if each eye is presented with a "flat" view of the same scene, but the two views differ in a manner similar to that which results from the difference in viewing angle in a "real" scene, the visual system will perceive depth in the image. This is the principle underlying stereoscopic viewers or displays, which are arranged so as to present "left-eye" and "right-eye" images separately to the two eyes.

However, this parallax effect is not the only means through which we perceive depth visually. Some people have only one working eye, and yet still function well in situations requiring an understanding of depth; they are able to compensate through reliance on these other cues. (There is also a small percentage of the population who have functional vision in both eyes, and yet do not perceive depth through the normal process. In these cases, the eye/brain system, for whatever reason, never gained the ability to synthesize depth information from the two different views. Such people often do not realize that their deficiency exists at all, until they are unable to see the "3-D" effect from a stereoscopic display or viewer.) Depth is also perceived through the changes required to focus on nearby vs. distant objects, from differences in the rate at which objects are passing through the visual field (rapidly moving objects are seen as being closer than slower or stationary objects, in the absence of other cues), and, curiously, through delays in processing a given image in one eye relative to the other. (This latter case is known as the *Pulfrich effect*, and may be produced simply by changing the luminance of the image presented to one eye relative to the other.)

2.7 Temporal Response and Seeing Motion

Our eyes have the ability to see motion, at least up to rates normally encountered in nature. This tells us that the mechanisms of vision work relatively quickly; it does not take an unreasonable amount of time from the moment a given scene is imaged on the retina, the receptor cells respond to the pattern of light making up the image, the impulses are passed to the visual centers of the brain, and the information interpreted as the sensation we call vision. However, this action is not infinitely fast, nor is motion perceived in quite the way we might initially think.

Clearly, the perception of motion is going to be governed by how rapidly our eyes can process new images, or changes in the visual information presented to them. It takes time for the receptors to respond to a change in light level, and then time to "reset" themselves in order to be ready for the next change. It takes time for this information to be conveyed to the brain and to be processed. We can reasonably expect, then, that there will be a maximum rate at which such changes can be perceived at all, but that this rate will vary with certain conditions, such as the brightness or contrast of the changing area relative to the background, the size of the object within the visual field, and so forth.

We also should understand that the eye/brain system has evolved to track moving objects – to follow them and fixate upon them, even while they are moving – and how this occurs. Obviously, being able to accurately follow a moving object was a very important skill for creatures who are both trying to be successful as hunters, and not being successfully hunted

26 THE HUMAN VISUAL SYSTEM

themselves. So we (and other higher animals) evolved the ability to predict the path of a moving object quite well, as is demonstrated each time one catches a ball. But this does not mean that the eye itself is tracking these objects via a smooth, fluid motion. This would not work well, due to the fact that the receptors do take some finite time to respond as mentioned above. Instead, the eye moves in short, very rapid steps – called *saccades* – with the sense of vision effectively suppressed during these transitions. The eye captures a scene, moves slightly "ahead" such that the moving object will remain fixed within the field, then stops and captures the "new" scene. In one way, this is very similar to the action of a motion picture camera, which captures individual still images to show motion. In fact, it is practically impossible for one to consciously move their eyes in a smooth manner; almost invariably, the actual motion of the eye will be in a series of quick, short steps.

The temporal response of vision affects display system design primarily in two areas – ensuring that the display of moving objects will appear natural, and in making sure that the performance of certain display types (which do not behave as constant-luminance light sources) is acceptable. The term *critical fusion frequency* (CFF) is used to describe the rate at which, under a given set of conditions, the eye can be "fooled" into perceiving motion (from a series of still images) or luminance (from a varying source) as "smooth" or "constant."

Flicker has always been one of the major concerns in the design and use of electronic displays, primarily because the dominant display type for years has been the cathode-ray tube, or CRT. CRT operation is discussed in depth in Chapter 4, but the one important aspect at this point is that the CRT behaves as a very inconstant light source; in effect, the image is displayed briefly and at high brightness, and then fades rapidly almost to zero light output before again being shown or *refreshed*. If this process is not repeated often enough, the display appears to be rapidly flashing, an effect with is very annoying and fatiguing for the viewer. The key question, of course, is how often the refresh must occur in order to avoid this appearance – what is the critical fusion frequency for such a source?

The prediction of the CFF for displays in general is a fairly complex task. Factors affecting it include the luminance of the display in question, the amount of the visual field it occupies, the frequency, amplitude, decay characteristics, etc., of the variation in luminance, the average luminance of the surrounding environment, and of course the sensitivity of the individual viewer. Contrary to a popular misconception, display flicker is generally not the result of a "beat frequency" with flickering ambient lighting (the most common form of this myth involves fluorescent lights); flickering ambients can result in modulation of the contrast ratio of the display, but this is usually a relatively minor, second-order effect. The overall level of the ambient lighting does affect flicker, but only because it is the perceived brightness of the display relative to its surroundings which is important. (Of course, exactly *how* important this is depends on the amount of the visual field occupied by both the display and the surroundings.)

The mathematical models used to predict flicker come in large part from work done by Dr. Joyce Farrell and her team at Hewlett-Packard Laboratories (working with researchers from the University of California, Berkeley) in the 1980s [1,2]. This work became the basis for several standards regarding display flicker, notably the International Standards Organization's ISO-9241-3 [3] set of ergonomic standards for CRT displays. A simplified form of the analysis, using assumptions appropriate for a typical CRT display in an office environment (specifically, a typical phosphor response with the display occupying about 70° of the visual field, in diagonal measurement), leads to an estimation of the CFF as a function of display luminance, as given in ISO-9241-3, of

Luminance (cd/m^2)	CFF (mean) (Hz)	SD	CFF (95%) (Hz)
25	59.504	**5.71**	68.925
50	64.802	**5.28**	73.514
75	67.901	5.53	77.026
100	70.100	**5.78**	79.637
125	71.806	6.07	81.821
150	73.199	6.36	83.693
175	74.377	6.64	85.333
200	75.398	**6.93**	86.833
225	76.298	7.1	88.013
250	77.104	7.27	89.099
275	77.832	7.44	90.108
300	78.497	**7.6**	91.037
325	79.109	7.78	91.946
350	79.676	7.95	92.793
375	80.203	8.12	93.601
400	80.696	**8.29**	94.375

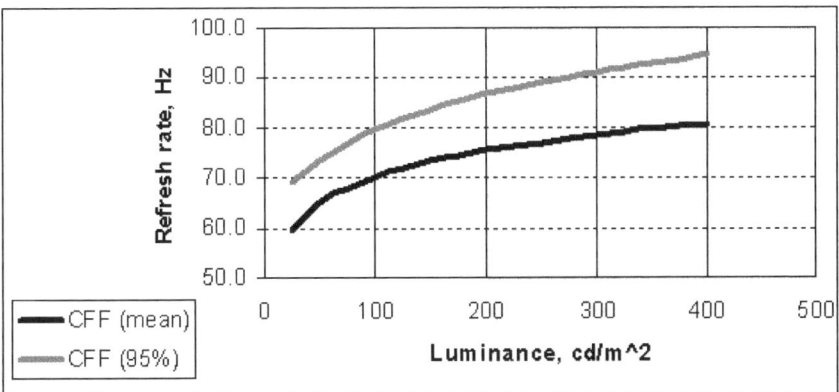

Figure 2-10 Critical flicker-fusion frequencies (CFF) given by the ISO-9241-3 formula for a range of display luminance values. This calculation assumes a display occupying 70° of the visual field (diagonal measurement). Figures are given for both the mean CFF, and the CFF for the 95th percentile of the population, calculated as $CFF_{(mean)} + 1.65 \times SD$ for the standard deviation values listed. The SD values in boldface are from the ISO-9241-3 standard; the remainder were derived via a linear interpolation. Note that these CFF calculation apply only to a CRT display, or a similar display technology in which the actual duration of the image is in reality relatively short compared to the refresh period. Such calculations do not apply to such types as the LCD, in which the display elements are illuminated to nearly their full intended value for most if not all of the frame time.

$$\text{CFF}_{(\text{mean})} = 34.9 + 17.6 \times \log(L_t) \quad (\text{Hz})$$

where L_t is the display luminance in cd/m². The distribution of CFF for the entire population has been shown to be essentially Gaussian, so to this mean one must add the appropriate multiple of the population's standard deviation in order to determine the frequency at which the display would appear "flicker-free" to a given percentage of the population. For example, the frequency at which the display would appear flicker-free to 95% of the population would be found by determining the CFF based on the display luminance, and then adding 1.65 times the standard deviation at that luminance. Note that these formulas have been based on assumptions regarding both display luminance and size, and the average viewing distance, which correspond to typical desktop-monitor use. The above formula suggests that, for a CRT-based computer display of 120 cd/m² luminance, and used at normal viewing distances, the refresh rate should be set to at least 71.5 Hz to appear flicker-free to half the population (this is the mean CFF predicted by the formula), and to not less than 81 Hz to satisfy 95% of the viewers. This is very typical for the desktop CRT monitor, and similar calculations have led to 85 Hz becoming a de-facto standard refresh rate to satisfy the "flicker-free" requirement of many ergonomic standards. A graph of the result of the above formula for mean CFF vs. luminance is shown in Figure 2-11, along with the standard deviations for inter-individual differences as established by the ISO-9241 standard. (Television, while operating at higher typical luminances, can get away with lower refresh rates since the display typically occupies a much smaller portion of the visual field than is the case with a desktop monitor.)

The update rate required for the perception of "smooth" motion is, fortunately, similar to that required for the avoidance of flicker, and in general is even lower. It is affected by many of the same factors, although one important consideration is that viewers on average tend to accept poorer motion rendition more readily than flicker. Acceptable motion can often be realized with an image update rate of only a few new images per second. For example, most "cartoon" animation employs a rate of between 10 and 24 new frames per second. The standard for the theatrical display of motion pictures is 24 frames/s in North America[2] (25 frames/s is the more common rate in Europe). Finally, television systems, which are generally seen as providing very realistic motion, use a rate of 50 or 60 new images per second. (This is not what is meant by "frame rate" in television terminology, as can be seen in Chapter 7.) This is, of course, very close to the refresh rates (60–85 Hz) generally considered to be "flicker-free" in many display applications.

While the rates required for good motion portrayal and a "flicker-free" image are similar, some interesting problems can arise when these rates are not precisely matched to each other. Examples of situations where this can occur are common in the computer graphics field (where new images may not be generated by the computer hardware at the same rate as that at which the display is being refreshed), and in cases of mixing systems of differing standard rates. An example of the latter is the display of film-sourced material on television; in North America, for instance, films are normally shot at 24 frames/s, while television uses a refresh rate of roughly 60 Hz. To accomplish this, a technique called "3:2 pulldown" is used. One frame of the film is shown for three refreshes of the television display ("fields"), while the

---

[2] However, it should be noted that motion pictures employ a technique known as "double shuttering", in which each frame of the film is shown twice before advancing to the next. This is done in order to raise the flicker component to twice the frame rate, and thus minimize the perception of flicker by the audience.

TEMPORAL RESPONSE AND SEEING MOTION 29

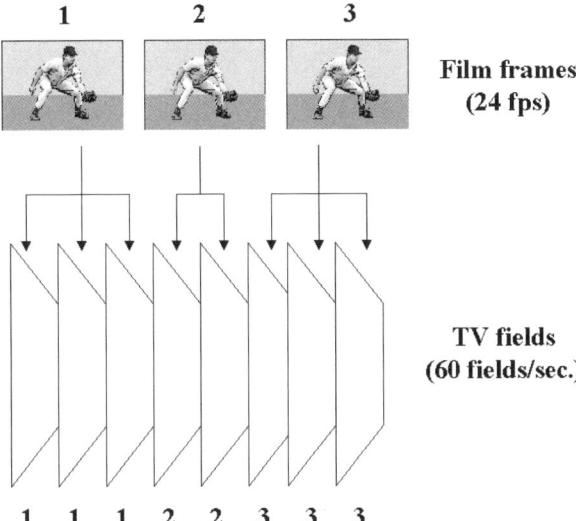

Figure 2-11 To show standard motion pictures (shot at 24 frames/s) on US standard television (approx. 60 fields/s), a technique known as "3:2 pulldown" is used. However, the uneven duration of the original frames, as seen now by the viewer, can result in certain objectionable motion artifacts.

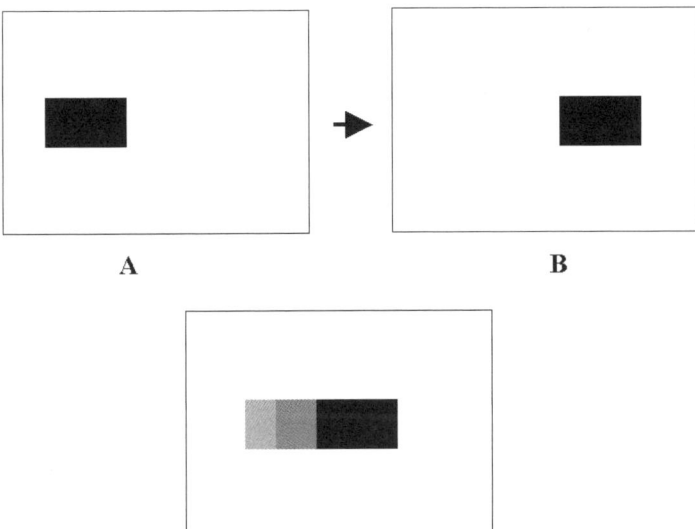

Figure 2-12 Effect of mismatched refresh and update rates. In this example, we assume that the display is being refresh 60 times per second; however, new images are being created only 20 times per second. This results in frame A being displayed for three refresh cycles, followed by frame B for the next three. The visual effect is simulated in the image at the bottom. Since the eye "expects" smooth movement, the center of the field of view moves slightly along the expected track of the moving object – but since the object in question has not actually moved for two out of every three displayed frames, the appearance is that of a moving object with "ghosts" or "shadows" resulting from the eye motion.

next appears for only two (Figure 2-11). This results in the frames of the film being unequal in duration as displayed, which can result in certain motion artifacts (known as "judder") as seen by the viewer.

The problems here again have to do with how the eye/brain system responds to moving objects. Again, the motion of the eye is not smooth – it occurs in quick, short saccades, based in large part on where the brain *expects* the object being tracked to appear. If the object does *not* appear in the expected position, its image now registers on a different part of the retina. A curious example of this may be seen when the image update rate is related to the display's refresh rate but is not the same. If, for instance, the display is being refreshed at 60 Hz, but only 20 new images are being provided per second, the object "really" appears in the same location three times before moving to its next position. The visual system, however, since it is expecting "smooth" motion, moves slightly "ahead" in the time of those two intermediate display refreshes. This results in the "stationary" image being seen by slightly different parts of the retina, and the object is seen as multiple copies along the direction of motion (Figure 2-12). In many applications, then, the perception of smooth motion will not depend as much on the absolute rate at which new images can be generated (at least above a certain minimum rate), but rather on making sure that this rate is kept constant and is properly matched to the display rate.

2.8 Display Ergonomics

Our desire, of course, is to produce display systems which are usable by the average viewer, and a large part of this means assuring that undue effort or stress is not required to use them. The field of properly matching machines to the capabilities and preferences of human beings is, of course, *ergonomics*, and the ergonomics of display systems has been a very important field in the past few decades. Many of the various international regulations and standards affecting display design have to do at least in part with ensuring proper display hardware and displayed image ergonomics.

Unfortunately, these factors were not always considered in the design and use of electronic displays, owing to a poor understanding of the field by early display system designers. This is not really the fault of those designers, as the widespread use of electronic displays was very new and the ergonomic factors themselves not yet researched in depth. However, today we have a far better understanding of these effects, and those wishing to implement successful display systems are well advised to be familiar with them. Not only will this lead to a product more acceptable to its intended users, but it compliance with the various standards in this area is often mandatory for even being able to sell the product into a given market.

Besides the standards for flicker already mentioned, items commonly covered in ergonomic guidelines or requirements include minimums and maximums for luminance and contrast, minimum capabilities for positioning the display screen (such as horizontal and vertical tilt/swivel requirements and minimum screen height from the work surface), character size and readability, the use of color, positional stability of the image (e.g., freedom from "jitter"), uniformity of brightness and color, and requirements for minimizing reflections or "glare" from the screen surface. A summary of specifications regarding some of the more important of these, from the ISO 9241-3 standard, is given in Table 2-1.

Table 2-1 Summary of ISO-9241-3 Ergonomic Requirements for CRT Displays

Item	ISO-9241-3 Ref.	Requirement
Design viewing distance	6.1	Min. 400 mm (300 mm in some cases)
Design line of sight angle	6.2	Horizontal to 60 deg. below horizontal
Angle of view	6.3	Legible up to 40° from the normal to the surface of the display
Displayed character height	6.4	Min. 16 minutes of arc; preferably 20–22
Character stroke width	6.5	1/6 to 1/12 of the character height
Character width/height ratio	6.6	0.5:1 to 1:1 allowed; 0.7:1 to 0.9:1 preferred
Between-word spacing	6.11	One character width (capital "N")
Between-line spacing	6.12	One pixel
Display luminance	6.15	35 cd/m² min
Luminance contrast	6.16	Minimum 0.5 contrast mod; minimum 3:1 contrast ratio
Luminance uniformity	6.20	Not to exceed 1:7 to 1, as measured from the center to the edge of the display screen
Temporal instability (flicker)	6.23	Flicker-free to at least 90% of the user population
Spatial instability (jitter)	6.24	Maximum 0.0002 mm per mm viewing distance, 0.5–30 Hz

Note: This table is for example only; the complete ISO-9241-3 standard imposes specific measurement requirements and other conditions not detailed here.

References

[1] "Predicting Flicker Thresholds For Video Display Terminals" – J.E. Farrell, Brian L. Benson, and Carl R. Haynie. *Proceedings of the Society For Information Display*, Vol. 28, No. 4, 1987 pp. 449–453
[2] "Designing Flicker-Free Video Display Terminals" – Joyce E. Farrell, Evanne J. Casson, Carl R. Haynie, and Brian L. Benson. *Displays*, July, 1988.
[3] International Standards Organization Document No. 9241, Visual Display Terminals (VDTs) Used For Office Tasks – Ergonomic Requirements – Part 3: Visual Displays, 1991.

3

Fundamentals of Color

3.1 Introduction

The theory of color – how we see it, how to use it, and how it may be created, analyzed, and represented – truly deserves an entire book, rather than just a single chapter, and there are of course any number of excellent texts on the subject available. (Several of these are listed in the bibliography of this book.) However, color is also such an important factor in electronic displays, and especially in terms of understanding many of the constraints which must be considered in designing an adequate display interface, that we would be remiss not to at least attempt coverage of the fundamentals here. The reader is cautioned that this will by necessity be a superficial and in some aspects simplified treatment, and reference to those works dedicated to the subject is highly recommended if one is to gain a complete understanding of the subject.

The first, and possibly the most fundamental, understanding that one must gain in a study of color is that color does not really exist; it is not, contrary to what would be indicated by lay usage of the term, a fundamental physical property of any object. Instead, color is a perception. It truly exists only in the mind of the viewer, and is simply the response of the eye/brain system to stimulation over an extremely narrow (slightly less than one octave) range of electromagnetic (EM) radiation. Visible light is fundamentally no different than any other EM wave; the fact that humans are equipped to directly perceive this particular band of the spectrum does not alter that. The perception that we call color, therefore, actually results from the interaction of a rather large number of factors, among them the spectral make-up of the illuminating light source, the reflectivity of the object in question over that same range, the sensitivity of the viewer, and so forth. Typically, few if any of these factors behave in a nice, regular, linear manner, and so color and its behavior under varying conditions becomes a rather complex thing to analyze.

As was discussed in the previous chapter, human color perception abilities come from the fact that we have multiple types of light receptors within our eyes, each of which has its own unique sensitivity curve across the visible spectrum. This gives us a good place to start in our discussion of how best to represent color in electronic display systems.

34 FUNDAMENTALS OF COLOR

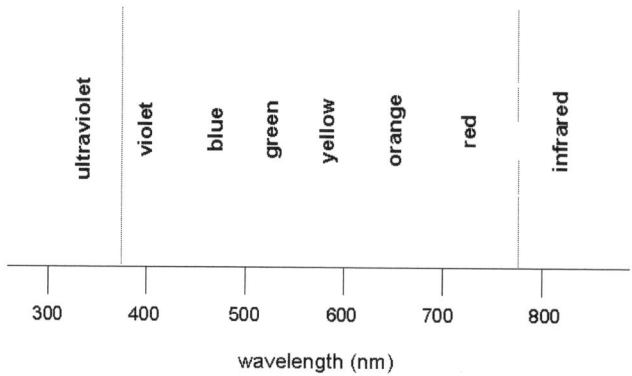

Figure 3-1   The visible spectrum.

3.2 Color Basics

The slice of the EM spectrum referred to as visible light covers a range of wavelengths from approximately 770 nm at the low-frequency end, to around 380 nm or so at the high (Figure 3-1). In terms of the common names for the colors of the spectrum, this range is from the deep reds to the blues, respectively; the common grade-school mnemonic for the colors of the rainbow ("Roy G. Biv", for red, orange, yellow, green, blue, indigo, violet) in fact labels the spectrum from the low-frequency end to the high. But as noted above, these terms actually describe perceptions. There is no real physical difference, other than frequency, between a light source at 770 nm and one at 560 or 390 nm. We perceive these as markedly different colors only because our brains interpret and present the information collected by four different types of light receptors in the eye – the "rods" – which have a relatively flat response across the spectrum, and the three types of "cone" cells which have very different and irregular response curves. The relative sensitivities of the four types are discussed in Chapter 2.

With the sensitivities of those receptors known, we have identified one of the three major factors which determine the perception of color. The other two are, of course, the characteristics of the illuminating light source, and the reflectance or transmission characteristics of the object being viewed. These combine to give the resultant total response of a given type of receptors to a stimulus:

$$\text{(Output of receptor } r) = \int_{\lambda=380\text{ nm}}^{\lambda=770\text{ nm}} L(\lambda) R(\lambda) S_r(\lambda) d\lambda$$

In other words, we must multiply the spectral distribution of the light source energy (L) by the reflectivity characteristics (R) of the object in question, and by the sensitivity of the receptor type in question (S_r), over the full spectral range of visible light, and integrate the response to determine the "output level" of those receptors given this stimulus. The brain synthesizes the responses from all of the various types of receptors to produce the perception that we call color. But note that this perception is basically obtained from a few parameters –

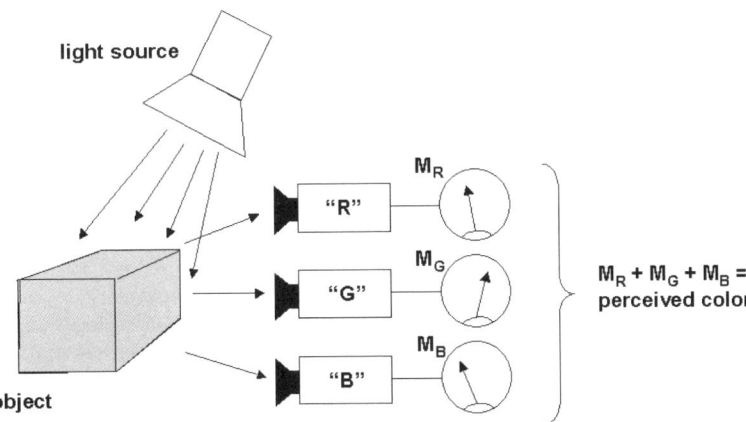

Figure 3-2 A model for the perception of color. In human vision, the perception of color may be viewed as the summation of the responses, shown here as "meter readings" (M_x), from the three types of color receptors (cones) in the eye. The values obtained from each "meter" depend not only on the reflectance characteristics (across the visible spectrum) of the object being viewed, but also on the spectral characteristics of the illuminating light source and the sensitivity curves of the three receptors. It is tempting to refer to these receptors as "red," "green," and "blue" (as has been done in this model, but in reality the response curves of the actual receptor cells in the human retina are fairly broad (see Chapter 2).

what might be seen as the "meter readings" from each of the different receptor types (Figure 3-2) of different wavelengths matters, not how much of each is reflected by the object being viewed, or the specific response of the receptor at any given discrete wavelength. It is the small set of values which result from combining all of these, across the visible spectrum, per the above formula. This has several interesting implications.

Objects with very different reflectivity curves might appear to be the same color, if illuminated by a light source of the proper spectral profile (Figure 3-3); such objects, or rather their spectral reflectivity curves, are said to exhibit *metamerism* under those lighting conditions. Similarly, two objects supposedly of the same color might look radically different if illuminated by different light sources, or a single object could appear to be different colors depending on the nature of the light source at any given moment. Further complicating the situation is the fact that even the sensitivity of a given individual viewer is not a constant, but varies with the light level, other objects and backgrounds in the field of view, and of course the viewer's health and age. Again, we are reminded that color is a perception, not by any means a physical property. If, for a given viewer, two visual stimuli result in the same set of outputs from the various color receptors – for whatever reason – then they will be perceived as being the same color.

This has a profound impact on how color can be achieved in electronic display devices. Edwin Land (the inventor of the Polaroid camera) showed that varying amounts of light from two or three light sources at discrete wavelengths, viewed in sufficient close spatial proximity, will be perceived as a single "intermediate" color – a color which, in the supposed "color space" describing the range of possible perceived colors, exists "between" the colors of the actual sources. For example, if a red source and a green source are both directed to a uniformly reflecting ("white") surface, the surface will appear yellow. There is, of course, no

36 FUNDAMENTALS OF COLOR

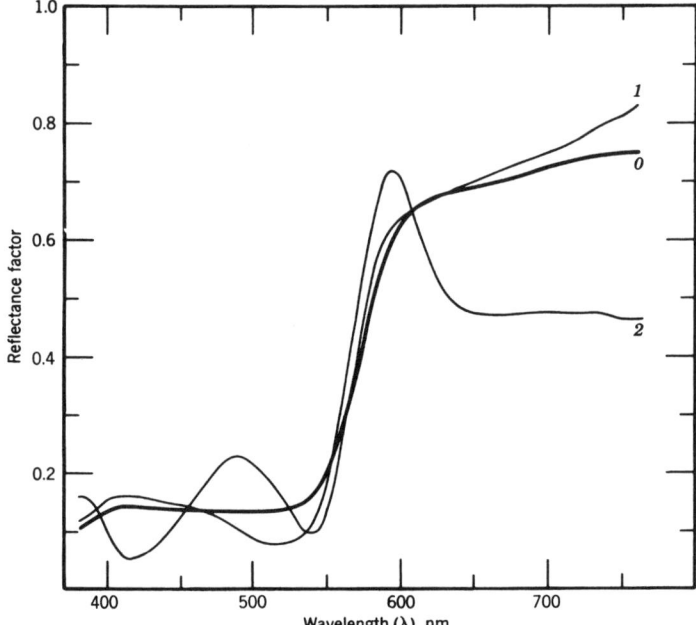

Figure 3-3 *Metamerism.* This graph shows the reflectance curves of three objects which, despite having clearly different reflectance characteristics, will be perceived as being the same color when viewed under the proper illumination (in this case, a light source conforming to the "D_{65}" illuminant specification, which is a standardized "daylight" white.) From Wyszecki and Stiles, *Color Science: Concepts and Methods, Qualitative Data and Formulae* (2nd edition, 1982; used by permission).

light energy being produced or received at a wavelength corresponding to "yellow" – but the important thing is that the receptors of the eye are being stimulated to produce the same outputs as they would if exposed to a "true" yellow source.

This phenomenon goes beyond simply combining two different sources to produce the perception of a third, intermediate color. Since there are three types of receptors responsible for distinguishing color in the human eye, we might expect that selecting the proper set of three sources, and combining these in varying amounts, could result in the perception of any desired color. This is, in fact, how "full-color" images are produced in electronic displays – through the combination of three *primary color* sources. The primaries for this purpose are generally referred to as red, green and blue, or "RGB" (although there may be considerable variation in the specific colors of each primary between any two real-world displays). These are the *additive* primaries – those that are used to create color through the addition of light. Those involved in the print media or artistic endeavors typically learn a different set; sometimes, these are given as "red, yellow, and blue", but more properly are labelled magenta, yellow, and cyan, respectively. This is the subtractive primary set, those colors which are used when dealing with the absorption of light – as in printed or painted pigments. (These are also properly referred to as the "CMY" set, more commonly seen as "CMYK." The "K" represents black, which is generally added to the basic three as practical printing devices cannot produce an acceptable black from the CMY primaries alone.) The use of two very different sets of "primary colors" is the source of considerable confusion and misunderstanding between engineers and artists!

However, the above turns out to be an oversimplification of the way we actually perceive color and the range of colors which can be detected. Due to the details of how color vision works, it turns out to be physically impossible to realize a color display based on a limited, finite set of discrete primary colors that can reproduce the *entire* range of possible colors. The reasons for this will become clearer as we look at how color is represented and analyzed, mathematically and in more detail. At this point, however, we should take some time to look at how the effects of various light sources, or illuminants, can be handled and specified.

3.3 Color Spaces and Color Coordinate Systems

The term *color space* refers to a three-dimensional model covering the possible range, in terms of both color and brightness, of light that can be perceived by human vision. Since the eye possesses three types of receptors, in terms of distinguishing color, it should not be surprising that a three-dimensional space is required to cover the range of possible color perceptions. Before looking into the color-specification systems in common use within the industry, it will be useful to look at the question of color from a more intuitive perspective.

When speaking of color in everyday speech, it is common to use words like "red" or "green" or "purple" – which are simply names for what in color-theory terminology is called *hue*. Hue is the property which most closely corresponds to the wavelength of the light – as you move through the visible spectrum, you are changing hue. When used as part of a system to describe any arbitrary color, however, we must extend the concept of hue to include colors which cannot be defined as a single unique wavelength. Combinations of two light sources at the ends of the spectrum – red and blue – are perceived as various shades of purple, a color which cannot be generated as a single wavelength (or *monochromatic*, meaning "single color") source. One way to complete the concept of hue, then, is to view it as giving the position around the circumference of a circle, one in which most of the circumference corresponds to the visible spectrum. The remainder then covers the purples, those colors which appear "between" the blue and red ends (Figure 3-4). Hue in this model becomes the angular portion of a polar-coordinate system.

The next concept in our "intuitive" model of color relates to the "purity" of the color, or to what degree the color in question really does correspond to a single wavelength of light. If red and white paint is mixed, for example, the expected result is pink – a color in which red predominates, but (through the white) all wavelengths of light are present to some degree. The quality which has varied in this case is called *saturation*, a measure of where the color in question is between being a purely monochromatic shade (100% saturation), and white (zero saturation, all wavelengths present in equal amounts). If saturation is added to the "circle" model with hue, it may be represented as the radial distance out from the center. This revised model, with white in the center and the pure (fully saturated) colors located around the circumference, is shown in Figure 3-5.

The last question to be addressed, is the "brightness" of the color – the difference between, for example, a bright, vivid red and the dull shade of a brick. The intuitive concept of brightness becomes, in standard terminology, *lightness* – or, more commonly, *value*. Adding value to the two-dimensional hue circle is achieved through the addition of an axis perpen-

38 FUNDAMENTALS OF COLOR

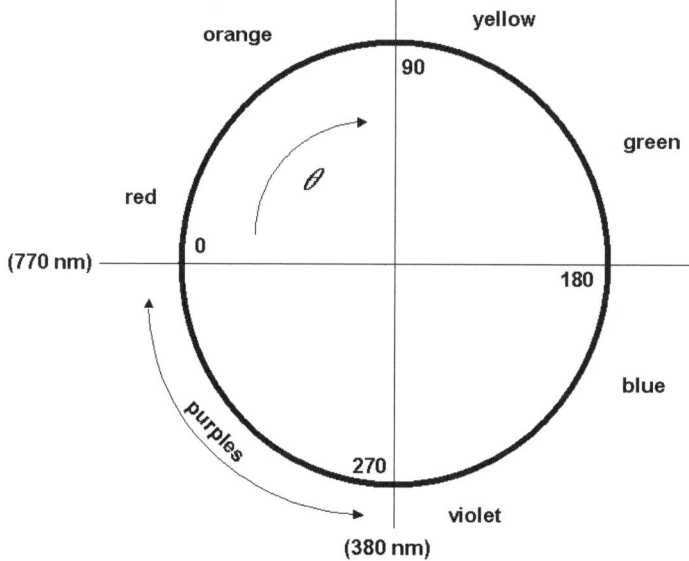

Figure 3-4 The beginnings of a system for specifying color numerically. Here, the wavelengths of the visible spectrum (770–380 nm) have been mapped to an angular measurement clockwise around a circle, with zero degrees arbitrarily set to equal 770 nm wavelength, and 270° equally arbitrarily set to correspond to 380 nm. The remaining quarter of the circle corresponds to the purple colors, those hues which do not correspond to any single wavelength of light, but rather are perceived when the eye is presented with varying amounts of red and blue light.

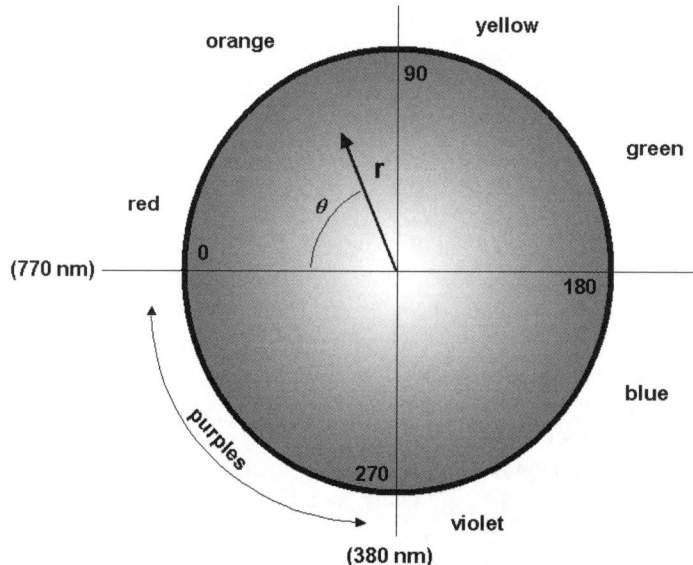

Figure 3-5 Adding saturation to the color model. In this diagram, saturation, or the "purity" of the hue, is indicated by the radial distance r outward from the center. Points on the circumference of the circle now represent "pure" colors, i.e., those which may be represented by a single wavelength, while the closer a point is to the center, the closer it is to white (all wavelengths present equally).

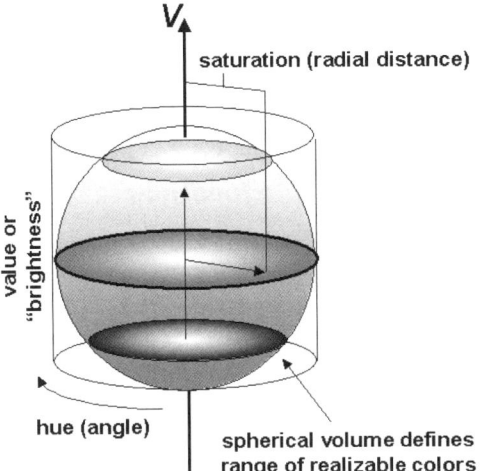

Figure 3-6 The completed HSV color model. At this point, a third dimension, labelled V (value) has been added as the axis of a cylindrical coordinate system; the two-dimensional, circular space defined earlier now appears as a cross-section of the full HSV space. However, the range of realizable colors does not fill the cylindrical space thus defined. This restriction has been (arbitrarily) shown here as a spherical volume within the full HSV cylinder, and results from the fact that color perception is greatly restricted (and in the extreme, fails completely) at very high and very low levels of perceived brightness. As will be seen, this simplistic model suffers from being *perceptually non-uniform*; that is to say that equal-distance translations within the defined space do not correspond to color changes perceived to be equal in magnitude by a normal observer.

dicular to the plane of the circle, such that the complete *HSV* model represents a cylindrical coordinate system (Figure 3-6).

Moving along the axis itself, at the center of this volume, represents changes between white and black – with no "color" in the common sense of the word at all. Moving out from the axis increases saturation – "adding color" to the white or gray, with the color becoming "purer" as the distance from the axis increases. And finally, moving around the axis changes the hue – we can move through the spectrum from red to yellow to green to blue, and then back through the purples to red. Note that the range of possible colors does not occupy the full cylinder defined by these three values, but instead appears in this case as a sphere within it. This is our first, crude attempt to account for the effect of luminance – or value, in this model – on the color sensitivity of human vision. Remember that in low-light conditions, color vision ceases to function. Similarly, at very high brightness, colors cannot be discriminated as the receptors "overload." This gives the range of realizable colors, within this HSV color space, its spherical shape – the difference between "white" and "black," and the fully saturated colors, decreases as we approach the extreme ends of the "lightness" axis.

While this simple HSV model provides an easy-to-use means of identifying color, which corresponds well to our everyday concepts, it is not very useful as a tool for precise colorimetry. In order to develop a better model, we must first develop a more accurate definition of exactly how the human eye responds to the visible spectrum of light. A standard model for the sensitivity of the eye's three types of color receptors, referred to as the "Standard Observer", was defined in 1931 by the Committee International de l'Eclairage (CIE, or in English, the International Color Committee). This model defines three sensitivity curves as func-

40 FUNDAMENTALS OF COLOR

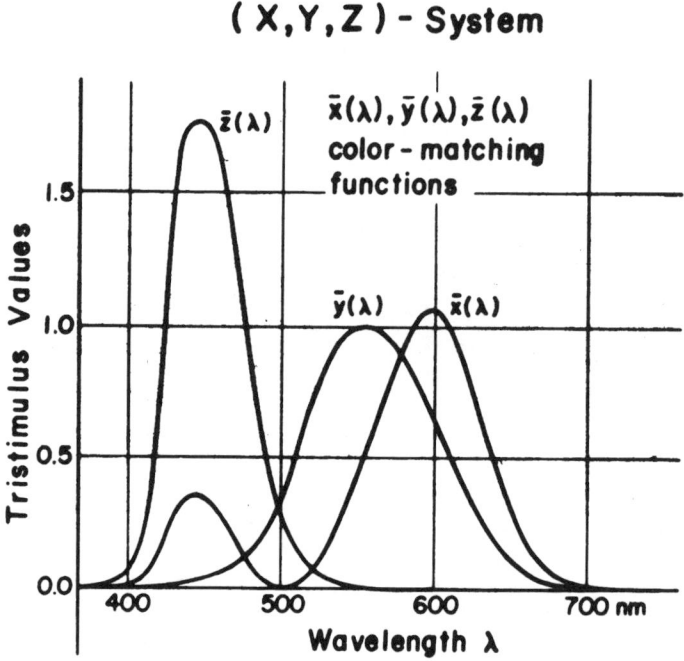

Figure 3-7 CIE color-matching functions. The three functions $x(\lambda)$, $y(\lambda)$, and $z(\lambda)$ are themselves derived from standardized visual sensitivity functions ($r(\lambda)$, $g(\lambda)$, and $b(\lambda)$), but avoid certain practical difficulties in those functions (such as negative responses in some portions of the spectrum). Integrating the response per these functions (the product of the function itself and the spectral distribution of the light being viewed) over the visual spectrum gives the tristimulus values, *XYZ*. Figure from Wyszecki and Stiles, *Color Science: Concepts and Methods, Qualitative Data and Formulae* (2nd editon, 1982; used by permission).

tions of wavelength across the visible spectrum: $x(\lambda)$, $y(\lambda)$, and $z(\lambda)$, for the "blue," "green," and "red" receptors, respectively (Figure 3-7). Note that the curves of this model do not correspond directly to the actual sensitivity curves of the cones of the eye; they have been modified somewhat due to practical concerns of the model. These are properly referred to as the *color-matching functions* of the CIE model, as they were derived through experiments in which observers were asked to match the colors of various sources. There are actually two sets of "Standard Observer" curves: the "2 degree" and the "10 degree" observer. These names refer to the area of the visual field covered by the test color in the color-matching experiments; of these, the "1931 CIE 2° Standard Observer" set is by far the more commonly used, although the difference between them is generally not important except in the most serious color work.

The CIE color-matching functions lead directly to a space defined by the CIE *tristimulus values XYZ*, which are simply measures of the integrated response to a given light source by receptors of these sensitivities. The tristimulus values for a given color certainly provide an unambiguous means of defining any color, but they are not often used. One reason for this is the fact that, while *X*, *Y*, and *Z* can each be considered as "primaries," this set is not itself physically realizable – they exist as mathematical constructs only, and lie outside the range of real-world colors. A much more useful means of expressing color was also defined by the CIE, based on the *XYZ* values. This is the *Yxy* color space, among the most widely used in electronic display work.

In this model, only the *Y* of the original *XYZ* set is retained; as the original functions were defined, the *Y* value corresponds to "lightness,", or the perceived brightness of the source. In the absence of "color" in the common sense of the word, *Y* can be thought of as defining the level of brightness in a "black and white" view of the world. The remaining two values, *x* and *y*, may be calculated from the *X*, *Y*, and *Z* values as

$$x = \frac{X}{X+Y+Z}, \quad y = \frac{Y}{X+Y+Z}$$

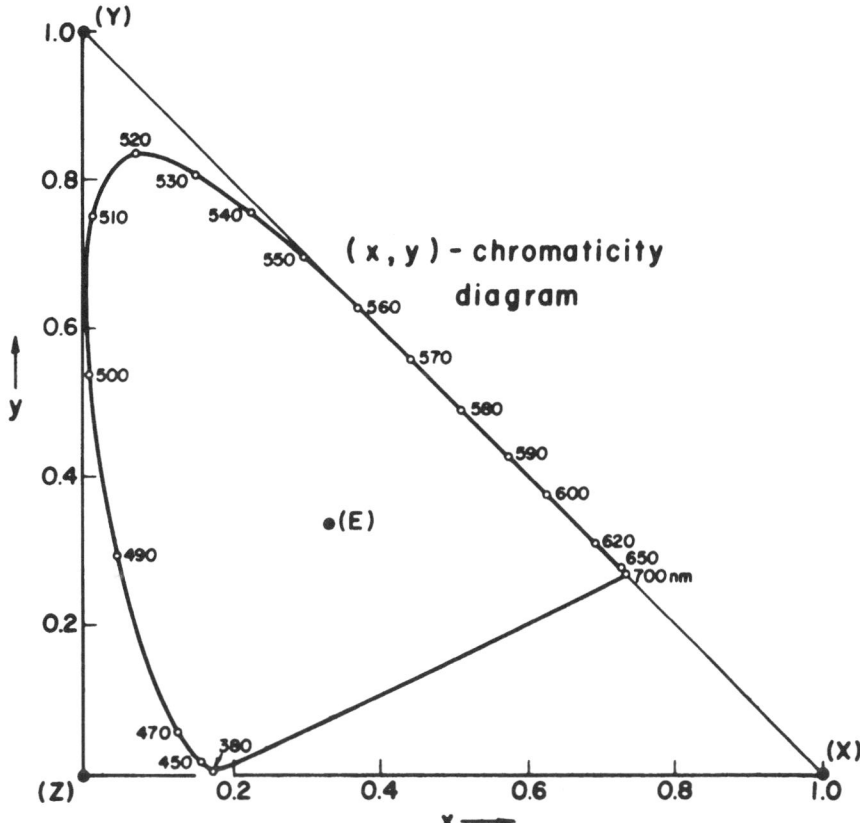

Figure 3-8 CIE *xy* chromaticity diagram. This chart is a two-dimensional slice of the three-dimensional *xyz* space derived from the CIE color-matching functions. The "pure" single-wavelength colors are located around the curved perimeter of the area (the numbers around this line are wavelengths, in nm); the straight line across the bottom, rising slightly from the lower left corner, is the limit of the region of purple shades obtained by various combinations of blue and red light. Whites are roughly at the center of the diagram, with the point labelled "E" being the so-called *equal energy white*, at $x = 0.3333$, $y = 0.3333$. As will be seen, the *xy* coordinates are a very popular means of specifying colors, but still this space still suffers from being perceptually non-uniform. Used by permission from John Wiley & Sons.

42 FUNDAMENTALS OF COLOR

These are actually two of a set of three chromaticity coordinates, with the final coordinate (z) derived from the Z value in a similar manner (which also results in $z = 1 - x - y$). However, the full xyz set is rarely used. The x and y values, however, define a two-dimensional space which is very useful in visualizing various aspects of color relationships.

Specifying the CIE (x,y) coordinates along with the luminance (Y) provides a very easy and practical means of describing the appearance of a wide range of light sources. The standard CIE *xy chromaticity diagram*, based on the 1931 2° standard observer, is shown in Figure 3-8. (Note that the xyz coordinates, given as such, refer specifically to the 2° Observer; the corresponding set based on the 10° values are properly identified as x_{10}, y_{10}, and z_{10}.)

The CIE xy diagram shares several characteristics with the two-dimensional slice of the HSV space presented in Figure 3-6. First, we again have the line of purely saturated colors around the curved periphery – moving around this edge again represents a change in hue. (Unlike the HSV model, however, the line between the extreme ends of the visible spectrum – the range of purples between red and blue – appears as a straight line. The reason for this will be clear shortly.) Whites are roughly in the center of the diagram, so again we see saturation increasing as one moves from the center to the periphery. (If any point can claim to be the "center" of this diagram, it might be the "equal energy" white point, the color of a flat spectral power distribution; this is at $x = 0.333$, $y = 0.333$.) Finally, we should again note that the xy diagram is just one slice through a three-dimensional color space. Perpendicular to the center of the diagram is where the Y axis can be imagined, representing luminance or "lightness". As with the HSV space, the two-dimensional diagram is the cross-section at the "widest" part of the space. The range of perceivable colors, and therefore the relative cross-sectional area, decreases at high and low luminances.

The Yxy system, and the related xy diagram, remain very widely used for specifying and plotting colors, despite a major shortcoming which is discussed shortly. There are several other color coordinate or color space definitions in common use, but before moving to those we can use this one for introducing several key concepts.

3.4 Color Temperature

A common means of indicating the characteristic spectral distribution of many light sources is to state their *color temperature*. The concept of stating color as a temperature comes from the fact that physical bodies radiate energy proportional to their temperature (when this radiation is visible, the object is in a state of *incandescence*). A *black body* is a theoretical construct that absorbs all light that falls upon it, and so its color depends completely on such emissions in the visible range. A color temperature, then, is simply the temperature (normally given in degrees Kelvin) at which such a body would be radiating light of a given color, or more precisely of a certain spectral power distribution. This concept is revisited in the next section, but for now we can note that increasing color temperature corresponds roughly to what we know from actual objects heated to incandescence. At the lowest temperatures at which such objects (a heated ingot of metal, for instance) begin to glow, they appear to be a dull "brick" red. As the temperature increases, the color of the object changes through brighter reds, to orange, yellow, and ultimately the object becomes "white hot." Beyond a certain temperature (somewhere above 30,000–40,000 K, far above the color temperature of any light source commonly encountered in practice), further increases in temperature produce no visible change in color; the black body is already emitting uniformly across

COLOR TEMPERATURE   43

the visible range, and appears a very bluish-white. We can plot the various colors that are specified by the color temperature system above onto the CIE *xy* chart (Figure 3-9). These lie along the line called the *black-body locus* (or *black-body curve*), which as expected starts in the very deep red and curves through the orange and yellow regions before passing through the central range of whites. We can now see that even a theoretically "infinite" color temperature still lies in the bluish-white region. It could not become a purer blue, as even an infinitely hot body would still be radiating across the lower-frequency portions of the spectrum.

Color temperature is often used to give at least an approximate idea of the color of a light source, especially to discriminate between the various "white" sources which are commonly encountered. A more precise description of these sources must wait until we have discussed a means for numerically specifying color in general, but we can at this point give some general indication of how color corresponds to the color temperature value. Most artificial light sources, such as incandescent lights and fluorescent tubes, produce a distinctly reddish light, with color temperatures in the range of 2500–4000 K. "Daylight" white, the color of sunlight as reflected from a diffuse, spectrally "flat" surface, is about 6000–6500 K, depending on conditions. This is very close to what we would perceive as a "true" white, an absence of any

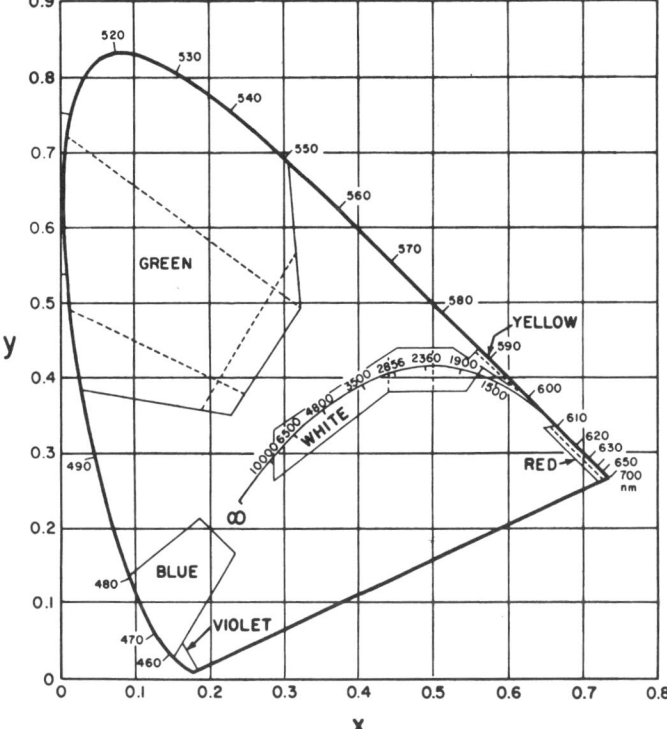

Figure 3-9   Color temperature on the CIE *xy* diagram. The curved line passing through the middle of the space is the *black-body curve*, which is the locus of points describing the color emitted by a theoretical "black body" at various temperatures. The numbered points along this curve give the corresponding temperature (in degrees Kelvin). Note that the curve ends in the bluish-white region of the color diagram; there is a point at which further increases in the black body's temperature do not change its radiation characteristics within the visible spectrum. From Wyszecki and Stiles, *Color Science: Concepts and Methods, Qualitative Data and Formulae* (2nd edition, 1982; used by permission).

discernible hue. The sun viewed directly, with an increase in the blue components provided by the scattering effects of the sky, appears as a much bluer white, with a color temperature as high as 10,000–12,000 K. Note that these are given as general ranges only; for most of these sources, there is simply too much dependence on a large number of variable conditions to be able to give a precise value.

3.5 Standard Illuminants

Since it is apparent that the color of a given object depends not only on the characteristics of that object itself but also on the light source, it is very helpful to have standards defined that can represent a range of commonly encountered light sources. Such definitions remove one source of variation from calculations or measurements of color, allowing us to focus more on the characteristics of and differences between objects. These definitions are referred to as *standard illuminants*, and are expressed as power distributions covering the visible light spectrum and sometimes beyond. (Definitions beyond the standard "visible" range are often useful, as energy outside this range – particularly at the high-frequency, or ultraviolet, region – can affect the perceived color of objects through phenomena such as fluorescence.) Standard illuminants have been defined to approximate a number of common real-world light sources, such as sunlight, incandescent lighting, fluorescent tubes, etc.. Besides standardizing the characteristics of sources of illumination, the colors of these sources are also commonly referenced as the standard "white" used in various applications. Some common standard illuminants and their characteristics are shown in Table 3-1.

As might be expected, since these standards are intended to represent common sources such as daylight and incandescent lighting, they are also very commonly associated with a particular color temperature. For example, the most common standard for "daylight white" is Illuminant D_{65}, which is approximately a 6500 K white. But it should be clear that not all colors can be identified through the color temperature system – there is no temperature at which a black body will produce a pure green light, for instance. In order to be able to specify and analyze color objectively, we need a numeric model which is capable of covering all perceivable colors.

Table 3-1 Standard illuminants[a]

Illuminant	CIE (x,y) coordinates		Color temperature	Comments
A	0.4476	0.4075	2854 K	Incandescent (tungsten) light
B	0.3840	0.3516	4874 K	Direct sunlight
C	0.3101	0.3162	6774 K	Indirect noon sunlight
D_{50}	0.3457	0.3586	5000 K	Common standard for publishing and document editing/preview applications; "paper" white.
D_{65}	0.3127	0.3297	6504 K	"Daylight" white; common standard white point for video applications.
E	0.3333	0.3333	5500 K	"Equal energy" white point

[a] It is important to note that, while each illuminant has a specific perceived color (as given by the color coordinates and color temperature specifications here), the actual specification of the illuminant is given as a standardized spectral power density curve

3.6 Color Gamut

Having a two-dimensional representation of color space such as the *xy* diagram also permits us to look at the idea of *color gamut,* which is simply the range of possible colors that may be produced by a given color display device. As mentioned earlier, practically all color displays function by varying the intensities of three primary-color images (for additive-color displays, usually simply red, green, and blue). Plotting the locations of these primary colors on a chromaticity diagram defines a triangular area covering the range of colors which can be produced from that set – the color gamut of the display. An example, using the chromaticity coordinates of a typical set of color CRT primaries, is shown as Figure 3-10. This also demonstrates why is it physically impossible to produce a practical display, based on a finite set of primaries, which can duplicate any color within the range of human vision. It is impossible, due to the curvature of the perimeter of the color space, to locate any three (or even four, five, or six) primaries such that the perimeter they define fully encloses the complete space. The only way to do this would be to locate at least one point outside the limits of the diagram, and the "outside" area represents physically unrealizable, "supersaturated" colors.

Still, the appearance of this color gamut plot, with real-world primaries, is somewhat misleading. When seen on the *xy* diagram, we are immediately struck by the very large portion of the color space that lies outside of the gamut triangle; these areas represent colors that cannot be produced on this display, and this conflicts with what we expect from our experience with common color displays. Such displays appear to present a very realistic image, and

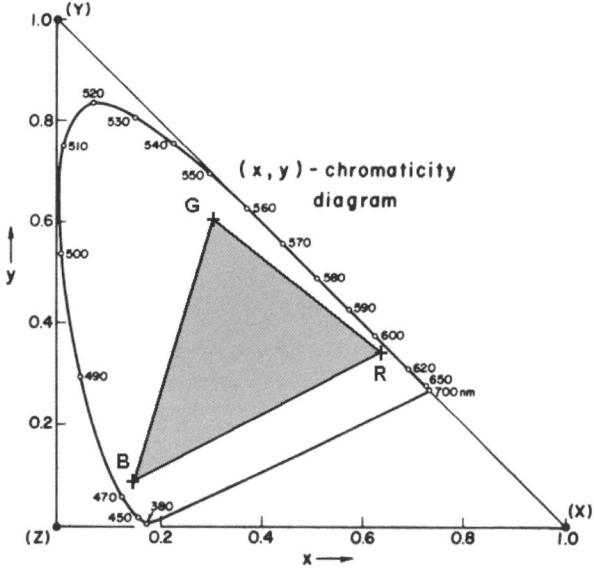

Figure 3-10 Color gamut. The shaded area shows the color gamut, or range of possible colors, for a typical electronic display; this area is defined by the locations within the color space of the three primaries used by the display, labelled here as R, G, and B. (The primary set shown here is that defined by the "sRGB" standard, with chromaticity coordinates of (0.640,0.330) for red, (0.300, 0.600) for green, and (0.150, 0.060) for blue.) Used by permission from John Wiley & Sons.

we have to ask how this can be so with so much of the colors of the "real world" unavailable. This is partially due to the eye/brain system's willingness to accommodate a less than ideal image; in other words, in the absence of a reference for comparison, we "think" we're seeing an image that is better than objective measurements would indicate. However, in this particular example, the color diagram we're using is also a part of the problem.

3.7 Perceptual Uniformity in Color Spaces; the CIE L*u*v* Space

Presenting color gamut information in this manner can be a very useful tool for getting an idea of the relative capabilities of displays or other imaging technologies, such as comparing the gamuts of emissive displays with a reflective/subtractive-color technology such as color printing. However, these plots can be misleading depending on the specific chromaticity coordinate system used. The problem here is that not all spaces, and especially not the *Yxy* space that we have been discussing, are *perceptually uniform*, meaning that they are not defined such that equal distances in any direction or in different areas of the space correspond to equal perceived changes in color.

To address this problem, other color spaces have been defined. Among the most popular is the CIE L*u*v* space, defined in 1976 through a rescaling of the *xy* coordinates and also applying a correction to the *Y* (luminance) value to produce a more perceptually accurate indicator of "lightness" for the situation being considered. The resulting two-dimensional diagram, corresponding to the original CIE *xy* chart, is formally referred to as the CIE 1976 Uniform Color Space (UCS) diagram, also known informally by the new axes identifiers, u' and v' (Figure 3-11). The u', v' coordinates are derived from the *x,y* coordinates (or the fundamental *XYZ* tristimulus values), as

$$u' = \frac{4X}{X+15Y+3Z} = \frac{4x}{-2x+12y+3}$$

$$v' = \frac{9X}{X+15Y+3Z} = \frac{9y}{-2x+12y+3}$$

The complete L*u*v* space is based on this coordinate system, plus the tristimulus values for an assumed "perfect reflecting diffuser" (a theoretical surface which reflects and diffuses uniformly across the entire spectrum) under the illuminant in question. The space is defined as

$$L^* = 116\left(\frac{Y}{Y_c}\right)^{1/3} - 16 \quad \text{when} \quad \left(\frac{Y}{Y_c}\right) > 0.008856$$

$$u^* = 13L^*(u' - u'_c)$$

$$v^* = 13L(v' - v'_c)$$

where Y_0, u'_0, and v'_0 are the values from the perfect reflecting diffuser.

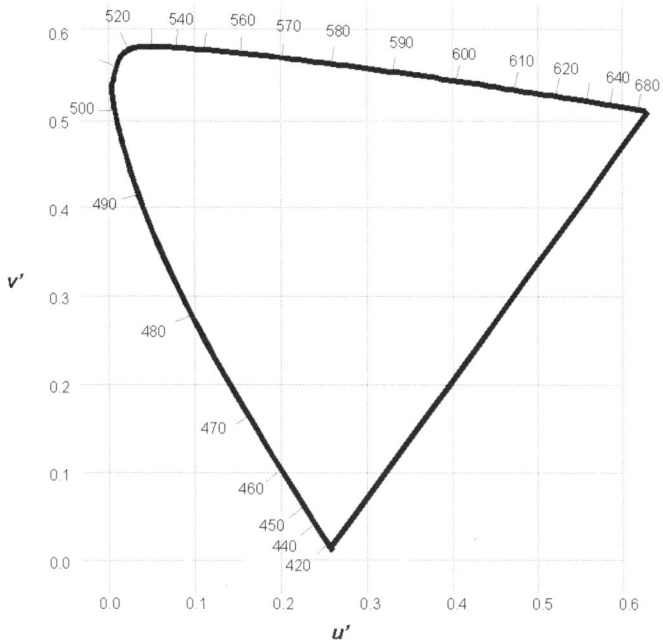

Figure 3-11 CIE u'v' chromaticity diagram. This is again a two-dimensional representation or cross-section of a three-dimensional space, derived from the original *xyz* coordinate system through a series of transformations which make the space *perceptually uniform*. This means that, to the standard observer, equal-distance changes within this space correspond to changes in the perceived colors that are of a similar degree.

Since the L*u*v* space is perceptually uniform, it is used as the basis of a very powerful means of expressing color differences and color uniformity, the ΔE^* (normally read "delta E star") specification. ΔE^* is calculated as the distance between two points in L*u*v* space as

$$\Delta E^*_{uv} = \sqrt{(L^*_1 - L_2)^2 + (u^*_1 - u^*_2)^2 + (v^*_1 - v^*_2)^2}$$

One of the most powerful uses of the ΔE^* metric in the display field is in measuring or specifying display uniformity, as it will permit combination of the traditional separate luminance and color uniformity requirements into a single perceptually accurate specification. If these are kept separate, a given display can quite easily be within specifications in terms of both luminance and color (usually stated as maximum permitted changes in *xy* coordinates) uniformity, and yet have visually objectionable overall non-uniformity.

All of the above is, of course, based fundamentally on the original 1931 2° Standard Observer definitions and *XYZ* tristimulus values. Should the 1964 10° Observer values be used, the identifiers are changed to u'_{10}, v'_{10}, etc.

Other perceptually uniform color spaces have also been defined. One of the more common is the CIE L*a*b* space, also defined in 1976 in a manner very similar to the L*u*v* space. However, the L*a*b* system is more commonly used for the specification of the color of objects (i.e., reflected light), rather than for color in light-emitting devices such as elec-

48 FUNDAMENTALS OF COLOR

tronic displays. It is rarely encountered in display work, where the *Yxy* and L*u*v* are by far the most common systems.

3.8 MacAdam Ellipses and MPCDs

One other graphical depiction of the non-uniformity of the *xy* diagram can be seen via *MacAdam ellipses*, which are experimentally derived groupings or small areas describing indistinguishable colors. In other words, the average viewer sees all colors within the area as the same color. The ellipses, then, define the distance corresponding to a "just-noticeable difference" (JND) in that particular area of the chart. (That they are referred to as "ellipses" is due to their elongated appearance only; they are not mathematically defined.) The size and shape of these areas when drawn on the *xy* diagram (shown in Figure 3-12, with the ellipses at roughly 10× their actual size, for visibility) dramatically demonstrate the perceptual non-uniformity of this space. The concept of a just-noticeable difference, or as it is sometimes called, a *minimum-perceivable color difference* (MPCD) is also used to define an incremental change or distance on a chromaticity diagram. For example, it is common to see the standard D_{65} illuminant referred to as "6500 K + 8MPCD", as its location does not lie exactly on the 6500 K point of the black-body locus.

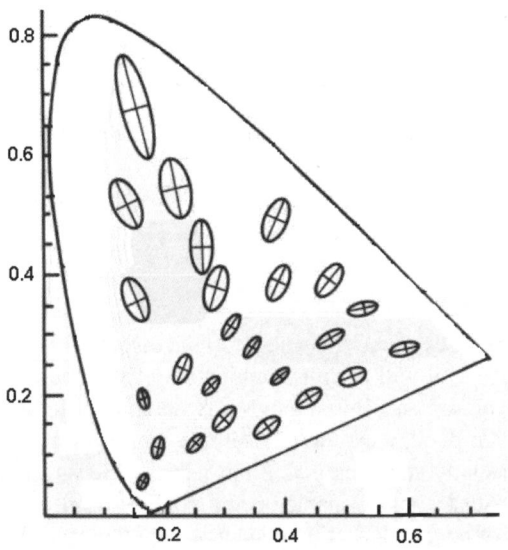

Figure 3-12 MacAdam ellipses, also known as "color ovals", sketched on the 1931 CIE *xy* diagram. These define areas (shown here 10× actual size) that define colors that will be perceived by the average viewer as the same; i.e., the boundary of each ellipse represents the locus of colors that are just barely noticeable as being different from the center color. Note that this shows the *xy* chart to be perceptually non-uniform, as the ellipses are not the same size in all parts of this space. Used by permission from John Wiley & Sons.

3.9 The Kelly Chart

One additional variation on the standard chromaticity diagrams should be mentioned before we move off this subject. In 1955, K. L. Kelly, of the US National Bureau of Standards (now NIST, the National Institute of Standards and Technology)., defined a set of standard names for regions of the chromaticity diagram. The Kelly chart, while not itself a new color space definition, does permit the approximate location of colors within the existing standard spaces to be quickly identified. Originally, of course, the Kelly chart was a variation of the standard CIE xy diagram, but since the $u'v'$ diagram is simply a rescaling of the xy space, the same regions can be mapped to that chart as well. The name "Kelly chart" is often misinterpreted as applying to the basic xy or $u'v'$ diagrams themselves, rather than simply the versions of these with the named regions overlaid.

3.10 Encoding Color

From the standpoint of one concerned with the supposed subject of this book – display interfaces – we still have not made it to the key question, which is how color information may be conveyed between physically or logically separate devices. As long as the image samples (*pixels*) were themselves simply one-dimensional values within a two- or three-dimensional sampling space, the interface could be expected to be relatively straightforward. Pixel data would be transmitted sequentially, in whatever scanning order was defined by the sampling process used, and as long as some means was provided for distinguishing the samples the original sampled image could easily be reconstructed. This is, of course, precisely how monochrome ("single-color", but generally meaning "luminance only") systems operate.

Introducing color into the picture (no pun intended) clearly complicates the situation. As has been shown in the above discussion, color is most commonly considered, when objectively quantified, as itself three-dimensional. The labels assigned to these three dimensions, and what each describes, can vary considerably (hue, saturation, value; red, green, and blue; X, Y, Z; and so forth), but at the very least we can expect to now require three separate (or at least separable) information channels where before there was but one. In the most obvious approaches, this will generally mean a tripling of the physical channels between image source and display, and along with this a tripling of the required – and available – total channel capacity. In many applications, this increase in required capacity or "bandwidth" is intolerable; some means must be developed for adding the color information without significantly increasing the demands on the physical channel, and/or without breaking compatibility with the simpler luminance-only representation. Television is the classic example of this, and the systems developed to meet these challenges in that particular case will be examined in detail in Chapter 9. In general, such systems are based on recognizing that a considerable amount of the information required for a full-color image is already present in a luminance-only sample. Therefore, the original luminance-only signal of a monochrome scheme is retained, and to this are added two additional signals that provide the information required to derive the full-color image.

These added signals are viewed as providing information representing the difference between the full-color and luminance-only representations, and so are generically referred to as "color-difference" signals. This term is also used to refer to signals that represent the difference between the luminance channel and the information obtained from sampling the image

50 FUNDAMENTALS OF COLOR

in any one of the three primaries in use in a given system. Another common term used to distinguish these additional channels is *chrominance*, which is somewhat inaccurate as a description but which leads to such encoding systems being referred to as *"luminance-chrominance"* (or "Y/C", since "Y" is the standard symbol for the luminance channel) types. Thus, we can generally separate color interfaces standards into two broad categories – those which maintain the information in the form of three primary colors throughout the system (as in the common "red-green-blue," or "RGB," systems), and those which employ some form of luminance/chrominance encoding. It is important to note, however, that practically all real-world display and image-sampling devices (video cameras, scanners, etc.) operate in a "three-primary" space. Use of a luminance/chrominance interface system almost always involves conversions between that system and a primary-based representation (RGB or CMY/CMYK) at either end.

One further division of possible color-encoding systems has to do with how the three signals are transmitted in time. If we consider the operation of an image-sampling device – a camera, for instance – which is intended to provide full-color information, we might view each sample point or pixel as being a composite of three separate samples. In the most obvious example, a camera is actually producing independent red, green, and blue values for each pixel. (In practice, this is most often achieved through what is in effect three cameras in one, each viewing the scene through a primary-color filter.) We can say that the complete image is made up of pixels with three values each (RGB pixels), or we could equally well view this as generating three separate images – one in red, one in green, and one in blue – which must

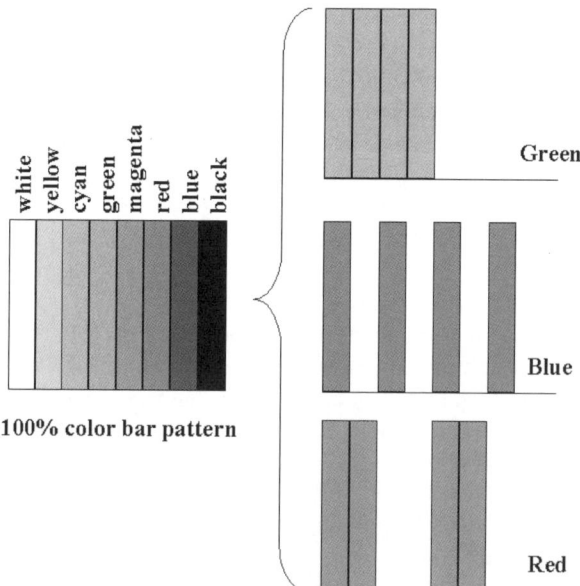

Figure 3-13 The separation of a color image into its primary components. Here, the standard color bar pattern (commonly used in television), consisting of various combinations of the three primaries (each at 100% value when used), is broken down into three single-color images or *fields*. This sort of *color separation* process is common in image capture, display, and printing systems, as these often deal with imagery through separate primary channels, sensors, and display devices. Recombining the three fields, in proper alignment, reproduces the original full-color image.

be combined to recover the image of the full-color original (Figure 3-13). This seems to be a trivial distinction, but in fact there are display technologies, as well as image-capture devices, which use each of these. Many display types provide physically distinct pixels which are, in fact composed of three separate primary-color elements. Other displays generate completely separate red, green, and blue images, and then present these to the viewer such that the desired full-color image is perceived. This leads to two very distinct methods for transmitting color data.

If the display is of the former type, and presenting the red, green, and blue information for each pixel simultaneously but through elements that are physically separated, it is said to be employing the *spatial color* type. The three primary-color images are presented such that the eye sees them simultaneously, occupying the same image plane and location, and so the appearance of a full-color image results. However, it is also possible to present the three primary-color images in the same location but separated in time. For example, the red image might be seen first, then the green, and then the blue. If this sequence is repeated rapidly enough, the eye cannot distinguish the separate color images (called *fields*), and again perceives a single full-color image. This is referred to as a *sequential color* display system, or sometimes as *field-sequential color*. Obviously, the interface requirements for the two are considerably different. In the spatial-color types, the information for the red, green and blue values for each pixel must be transmitted essentially simultaneously, or at least sufficiently close in time such that the display can recover all three and present them together. This is commonly achieved through the use of separate, dedicated channels for each primary color (Figure 3-14a). With a sequential-color display, however, all of the information for the each

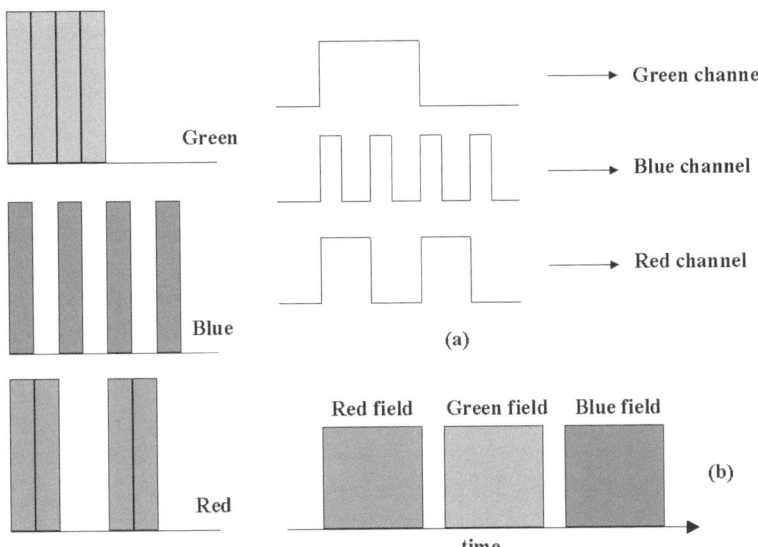

Figure 3-14 Transmission of the color fields. Many display systems employ spatial color techniques, or other types in which the three primary-color fields may be transmitted and processed simultaneously, in parallel (a). Some, however, employ a field-sequential method (b), in which the separate primary fields are transmitted in a sequence, often over a single physical channel. Used by permission from John Wiley & Sons.

52 FUNDAMENTALS OF COLOR

single-color field must be transmitted in the proper order, and only then will the information for the next color be sent. This can be implemented through any number of physical channels, but the fields themselves will be serialized in time (Figure 3-14b).

In the next chapter, the last of the "fundamentals" section of this book, an overview of the most popular current display technologies, and some promising future possibilities, are presented. With that review completed, we will have examined each of the separate components of the overall display system, and be ready to move to the main topic – the interfaces between these components. Of particular interest, however, will be how each display technology, and each of the various display-interface systems, handles the encoding of color, and the limitations on color reproduction that these methods create. Chapter 8, which deals with the details of the modern analog television broadcast standards, covers an especially significant development in terms of handling color electronically, in the form of the color-encoding schemes developed for the broadcast medium.

4

Display Technologies and Applications

4.1 Introduction

Having examined the workings of vision and color, it is now time to look at the other side of the human/display interface: the display itself. While the basic purpose of all displays is the same – to deliver visual information to the user – the means through which this is accomplished varies significantly among the various types and technologies of display devices. Quite often, as we will see, the display technology itself has a significant impact on the appearance of the image, and it is up to the designer of the complete display system to ensure that the appearance is satisfactory and meets the requirements of the application. In this chapter, we review some of the basics of the most popular display technologies as well as several new types, along with their unique characteristics and the typical applications and markets for each.

Organizing the various display technologies into a manageable hierarchy may be done in several ways; one such attempt is shown in Figure 4-1. Here, the various display technologies are organized first by whether the fundamental display device produces light – the emissive types – or if it instead functions by modulating externally produced light. However, in a very real and practical sense, the display world divides very neatly into just two categories for many applications: the cathode-ray tube (CRT) display, and everything else. The CRT display, now over 100 years old and still the dominant display technology in almost all non-portable applications, also has some very fundamental differences from practically all of the non-CRT types, and so is reviewed first and receives the lion's share of attention.

The non-CRT types further divide into subcategories based on their functional properties and the basic technology behind each. However, almost all of these share one characteristic which puts them in sharp contrast with the CRT display: they are typically fixed-format devices, meaning that the display itself comprises a fixed array of independently controllable picture elements, or *pixels*. (It is important to at all times keep in mind the distinction be-

54 DISPLAY TECHNOLOGIES AND APPLICATIONS

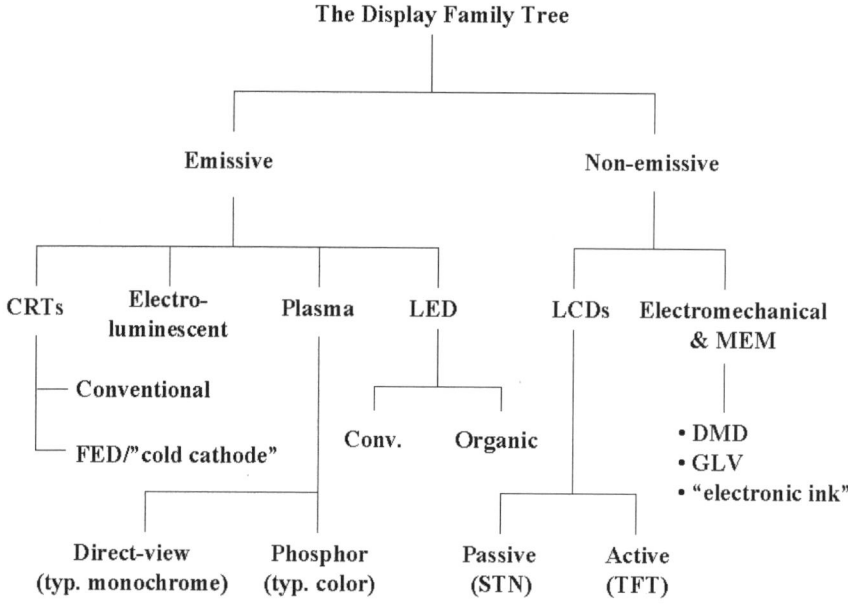

Figure 4-1 A hierarchy of display technologies.

tween "pixel" used in this sense – a physical structure in a fixed-format display device – and the meaning of the word as it was originally introduced in Chapter 1, as a point sample of an image.)

The non-CRT types are often referred to in general as "flat panel displays" (FPDs), since as least most of the direct-view technologies in this category do take that physical form. But this term is becoming less and less applicable as new types, such as the *microdisplay*, enter the market. A useful breakdown of these types begins with distinguishing whether or not the display device itself emits light (an emissive display), or produces images through controlling the light from an external source. The latter class further divides as we consider where the light source is relative to the display device and viewer, which tells how the display must interact with the light. Thus we have reflective displays (those which control light reflected from the display surface to the viewer), transmissive displays, and some technologies which are capable of operation in either mode. Further categorization in all classes is then by the fundamental operating principle or basic technology used in the construction of the display device.

Despite the very wide range of methods used in creating their images, practically all of the display types to be reviewed here have one basic factor in common: all present spatially two-dimensional imagery which is repeatedly transmitted to the display in a regular order. Almost without exception, unless the apparent orientation of the display device with respect to the viewer is altered through optical means, the transmission ordering used is the regular raster scan as presented in Chapter 1. In the typical raster scan, the information corresponding to the extreme upper-left pixel is sent first, and the scan progresses horizontally (to the right) from that point, and then each horizontal row of pixels is transmitted in similar manner from the top of the image to the bottom. There is no inherent advantage to this ordering over any other possible scheme, with the minor exception that progressing along the "long axis" of the display device generally will slightly reduce the transmission rates required. But once this standard was established by early CRT-based devices (and especially television), it remained the norm.

4.2 The CRT Display

The origins of the cathode-ray tube, or CRT, stretch back to the late 19th century, and the invention of the electric light bulb by Thomas A. Edison in 1879. In later experiments with this device, it was discovered that a current would pass from the heated filament to a separate plate within the bulb, if that plate was at a positive potential with respect to the filament (the thermionic emission of electrons, originally referred to as the "Edison effect"). Two other discoveries which soon followed paved the way for the CRT as the display we know today: first, in 1897, the German physicist Karl Braun invented what he called an "indicator tube", based on the recent discovery that certain materials could be made to emit light when struck by the stream of electrons in the Edison-effect bulbs. (Actually, at the time, the electron was not yet known as a unique particle; Braun and others referred to the mechanism that produced this effect as "cathode rays", thus originating the name for the final display device.) Later, in 1903, the American inventor Lee Deforest showed that the "cathode rays" (the electrons being emitted in these devices) could be controlled by placing a conducting grid between the emitting electrode (the cathode) and the receiving electrode (the anode), and varying the potential on this grid. Combined, these meant that not only could a beam of electrons – the "cathode ray" – be made to produce light, but the intensity of that beam could be varied by a controlling voltage. (DeForest had actually invented the first electronic component capable of the amplification of signals – what in America is called the "vacuum tube", but which in the UK took the more descriptive name "thermionic valve".) The modern cathode-ray tube is in essence a highly modified vacuum tube or valve, and is one of the few remaining practical applications of vacuum-tube technology.

A cross-sectional schematic view of a typical monochrome CRT is shown in Figure 4-2. The cathode, its filament, and the controlling grids and other structures form the electron gun, located in the neck of the tube. The plate, or anode, of the tube is in this case the screen itself, covering the entire intended viewing area (and generally somewhat beyond). This is the first obvious modification from the more typical vacuum-tube structure; the anode is located a considerable distance from the cathode, and is greatly enlarged. This screen structure comprises both the electrical anode of the tube, and the light-emitting surface. Typically, a layer of light-emitting chemicals (the phosphor) is placed on the inner surface of the faceplate glass, and is then covered by a thin, vacuum-deposited metal layer, most often aluminum. The aluminum layer both serves as an electrical contact, and protects the phosphor from direct bombardment by the electron beam. As the light emitted by the phosphor is proportional to the energy transferred from the incoming beam, this stream of electrons must be accelerated to a sufficiently high level. For this reason, the screen is maintained at a very high positive potential with respect to the cathode; usually several thousand volts as an absolute minimum for the smallest tubes, and up to 50 kV or more for the larger sizes. (The screen itself is not adequate to perform the task of accelerating the electrons of the beam; there is almost always an electrode at the anode potential as part of the electron gun structure itself, connected to the screen proper by a layer of conductive material on the inside of the CRT. Therefore, the screen at the face of the tube is more properly referred to as the *second* anode.)

56 DISPLAY TECHNOLOGIES AND APPLICATIONS

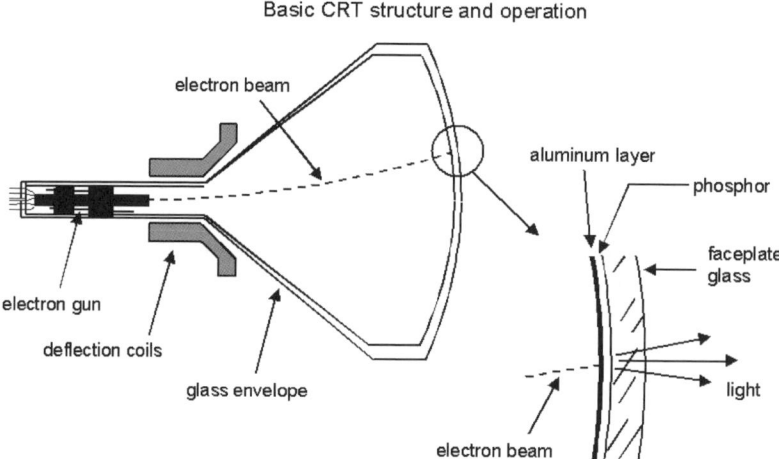

Figure 4-2 Basic monochrome CRT. In this, the oldest of current electronic display devices, light is produced when a beam of electrons, accelerated by the high potential on the front surface of the tube, strikes a chemical layer (the phosphor). The beam is directed across the screen of the CRT by varying magnetic fields produced by the deflection coils.

In order to be usable as a display device, it is not sufficient that the beam simply impact the screen and cause the phosphor to emit light. The beam must be constrained to produce a small, well-defined spot of light, and some means must be provided for directing that spot and varying its intensity in order to "write" images. Forming a distinct spot from the stream of electrons is achieved through the use of electrodes within the electron gun structure, which produce one or more lenses in the form of shaped electric fields. These work in a manner directly analogous to the action of normal optical lenses on light. To move the resulting spot across the screen, the beam is deflected using magnetic fields. Pairs of coils are mounted on the outside of the tube, around the funnel area, and produce fields which move the beam both horizontally and vertically. In order to produce the image using the raster-scan method described in Chapter 1, the waveform of the current in each pair of coils is a modified sawtooth, as shown in Figure 4-3. One pair of coils – usually those which move the beam along the horizontal axis – scans at a much higher rate than the other, producing the raster pattern as shown. These coils are normally constructed as a single component, referred to as the *deflection yoke*.

The intensity of the beam, and so the intensity of the light emitted by the phosphor at the screen, is varied by applying a signal corresponding to the desired luminance to the electrodes of the electron gun. Typically, this video signal is applied to the cathode (superimposed on to the DC bias voltage of the cathode), while the "grid" of the tube is tied to a variable DC supply; using cathode drive for a CRT generally results in more stable operation (vs. variations in the anode voltage, etc.). By driving the cathode more negative, the beam current (and so the intensity of the light produced) is increased; therefore, the video signal normally appears "inverted" at this point. The potential difference between the cathode and the fixed grid determines the point at which the beam current goes to zero, also referred to as the "cutoff" or "blanking" point of the CRT. The beam is normally cut off during those portions of

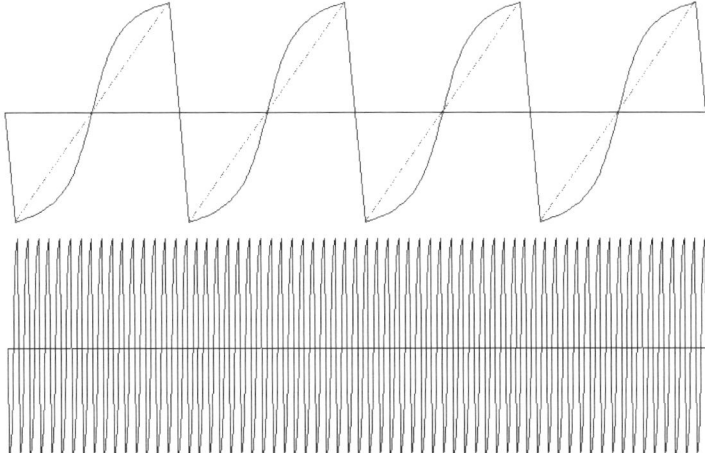

Figure 4-3 Deflection current waveforms. To produce the raster-scan pattern described in Chapter 1, one axis – usually the horizontal – is scanned at a rate much faster than the other. (In most CRT displays, the horizontal rate is hundreds or thousands of times that of the vertical.) Note that the waveform has a relatively slow portion, during which the active video is being displayed, followed by a rapid return of the beam to the other side of the screen (the "retrace period"). Also, the waveform in most cases is not a pure "sawtooth" shape (the dotted line in the above figure), but rather is modified into an "S" shape (somewhat exaggerated here for clarity). This compensates for geometric distortion that would otherwise result from the mismatch in the CRT screen radius and the radius of deflection of the electron beam.

the deflection cycle in which the scan is rapidly returned from one side of the tube to the other (or from the bottom of the scan to the top); these are the rapid transitions in the deflection waveform, and are commonly referred to as the *retrace* periods. The time during which the beam is kept in cutoff, or blanked, must be longer than the actual time required for the retrace, as the deflection coils must be permitted to complete the retrace and stabilize before beginning the next line or frame of the "active" scan.

This is a greatly simplified description of the operation of a monochrome CRT, but it will suffice for our purposes at this point. Before moving on to the finer details of CRT displays, we next consider how color operation is achieved in this technology.

4.3 Color CRTs

The earliest attempts at producing color displays using the CRT employed the field-sequential color technique described in the preceding chapter. Separate scans of the fields containing the red, green, and blue components of the color image are displayed in sequence on a CRT with a white phosphor, and viewed through a synchronized color filter wheel. This method can produce a high-quality color image, but is limited by the rate at which the color wheel can operate and its size. Obviously, large-screen color displays using this technique are not practical. (A more recent variant of this technique employs an electrically switched filter panel placed over the CRT, which permits both higher frame rates and larger screen sizes; however, the cost of the filter has restricted its use to only very high-end applications.) Another method of producing color in the CRT, again very seldom used in current designs, is

58 DISPLAY TECHNOLOGIES AND APPLICATIONS

to use a multi-layer phosphor screen, with each layer producing a different color of light. Higher electron-beam energies result in the excitation of more layers of the phosphor, producing a range of output colors. Such *beam-penetration* color CRTs, however, have a very limited color range, and clearly do not provide a means of varying both the color and the intensity of the light produced. Their use has been restricted to a very small range of industrial applications.

By far the most successful color CRT type is the now-common tricolor "shadow mask" CRT, introduced by RCA in the 1950s as part of their proposed color television system. This basic concept, embodied in a fairly wide range of different tube designs, is used in practically all color CRT displays to this day. In this type of color CRT, the phosphor screen is actually made up of a very large number of discrete dots, stripes, or other patterning of several different phosphors – one for each of the primary colors (typically red, green, and blue) to be produced by the tube. Examples of several different phosphor patterns are shown in Figure 4-4.

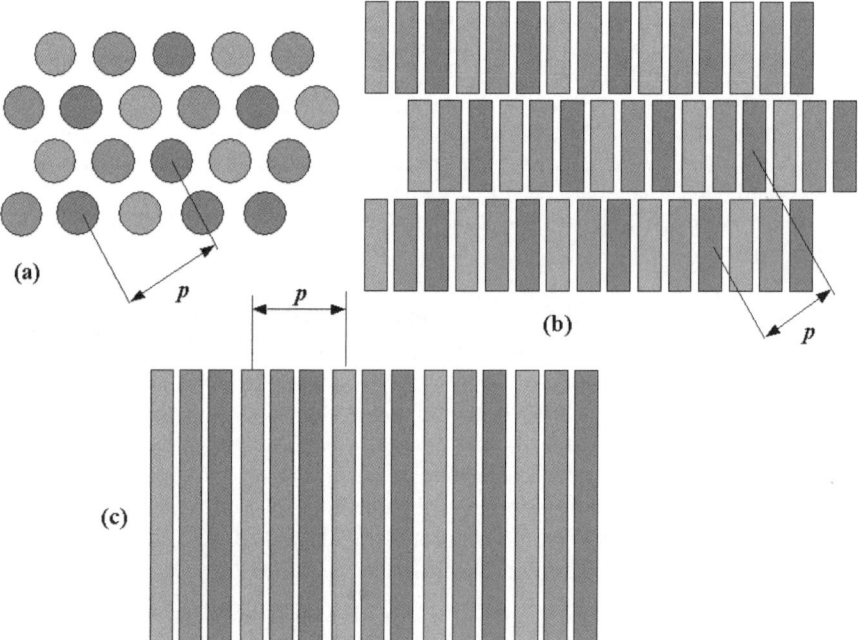

Figure 4-4 Color CRT phosphor patterns. The early color CRTs used three electron guns arranged in a triangle in the neck, and the phosphors of these "delta-gun" tubes were also arranged in triangular patterns (a). Later tubes, including most current designs, placed the three guns in a horizontal line, and the "in-line" types use a pattern of rectangular holes as shown in (b). Sony Corp. pioneered a different form of mask, in which the beams were isolated via a grille of flat metal strips under tension; the "aperture-grille" types, including Sony's own Trinitron™ tubes, employ a distinctive phosphor arrangement of continuous vertical stripes. In all types, the phosphor "dot pitch" (*p* in the above drawings) is defined as the center-to-center distance between adjacent phosphors of the same color; this measurement imposes a fundamental limitation on the resolution capabilities of the tube. However, note that the pitch is measured along a diagonal in the conventional shadow-mask types, and along the horizontal in the aperture-grille types, leading to figures that are not directly comparable in terms of their implied resolution limits.

The electron gun structure is also tripled – three individually controlled beams are produced and directed at the phosphor screen. In order to properly generate a full-color image, however, it should be apparent that each beam should be allowed to excite only one of the three phosphors, thus giving us in effect three independent displays in one – three images, one in each of the primary colors, which are overlaid in a single image area. (This is the first example of the spatial color technique mentioned in Chapter 3.)

The task of keeping the beams, and therefore the images, separated falls to a structure called the *shadow mask*. This mask is built into the CRT, located just behind the phosphor screen (and generally mounted onto the faceplate glass). Constructed of metal (originally iron or steel, but today more commonly an alloy with a lower coefficient of thermal expansion), the mask has numerous holes through which the beams can pass. In fact, there is one hole for each set of three phosphor dots on the screen; with the mask so close to the screen, however, only one dot is "visible" to any beam through the mask holes, depending on the angle at which the beam strikes the mask (Figure 4-5). By varying the relative angles at which the three beams strike the mask, then, each beam can be made to excite only the phosphor dots of the correct color, and the others are blocked or "shaded" from that beam by the shadow mask. This method of color separation comes at a price, however; the mask intercepts a high percentage of the beam current (up to 80% or even higher in some designs), and so this type of color CRT requires considerably higher beam energies – and so higher anode voltages – than a similar-sized monochrome tube, in order to provide comparable brightness.

This system clearly requires some fairly sophisticated control of the beam position and trajectory, but nevertheless gave us the first truly practical full-color CRT display. The three beams must be made to strike the same area at the same time, throughout the scanning process, so that only a single spot of the proper color is perceived by the viewer. This is referred

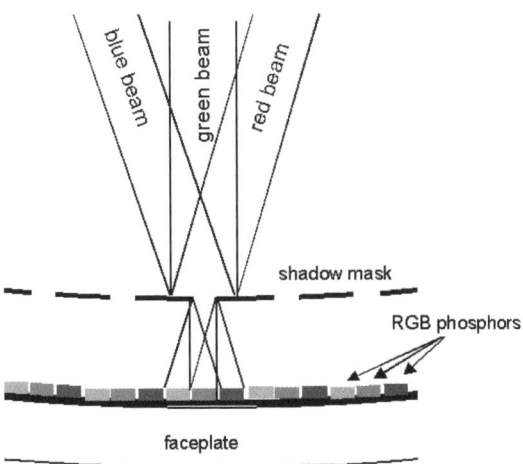

Figure 4-5 The action of the shadow mask. In a tricolor CRT, the shadow mask acts to separate the effects of the three electron beams on the phosphor dots of the screen. Each beam is made to intercept the mask at a slightly different angle, and so the holes of the mask expose only the desired color of phosphor to a given beam. This diagram is somewhat simplified, and not to scale; for one thing, the electron beams are typically larger in diameter than shown here, such that each may illuminate multiple phosphor dots of that color simultaneously.

to as the *convergence* of the beams, and the failure to bring the beams together at all times results in *misconvergence,* the appearance of primary-colored "fringes" around areas of otherwise correct color, or the visible "breakup" of fine details – such as thin lines – into three distinct primary-color images. Even if the beams come together at the proper point, it is still possible that they will not be properly separated by the mask, and one or more beams will then illuminate phosphor of the wrong color. This is referred to as a *beam mislanding* or *color purity* error, and results in areas of incorrect color being produced. Common problems which can result in this form of error include unwanted magnetic fields, either externally produced or resulting from the buildup of magnetism in the mask or other portions of the display, and/or the physical distortion or shift in location of the mask itself. Mask distortion due to heat is a very common source of beam mislanding; as mentioned above, the mask intercepts much of the energy of the electron beams, and so is prone to heating from this source. Localized expansion of the mask, as might occur when the image has a small, isolated area of high brightness, results in a problem called *doming*; this is what caused CRT manufacturers to switch to mask materials which have less thermal expansion, or to use other means of better controlling the position of the mask surface.

One such solution was pioneered by Sony Corp. in the 1960s. Instead of a conventional shadow mask, the Sony "Trinitron™" CRT uses an array of flat metal strips or wires, held under tension by a metal frame (the strips run vertically as seen by the viewer of the CRT). The slots between these strips act in the same manner as the holes in a conventional mask, this time with phosphors patterned in continuous stripes. This type of mask structure was named an "aperture grille" by Sony, to distinguish it from the conventional shadow mask, and for many years was unique to Sony tubes. (The aperture grille design has since been licensed to other manufacturers, notably Mitsubishi.) This design results in two visible distinguishing features for the aperture grille tubes. First, since the grille is under tension vertically, the faceplate may be made flat (or nearly so) along this axis, resulting in a cylindrical shape for the screen. Second, the individual strips of the aperture grille would, if not restrained, be subject to vibration and possible damage if the tube were subjected to mechanical shock. To prevent this, very thin *damper wires* are welded to the aperture grille, running horizontally across the strips. Since these wires do block a small amount of the beams, they are visible are very faint horizontal lines across the image. (Smaller tubes may have only one wire, while the larger ones generally have two.) At normal viewing distances and with most image content – large white areas are obviously the worst case – these artifacts are generally ignorable.

4.4 Advantages and Limitations of the CRT

As might be expected of a display device of such relative complexity, the CRT suffers from a number of limitations and shortcomings. However, it also represents a very mature technology, and one in which these problems have for the most part been addressed. This maturity also results in one of the CRT's major advantages over other technologies – it is by far the least costly alternative in a wide range of potential applications.

Another major advantage of the CRT display, from a functional perspective, is its ability to adapt quickly and easily to a wide range of image formats. Unlike most other display technologies, the CRT does not have an inherent, fixed format of its own. The number of scan lines in the raster is determined solely by the scanning frequencies used in the beam deflection system. Even in the case of the color CRT, the phosphor triads of the screen do not represent fixed pixels; there is neither a requirement, nor any mechanism provided, to ensure that the samples in the image in any way align with these. This has permitted the development of "multifrequency" CRT displays, capable of accepting a wide range of image formats. Other advantages of the CRT include a wide viewing angle, a reasonably wide color gamut, and the fact that it is an emissive display and so does not rely on any external light source.

Limitations or disadvantages of the CRT include the difficulty in obtaining an acceptable level of geometric distortion, linearity, and focus, plus the aforementioned color problems of misconvergence and color purity. Most of these are complicated by the fact that the distance from the electron gun (or more precisely, from an imagined "point of deflection" at which the beam is thought of as being "bent" onto the correct trajectory) to the screen varies considerably from the center of the screen to the corners. This means that there will be no one fixed focus or convergence correction which will be optimum for the entire screen, and also results in an inherent geometric distortion if a straightforward linear-ramp waveform were to be used for the deflection currents. All of these require compensation in the form of complex waveforms which correct the focus, beam position, etc., vs. the beam's position on the screen. In modern monitor designs, however, these are relatively easy to produce via digital synthesis.

More difficult to correct, however, are the CRT's inherent size, weight, and relatively high power requirements. Still, given its low cost and generally good performance across the board, the CRT remains the display of choice in a wide range of applications; the only display needs which have not somehow been addressed by this technology are limited to a relatively narrow range of portable systems, or a few specific environments (such as those with high ambient magnetic fields). CRTs have been built small enough (approx. 2.5 cm diagonal) to permit their use in such applications as viewfinders or head-mounted displays; they have also been used as direct-view displays, as in a monitor or television, up to 1 m in diagonal measurement and even higher. Special CRT types are also used as the basis of projection systems (to be discussed in more detail later in this chapter) providing images several meters or more across, and even more specialized relatives of the CRT are used as the basic element making up extremely large display systems for sports arenas and similar venues. After almost 100 years as a practical display device, the CRT is still in widespread use and appears likely to remain a significant portion of the overall display market for many years to come.

4.5 The "Flat Panel" Display Technologies

The vast majority of commercially viable, non-CRT display technologies may be referred to generically as "flat-panel" displays, or FPDs. These types have several characteristics in common. First, they are, as the name implies, physically flat and relatively thin devices, certainly with much less overall depth than the CRT. Second, they are fixed-format displays; they are in general composites, a matrix of simple display "cells" or "pixels", each producing or controlling light uniformly over a small area. Each of the individual cells of such displays is driven such that they correspond one-to-one with the samples or pixels of the image. This is a fundamental difference with the operation of the CRT, and accounts for much of the differences in both the appearance and the interface requirements for the two classes of display.

62 DISPLAY TECHNOLOGIES AND APPLICATIONS

Since these displays *do* consist of a fixed array of individual elements, it is natural that they be organized and driven in a manner analogous to the array of pixels resulting from the sampling of the image, as discussed in Chapter 1. In short, these elements are typically arranged and driven in a regular rectangular array of rows and columns. Each element may be accessed, or *addressed*, simply by selecting the appropriate row and column. This matrix-addressing scheme is common to practically all flat-panel display types. One result is that, unlike the CRT, which readily accepts a continuous video signal, the flat-panel types require that the video information be provided in discrete samples, corresponding to the discrete pixel structure. (This discrete structure has led some to refer to these types of displays as "inherently digital"; this belief is, however, in error. The fixed-format structure places additional demands on the timing or sampling of the incoming information, but does not necessarily require either digital or analog encoding of this information. This is covered in much more detail in Chapter 7.)

A simplified display employing a matrix of row and column electrodes is shown in Figure 4-6. In this case, we are using separate light-emitting diodes (LEDs) located at the intersection of the rows and columns; a given LED will light only when the proper row and column electrode pair is selected and driven. Thus, each LED may be considered as forming one pixel of this display. In most practical displays, however, the active display element will commonly be located between the electrodes, with the electrodes themselves carried on substrates defining the "front" and "back" (or "top" and "bottom") of the display device. As these *are* displays, which will either produce or transmit light at each pixel location, at least one set of electrodes must typically be constructed on a transparent substrate, usually glass. In many cases, the electrodes themselves are transparent, made from very thin but still suffi-

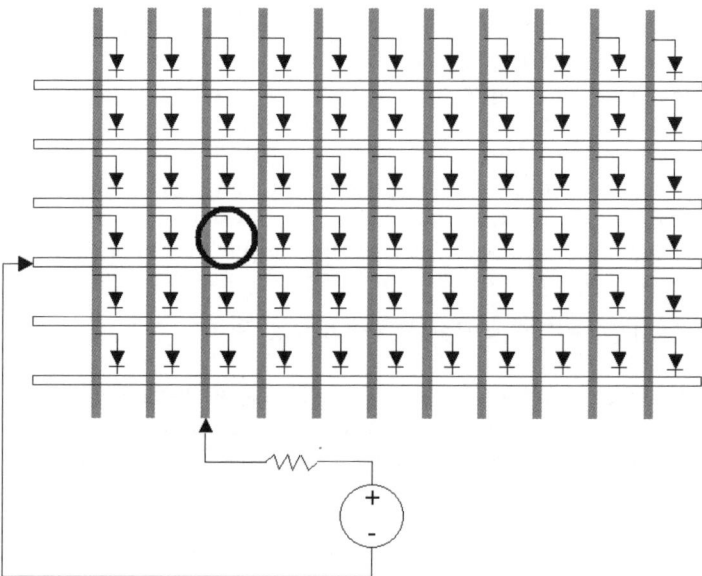

Figure 4-6 Row-column matrix addressing. In this simple display, the picture elements (LEDs) are located at the intersection of row (white) and column (dark) electrodes. Driving a given pixel requires simply connecting a source to the row and column electrodes that intersect at the desired location, as shown.

ciently conductive layers of metal film or metallic oxides. (One of the most common materials for this purpose is indium-tin oxide, generally referred to as "ITO".) Drive circuitry is located around the periphery of the panel, such that the electrodes may be selectively energized to drive the desired pixel. In operation, with the video information provided in the previously described raster-scan order, each pixel would be written to the column electrodes while the row electrode corresponding to that line is selected. However, this shows a potential problem with this simplistic drive scheme; each pixel will be active only during that brief period of time in which it is being driven, and will return to its inactive state as soon as the next pixel is selected. In a display with a large number of pixels, it becomes very difficult to drive each pixel long enough to ensure that it is at least seen as in a stable state.

One way to correct for this is to load the video information into a storage element, such as a shift register, and then write all of the pixels in a given row (or at least part of a row) at the same time. This permits each pixel to be driven for a line time, rather than a pixel time, before the next row must be written. With each pixel enabled for a longer time, the display can appear more stable (with higher brightness and contrast) to the viewer. There is still a limit to how far such a drive scheme can be extended, however; with increasing line counts, the "on" time for each pixel may again be decreased to the point at which the display would not be usable. Increasing the pixel count and/or the physical size of the display also leads to increased capacitance in the row and column electrodes, making it more difficult to drive them quickly. This is especially a problem with those display technologies which require relatively high voltages and/or currents.

The basic problem faced here is the same as in the CRT; any given area needs to be driven long enough, and/or with sufficient intensity, to register visually, and must repeatedly be driven or refreshed so as to create the illusion of a steady image. But increasing the pixel count (or the physical size of the screen) also increases the difficulty of achieving this. In the case of the CRT, the problem is partially ameliorated through the persistence of the phosphor; it continues to emit light for some period of time after the direct excitation of the beam is removed. Some of the flat-panel technologies provide similar characteristics (in some cases, exactly the same: phosphor persistence), but a more common solution is to design "memory" into each pixel. In other words, the panel will be designed such that each pixel, once addressed and driven, will maintain the proper drive level on its own. This is typically achieved by constructing a storage element, usually comprising at least a transistor and a storage capacitor, at each pixel location. Flat-panel displays employing such schemes are generically known as *active-matrix* displays, due to the active electronic elements within the pixel array itself, while the simpler system in which the electrodes drive the picture elements directly become the *passive-matrix* displays.

The fact that these displays are of a fixed pixel format is, again, one of the chief functional differences between this class and the highly flexible CRT. There are a certain fixed number of pixels in an FPD, and so it must *always* be driven at its "native" format. In order to use a flat-panel display in applications that traditionally have used the CRT – such as computer monitors – it is often necessary to add intermediary circuitry which will convert various incoming image formats to the single format required by the display. Such image-scaling may be done using a variety of techniques, some more successful than others in providing a "natural-looking" image. The FPD also, again unlike the CRT, may be restricted to a relatively narrow range of frame rates, requiring also that *frame-rate conversion* be provided for even if the input is of the correct spatial format. This can again result in differences in image appearance between the CRT and FPD displays, especially if moving images are to

64   DISPLAY TECHNOLOGIES AND APPLICATIONS

be shown. Finally, even if no spatial or temporal conversions of the input image are required for display on the FPD, the simple fact that its pixels are of a fixed and well-defined shape results in a significant difference in appearance between the image on an FPD-based monitor and the same image as seen on a CRT.

At this point, we review the fundamental operation of several of the more popular non-CRT technologies. While this is certainly not be an exhaustive, detailed description of all FPD operating modes, it should serve to give some idea of the wide and varied range of types which are offered under this general classification.

4.6 Liquid-Crystal Displays

By far the most common of the flat-panel display technologies is the liquid crystal display, or LCD. Now used in everything from simple calculator, watch, and control panel displays to sophisticated full-color desktop monitors, the LCD is almost synonymous with "flat-panel display" in many market at present.

Unlike the other flat-panel types to be reviewed here, the LCD is a non-emissive display. It acts only to modulate or switch an external light source, either as that light passes through the LCD (the transmissive mode of operation), or as the light is reflected from the LCD structure (operating in reflective mode). There are numerous specific means through which LCDs control light, but all operate in the same fundamental manner – the arrangement of molecules within a fluid is altered through the application of an electric field across the material. The effect on the light transmission or reflection may be through phase or polarization changes, the selective absorption of light, or by switching between scattering and non-scattering states.

Probably the most common operating mode, and certainly one of the most useful in explaining the basic of LC operation, is the *twisted-nematic* mode. Liquid crystals are so named because the molecules of the liquid tend to align themselves in ordered arrays, as in a solid crystalline substance. These materials are also generally organic compounds in which the molecules are relatively long and thin; for the purposes of analyzing their electro-optical behavior, they may be though of as extremely small rods in suspension in a fluid medium. In the nematic state, these molecules – the "rods" – align themselves in layers throughout the fluid, and such that those in adjacent layers tend to be oriented in the same direction. The molecules will also align themselves with fine physical structures in the substrate of the display. (In practice, these are created by physically rubbing a relatively soft layer of material deposited on top of the glass substrate, creating a very large number of very fine scratches, all aligned in the same direction.) If the liquid crystal material is placed between two such substrates, the tendency of the molecules to align themselves with those above and below, plus the tendency of the outermost layers to align with the "rubbing direction" of the substrate, a sort of helical arrangement of the molecules through the liquid crystal occurs, as shown in Figure 4-7a. This helix has the effect of twisting the polarization of light passing through it by 90°. If crossed polarizing layers are then placed on either side of this structure, light can still pass through by virtue of the polarization rotation.

In Figure 4-7b, however, the effect of placing an electric field across the material is shown. The LC molecules' tendency to form the helical structure described above can be overcome by a field of sufficient rubbing directions orthogonal to one another, the tendency of the molecules of a given layer to align with those above strength, and the molecules will

LIQUID-CRYSTAL DISPLAYS 65

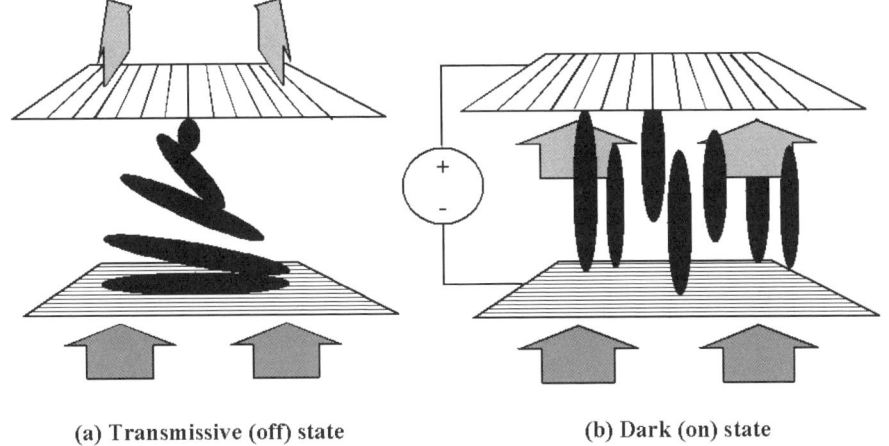

(a) Transmissive (off) state   (b) Dark (on) state

Figure 4-7 Basic twisted-nematic (TN) liquid-crystal operation. In the off state (a), with no electric field applied across the cell, the liquid-crystal molecules align with each other and with the "rubbing" direction on both substrates. With the substrates crossed as shown, the molecules then form helical structures; this is the *twisted-nematic* state. This helical structure is also optically active, and will twist the polarization of light by 90° as it passes through the cell. If the substrates also carry crossed polarizing layers, this polarization-rotation action will permit light to pass through the cell, and thus this example is transmissive in the off state. However, if an electric field is applied across the cell, the LC molecules will align with the field, destroying the helical structure and thus eliminating the polarization rotation. Thus, light polarized by the bottom polarizer will not pass the upper, and the cell appears dark. This change of state is completely reversible, simply by removing and applying the electric field, and so will form the basis for a practical display device.

then instead align themselves with the field. This destroys the helical structure, and with it the polarization rotation effect. Light that previously passed through the second polarizing layer is now blocked. Removal of the electric field permits the helical structure to re-form, and light once again will pass through. The transition between the two states is not especially abrupt, as may be seen in the graph of light transmission vs. applied voltage for a typical LC cell, in Figure 4-8. This gives the TN LCD the inherent capability of producing a range of intensities, or a "gray scale", although the shape of the response curve is less than ideal.

It should be noted at this point that the action described above depends solely on the magnitude of the electric field across the LC cell, not on its polarity; in other words, the liquid crystal display would operate as shown with the source connected in either direction. This turns out to be very important, as it was discovered early in the commercial history of LC displays that the display would be damaged if exposed to a long-term net DC voltage across the cells. This is due to

Many simple liquid-crystal displays are of the passive-matrix type. However, to provide sufficient contrast, the LC materials and cell design used for these result in relatively slow operation. This is necessary so that the individual pixels will remain in the desired state long enough between drive pulses, but makes this type ill-suited to applications requiring the display of rapid motion. Use of an active-matrix design enables faster response, and can result in an LCD suited to motion-imaging applications. Most LCD panels used in high-end applications, such as desktop monitors and notebook computers, are of the active-matrix type, also known as "TFT-LCD" (for "thin film transistor liquid crystal display"; the active com-

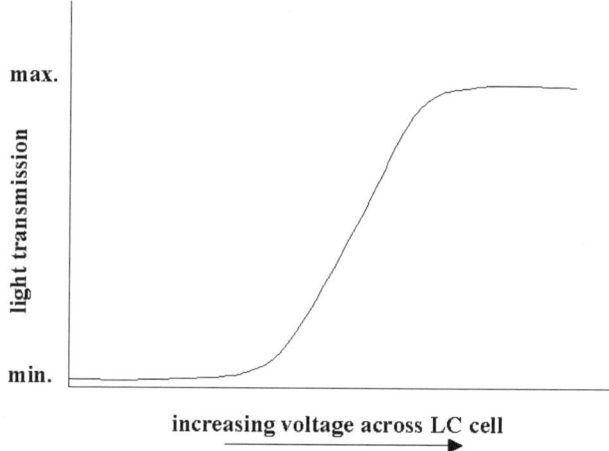

Figure 4-8 Idealized response curve of an LC cell. In this case, the cell has been designed to pass more light with the application of an electric field, the opposite of the case shown in Figure 4.7. Both types are in common use.

ponents are constructed via thin films deposited directly onto the display substrate). However, in addition to the added complexity of the active-matrix pixels, this type generally requires more power than the passive-matrix LCDs, making the passive-matrix often the more attractive choice in power-critical portable applications.

The simple TN-LCD also suffers from a limited viewing angle, meaning that the appearance of the display is optimum only through a certain limit range of angles, centered around a line roughly perpendicular to display surface. (It should be noted that in almost all practical LC displays, the direction of maximum contrast will not be precisely normal to the plane of the display.) This results from the nature of the electro-optical effect behind the operation of the display, which clearly functions best along the axis of the helical arrangement of molecules. Light passing through the structure at an angle does not experience the distinct change in transmission states, and so the contrast of the display falls off rapidly off-axis. This can be compensated for, to some degree, through the addition of optically active film layers on top of the basic TN panel, or through the use of different LC modes. In the passive-matrix types, the most common approach is to employ the "super-twisted nematic", or "STN" mode. Without going into unnecessary detail, this mode involves a 270° twist in the helical arrangement of the molecules, rather than the 90° of the standard TN, and provides both higher contrast and a wider viewing angle, along with a much sharper response curve.

Active-matrix LCDs may also use other LC modes rather than the simple TN (with or without compensating film) in order to obtain improved contrast and viewing angle. Two of the more common in current displays are the in-plane switching, or IPS type, and the vertical linear alignment (VLA) mode, both shown in Figure 4-9. These modes are not used in passive-matrix displays, due to their requirement for more complex pixel structures and/or higher power requirements, both of which are contrary to the low-cost/low-power aims of most passive-matrix designs. Both offer greatly improved viewing angle and response times over the conventional TN mode. However, the higher power requirement has limited their use to date to panels intended for desktop monitor or television applications (as opposed to notebook PC applications, which are of course more power-critical). More recently, both

LIQUID-CRYSTAL DISPLAYS 67

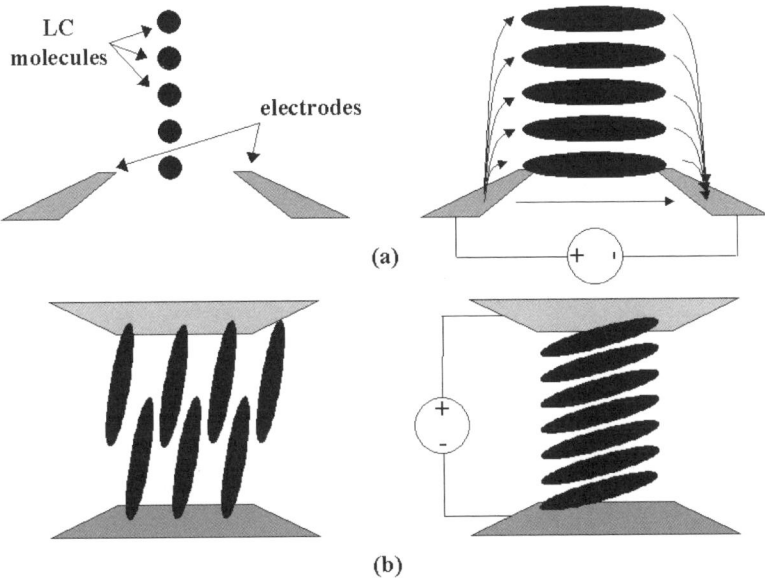

Figure 4-9 The in-plane switching (IPS) and vertical linear alignment (VLA) LC types. The IPS (a) uses LC molecules aligned in the same direction between coplanar electrodes, rather than the helical arrangement of the TN type. When an electric field is generated between the electrodes, the molecules rotate to align with the field. This 90° (approximately) rotation may also be used to control light passing through the cell, based on polarization. In the VLA type, the molecules are aligned vertically in the off state, but when the field is applied the alignment changes as shown.

types have evolved into "multi-domain" variants; these address color and contrast uniformity issues in the original IPS and VLA types, which resulted from the fact that the LC molecules do not actually swing exactly 90° between states as shown in Figure 4-9. The multi-domain solution is shown in Figure 4-10, using the vertically aligned type as an example. In this approach, the display area is broken into many small areas, each with a different orientation of the LC molecules as shown. When viewed at a normal distance, the color errors introduced by each domain, as viewed from a given angle, cancel each other and the display appears uniform on average.

Both active- and passive-matrix designs may be used in either transmissive or reflective displays. Transmissive-mode displays most often incorporate an integral "backlight" structure, as shown in Figure 4-11a. The light source itself may be one or more small fluorescent tubes (most often of the cold-cathode fluorescent, or CCFL, type), LEDs, or an electroluminescent panel. To provide acceptable brightness uniformity, some type of diffusing layer is generally also included. The backlight, along with the additional power supply generally required to drive it, again increases the complexity and cost of the complete display system, and so may limit the applicability of such displays to relatively high-end applications. In the reflective LCDs (Figure 4-11b), ambient lighting is used to view the display; rather than a backlight, a reflective layer is placed "behind" the LC panel (as seen by the viewer). Due to the light losses involved in two passes through both polarizing layers and the LC material itself, reflective displays generally provide poor contrast compared to their backlit transmissive counterparts, but still are often the preferred choice where low power con-

68 DISPLAY TECHNOLOGIES AND APPLICATIONS

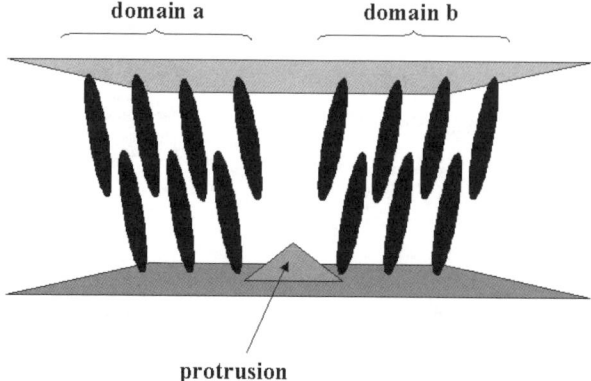

Figure 4-10 Multi-domain VLA. To compensate for the non-uniformity of the display if the VLA mode is used, the display may be divided into multiple small domains (at least two per pixel) which differ by having opposite *pretilt angles*. This is achieved by adding small protrusions to the lower substrate; the opposing domains result in a uniform appearance when viewed together.

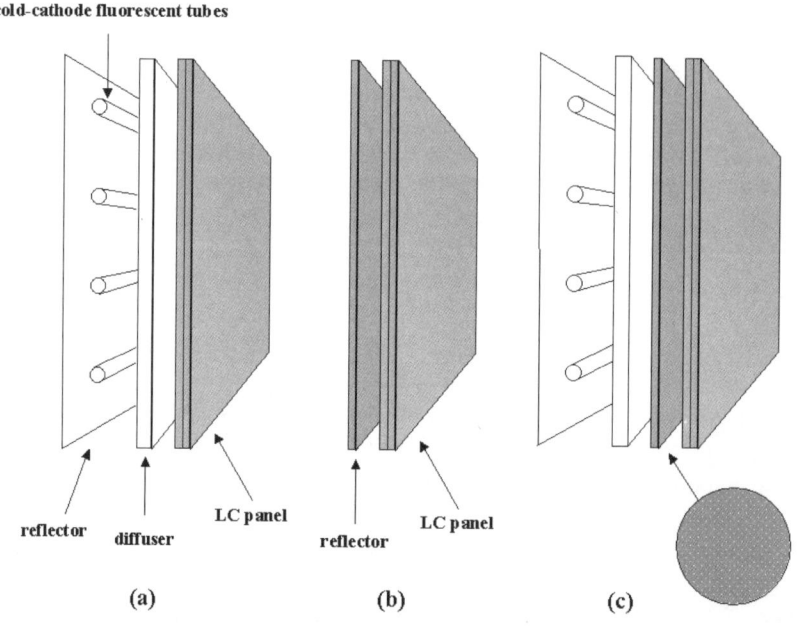

Figure 4-11 Transmissive, reflective, and transflective LC displays. In the transmissive type (a), the light source (*backlight*) is located behind the LC panel itself, and is typically comprised of fluorescent tubes, an electroluminescent panel, or LEDs. In the reflective type (b), common in low-power applications, ambient light is used, passing through the panel from the front and then being reflected back through via a reflective surface behind the panel. The transflective type (c) is a compromise, combining elements of both (a) and (b). Primarily used in the reflective mode, the reflector is made to pass some light from a backlight (usually by making the reflector from a mesh-type material), which is turned on only when insufficient ambient light is available.

sumption is of paramount concern. A hybrid type, the *transflective* display (Figure 4-11c), typically adds a limited-use backlight to a normally reflective display, to enable occasional use in low-ambient-light environments.

Making the LCD into a full-color display is conceptually very simple. With the exception of certain LC modes which involve wavelength-specific effects, this type of display has little or no inherent color, instead passing or reflecting an external light source essentially unchanged. In order to make a full-color display, then, all that is required is the addition of color filters over the LC cells, and the use of a white light source. Various pixel layouts have been used in the design of color panels, but one typical arrangement is simply to place three complete pixel structures – now becoming the three primary-color sub-pixels – into a single square area that is now the complete full-color pixel. Besides the additional complexity in the panel design (which now has at least three times as many "pixels" as in a monochrome panel of the same format), the fabrication and alignment of the color filter layer adds considerable cost to the display.

An alternative method of producing a color LC display is to employ three stacked panels with filter layers corresponding to the *subtractive-color* primaries (cyan, magenta, and yellow). As a reflective display, this permits full-color operation by selectively absorbing these primary colors.

The term "LCD" covers a wider range of specific technologies than any other of the flat-panel types. There are a very wide range of liquid-crystal types and operating modes which have not been covered in detail here, with varying advantages, disadvantages, and unique features. Some provide very high contrast; some provide bistability, and with it the ability to retain an image even after electrical power is disconnected from the display. However, LCDs have until very recently generally been limited to small-to-medium sized applications; from roughly 2.5 cm (1 inch) (or less) diagonal up to perhaps 63 cm (25 inches) at the upper end. The larger sizes are almost exclusively the domain of the active-matrix types, and the size is for the most part limited by the ability of manufacturers to process sufficiently large panels while maintaining acceptable uniformity and defect counts. There has, however, been some success demonstrated in *tiling* LCDs, using panels specifically designed to be placed adjacent to one another in order to form a much larger complete display system.

4.7 Plasma Displays

Plasma displays are closely related to the simple neon lamp. It has long been known that certain gas mixtures will, if subjected to a sufficiently strong electric field, break down into a "plasma" which both conducts an electric current and converts a part of the electrical energy into visible light. This effect produces the familiar orange glow of the neon lamp or neon sign, and can readily be used as the basis of a matrix display simply by placing this same gas between the familiar array of row and column electrodes carried by a glass substrate (Figure 4-12a). A dot of light can be produced at any desired location in the array simply by placing a sufficiently high voltage across the appropriate row-column electrode pair. The plasma display panel, or PDP, is clearly an emissive display type, in that it generates its own light. However, unlike the CRT, there is no easy means of controlling the intensity of the light produced at each cell or pixel. In order to produce a range of intensities, or a *gray scale*, plasma displays generally rely on temporal modulation techniques, varying the duration of the "on" time of each pixel (generally across multiple successive frames) in order to provide the appearance of different intensities.

Plasma displays may use either direct current (DC) or alternating current (AC) drive; each has certain advantages and disadvantages. The DC type has the advantage of simplicity, both in the basic structure and its drive, but can have certain unique reliability problems owing to the direct exposure of the electrodes to the plasma. In the AC type, the electrodes may be covered by an insulating protective layer, and coupled to the plasma itself capacitively. This results in an interesting side-effect; residual charge in the "capacitor" structure thus formed in a given cell of the AC display in the "on" state pre-biases that cell toward that state. Even after the power is removed, then, the panel retains a "memory" in those cells which were on, and the image can then be restored at the next application of power to the panel.

Color is achieved in plasma panels in the same way as in CRTs; through the use of phosphors which emit different colors of light when excited. In color plasma panels, the gas mixture is modified to optimize it for ultraviolet (UV) emission rather than visible light; it is the UV light that excites the phosphors in this type of display, rather than an electron beam. A typical AC color-plasma structure is shown in Figure 4-12b; note that in this type of display, barriers are built on or into the substrate glass, in order to prevent adjacent sub-pixels from exciting each others' phosphors. Again, due to the difficulty of directly modulating the light output of each cell, temporal modulation techniques are used to provide a "gray scale" capability in these displays.

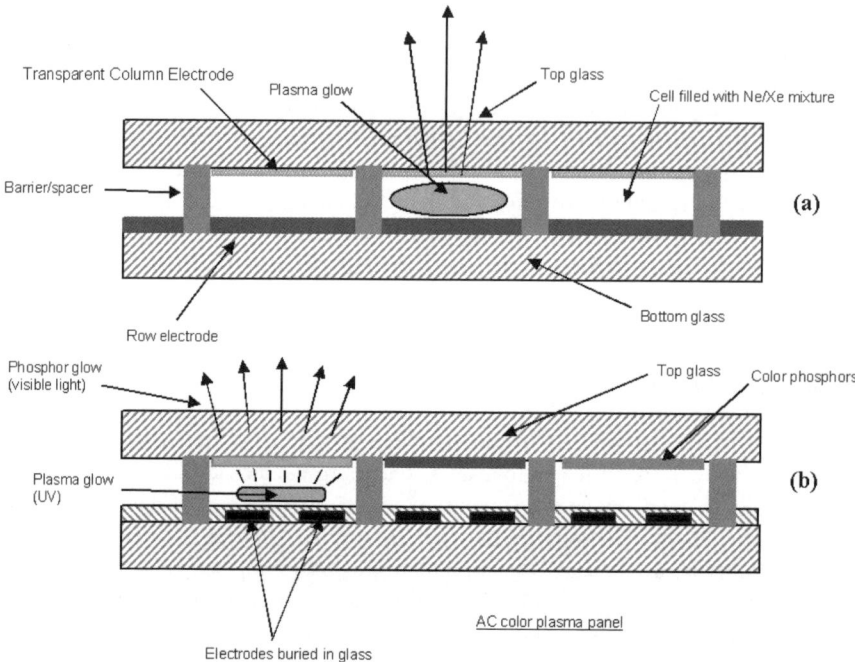

Figure 4-12 Plasma displays. In the typical monochrome plasma display panel (PDP), light is produced as in a neon bulb – a glowing plasma appears between the electrodes when a gas mixture is subjected to a sufficiently high voltage across them. In a color plasma panel, shown here as an AC type, the gas mixture is optimized for ultraviolet emission, which then excites color phosphors similar to those used in CRTs.

The fundamental mechanism behind the plasma display panel generally requires much higher voltages and currents than most other "flat-panel" technologies; the drive circuitry required is therefore relatively large and robust, and the structures of the display itself are larger than in other technologies. Owing to these factors, plasma displays have in practice been restricted to larger sizes – from perhaps 50–125 cm (20–60 inches) diagonal – and relatively low pixel counts. For this reason, plasma technology has not enjoyed the high unit volumes of other types, such as the LCD, but has seen significant success in many larger-screen applications such as television and "presentation" displays. The plasma display, especially in its color form, competes well against the CRT in those applications where, for reasons of space restrictions or environmental concerns, the much higher cost can be justified.

4.8 Electroluminescent (EL) Displays

Probably the simplest display, at least conceptually, is the electroluminescent or "EL" panel. Phosphor materials, in some cases identical to those used in the more common CRT, will glow not only when struck with an electron beam but also when subjected to a sufficiently strong electric field. Therefore, placing these materials between electrodes in the now-common row and column arrangement can produce an emissive display with an attractively wide viewing angle. In order to increase the light output, the rear substrate can be made reflective, although this can reduce the display contrast as it also reflects incoming ambient light. Another common design is to place a black (light-absorbing) layer at the rear of the structure, and/or place a circular polarizing layer on the front surface, both done in order to increase the display contrast . The circular polarizer will pass light produced by the display, but ambient light entering the panel and reflecting off the rear surface will not exit the panel due to the reversal of polarization occurring upon reflection.

Like the plasma displays, EL panels have been produced in both DC- and AC-drive versions, and are further classified by the nature of the electroluminescent layer (thin-film or powder); they can also employ either a passive-matrix or active-matrix drive scheme. Thin-film AC EL panels are currently the most popular commercial type. The technology does provide luminance control, although in this case it is by either varying the refresh frequency, or through the use of pulse-width modulation or other temporal techniques.

To date, EL technology has not seen the commercial success of the LCD or plasma types, and the use of this type of display has been for the most part restricted to certain industrial or military applications where the inherent ruggedness of the panels make them attractive. EL displays suffer from the need for high drive voltages and until recently relatively low luminance and contrast. The most common monochrome EL material (zinc sulfide, with manganese as an activator) produces a yellowish-orange light. Full-color EL panels have been very slow in coming. There are two options for producing color displays in this technology: first, a panel can be constructed using a white-emitting phosphor, and color filters applied over it as in the LCD case. The other option is to pattern individual sub-pixels of red-, green-, and blue-emitting phosphors, to form full-color triads as in the case of the CRT or color plasma types. Both of these have had their problems; the color-filter approach suffers from not having a sufficiently bright white-emitting phosphor available to tolerate the luminance reduction which comes from the filter layer. Using three separate phosphors, one for each primary, has resulted in workable full-color EL panels, but to date the blue phosphors used have not provided sufficient light output.

72 DISPLAY TECHNOLOGIES AND APPLICATIONS

At the present time, EL technology appears to be in danger of being relegated to certain niche markets and applications, and potentially bypassed altogether due to advancements in other technologies. However, the history of EL development has been one of periods of rapid progress separated by times of relative stagnation, and it would be premature to count EL out just yet.

4.9 Organic Light-Emitting Devices (OLEDs)

A relatively recent development, just now coming to the market in commercial products, is the organic light-emitting device, or OLED. As a matrix display, the OLED panel functionally most resembles the EL types – a layer of light-emitting material placed between the electrodes which define the pixel array. However, unlike EL, the OLED materials operate at a much lower voltage, approximately in the same range as is used in the liquid-crystal types. Further, high-brightness OLED materials have already been demonstrated in all colors, including white, and so full-color operation is relatively easy to achieve. OLED displays promise a combination of the best of both the EL and LCD technologies: an emissive display with a wide viewing angle, good brightness and contrast, and yet with relatively modest power requirements and low operating voltages. Electrically, the OLED's drive requirements are so similar to those of the LCD that it is expected that many LCD production lines could be converted to OLED production in a reasonably straightforward and economical manner.

While the OLED display structure superficially resembles EL, in that an emissive layer is located between the row and column electrodes, the actual OLED structure is somewhat more complex. The basis for the OLED is a layered structure of organic polymer semiconductors, arranged so as to produce light through a mechanism similar to that of the ubiquitous light-emitting diode. A typical OLED structure is shown in Figure 4-13.

Active-matrix OLED panels have already been shown to be practical alternatives to the TFT-LCD, with the potential to compare favorably with that technology in terms of both cost and performance. OLEDs are expected to compete with the LCD in practically all current

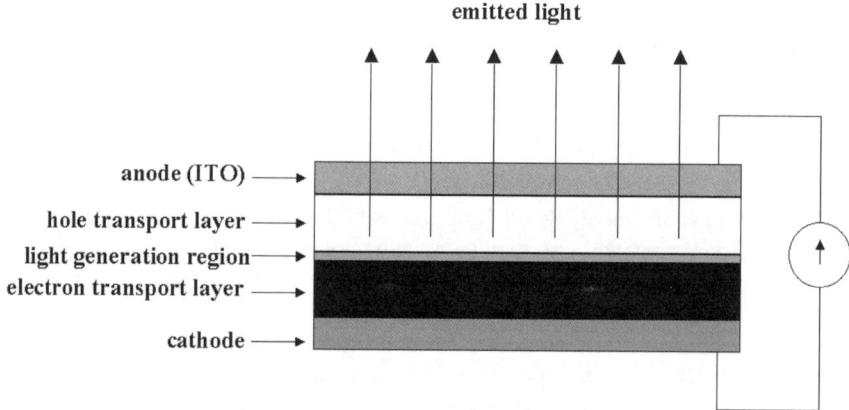

Figure 4-13 The structure of an OLED. Not shown are the glass substrates between which these layers would be built. Note that the OLED, unlike the LED, is a current-driven, rather than voltage-driven, device.

LCD applications, from small calculator, PDA, cell phone, and similar displays, through notebook computer displays and panels intended for desktop monitors. The materials also appear to be a reasonable choice for use with polymer substrates, raising the possibility of low-cost, flexible color displays. Further, the OLED technology may scale to larger panel sizes than has been possible with the LCD, making it a potential alternative to plasma in at least the lower end of the size range covered by that technology. OLEDs are still in their infancy in terms of commercial development, but definitely seem poised to take a significant share of the worldwide display market in the near future.

4.10 Field-Emission Displays (FEDs)

For almost as long as there have been CRT displays, there have been those who have tried to construct a version of this technology – a display which uses electron beams to stimulate phosphors – in flat-panel form. Probably the closest commercial technology until recently, in terms of being analogous in operation to the CRT, has been the color plasma panels, which, as noted above, use UV light to excite the phosphors. The main problem with translating the CRT to a true "flat panel" display has always been the source of the electron beam; the conventional CRT uses a heated cathode along with extremely high accelerating voltages. Besides using a considerable amount of power, this approach does not readily lend itself to incorporation in a thin display. It should be noted that several manufacturers have attempted to make flat, thin (or at least thinner than normal) CRTs using more-or-less conventional heated-cathode electron sources, with varying degrees of success, but a true flat-panel equivalent to the CRT display required the development of a "cold" source of electrons which would operate at lower voltages.

This essentially defines the distinguishing feature of a class known as *field-emission displays*, or FEDs. These are display devices which produce light via phosphors, again excited by streams of electrons, but unlike the CRT the electrons originate from emitters which do not require heating above ambient levels. The term "cold-cathode CRT" has also been used to refer to this class of display.

There have been several different approaches to the problem of designing a practical electron emitter for these devices. The quantitative measure of the ease with which electrons may be driven off (or, from a different perspective, extracted from) a given surface, material, or structure is the *work function,* which may be expressed as either the potential required to cause electron emission (the *work function potential*) or the equivalent energy requirement in joules or electron-volts. In these terms, then, what is needed is a practical emitter design with a sufficiently low work function so as to permit adequate electron emission at ambient temperatures.

In a conventional, heated CRT cathode, achieving the required work function level is generally done through the use of certain materials to form the actual emitting surface; a common example is barium oxide, a layer of which is applied to the "top" surface of the metal CRT cathode. This is in general not practical in a flat-panel device; not only so such materials fail to provide a low enough work function on their own, but there are increased requirements for emission uniformity over a relatively small area and the requirement that the emitter be capable of fabrication using available FPD processes. Therefore, other approaches are used for FEDs. Most rely on the fact that sharp edges, points, or similar structures are relatively easy places from which to extract or inject charge (owing to the concentration of charge in such regions, and the resulting concentration of electric fields there; this is similar to the principle behind the common lightning rod). Such structures are fairly easy to produce using conventional silicon-IC processing techniques, which themselves are readily adaptable to flat-panel display production.

Several emitter designs have been used in the development of FEDs. Each of these involve structures which are sufficiently small so as to permit multiple emission sites per pixel (or sub-pixel, in the case of a color display), in order to address the need for overall uniformity. The Spindt cathode (named for its inventor, Charles "Capp" Spindt, then of the Stanford Research Institute) use a conical emitter, formed through standard IC fabrication processes, as the source of the electrons, which pass through a hole in a surrounding conductive layer which acts as a control grid. A similar approach uses a long sharp edge as the emitting structure, again with structure acting as a control grid placed above and to either side of the emission site. Recently, *carbon nanotubes* have shown great promise as the electron emitters for field-emission displays. These are microscopic hollow filaments of carbon, which can be deposited on the display substrate so that many are oriented orthogonal (or nearly so) to the surface. The nanotubes are small enough, and packed densely enough on the substrate, so as to provide many emission sites per pixel or sub-pixel, and thus fulfill the requirement for uniform electron emission across the area to be illuminated.

Outside of the unique requirements for the emitters, the FED is very similar in basic structure to the other flat-panel types, most closely resembling the plasma display (especially in the color form). A cross-section of a typical color FED is shown in Figure 4-14. This display uses the familiar row-and-column addressing scheme, with drivers located on the periphery of the panel. Conventional CRT phosphors are placed on the inner surface of the front glass, and barriers are constructed between pixels and sub-pixels to isolate them from each other, and also often form the spacers between the front glass and rear substrate. FEDs have been designed in both low-voltage (up to several hundreds of volts potential difference between cathode and anode) and high-voltage (thousands of volts) forms; each is promoted as having

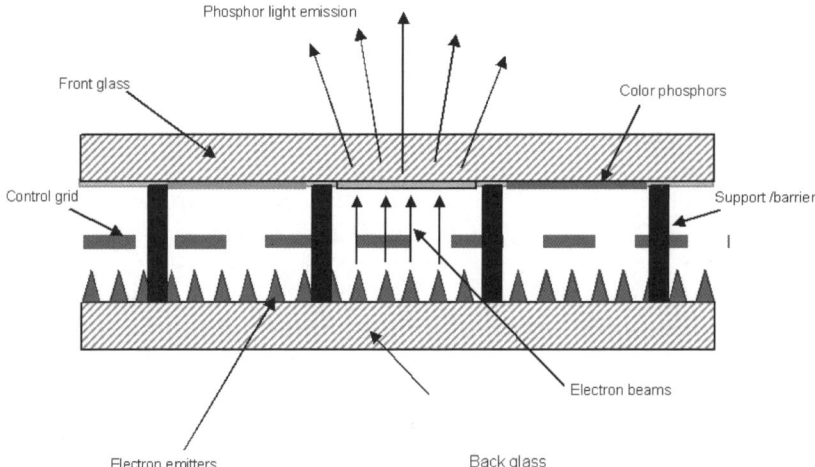

Figure 4-14 The structure of the field-emission display, or FED. This drawing is not to scale, especially with respect to the electron emitters. These are typically microscopic, and of sufficient quantity that thousands of individual emission sites may comprise a single color sub-pixel.

certain advantages. The tradeoff to be made is basically one of acceptable luminance at lower cathode currents, versus the complexities and costs involved with generating and controlling higher voltages.

As in the case of the OLED display, FEDs are just now entering the market commercially. These displays also promise high brightness and contrast at power levels and costs competitive with the TFT-LCD, and again offer the viewing angle and potential size advantages of an emissive display, requiring no backlight. FEDs do, of course, have some unique challenges, including the requirement for higher drive voltages and processes which to date have not been commonly used in smaller-size FPDs. Time will soon tell how successful these new FPD technologies are in the various display markets.

4.11 Microdisplays

Perhaps the ultimate marriage of flat-panel display and silicon IC technologies, *microdisplays* have recently opened numerous new opportunities for electronic displays. Essentially a display constructed on (or even *as*) an integrated circuit, this class covers multiple technologies sharing two main distinguishing features: they may be considered "flat-panel" displays, but they are of such a small size (generally under 5 cm (2 inch) diagonal, and often less than 2.5 cm (1 inch)) that they are not used in a conventional direct-view manner. Instead, microdisplays are either used to generate the appearance (a "virtual image") of a much larger display via magnifying optics, or the image of the display is projected onto a screen for viewing. Products using the former mode are often classed as "near-eye" applications, since the display device itself is physically located near the viewer's eye; example are camera or camcorder viewfinders, or so-called "eyeglass" or "head-mounted" display systems. As the basis for projection displays, microdisplays become the hearts of products competing with direct-view monitors, televisions, and much larger presentation display systems.

Microdisplays may be categorized into two broad groups: those which are essentially miniaturized versions of any of several of the conventional FP technologies, such as LCD or OLED displays, and those which employ micro-electro-mechanical (MEM) structures to control light. The former category is currently dominated by the liquid-crystal types, often referred to as liquid crystal on silicon (LCoS) microdisplays. These are exactly what the name implies: a liquid-crystal display constructed on top of a silicon IC. The IC is basically a slightly modified memory array, in which the individual memory cells form the storage and drive elements for the LC pixels. The most obvious modification involves a slight change to the LC process – each element in the array must be topped by a large pad of reflective metal, formed as the last metallization step in the IC processing, which acts as both the driven electrical contact for the LC cell and the light-reflecting "back" of the display. A typical LCoS microdisplay is shown in Figure 4-15. Outside of the silicon IC substrate, this display is virtually identical to its larger, direct-view cousin. An alignment layer is deposited on top of the IC's metal pads, a glass panel carrying the transparent upper electrode (ITO) is placed on top of the IC, and the cavity between glass and silicon filled with liquid-crystal material. (Note that only a single common electrode need be supplied by the glass, since the addressing of individual pixels is handled completely by the IC.) This example, as is the case with almost all LC microdisplays, is obviously a reflective display; polarized light enters through the glass, and is reflected from the metal pad at the "bottom" of each cell. The polarization of the light may either be altered by the LC or not, depending on its state, which provides the basis

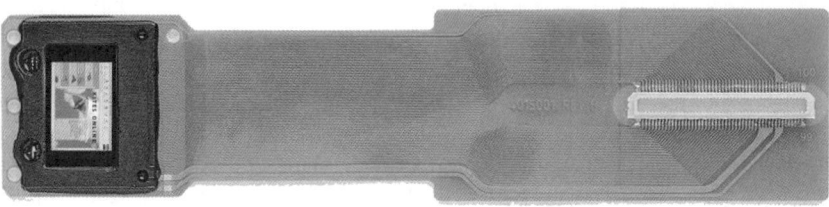

Typical LCoS microdisplay

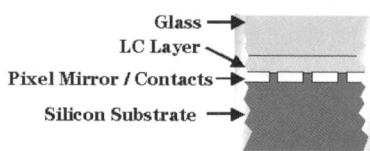

Figure 4-15 A typical liquid-crystal-on-silicon (LCoS) microdisplay. These devices are essentially LC displays built on top of a silicon IC, which provides both the lower electrodes (doubling as reflective surfaces) and the drive and interface electronics. Reflective microdisplays of this type may be used in both direct view (through magnifying optics, and then generally referred to as "near-eye" types) and projection applications. (*Illustration courtesy of Displaytech, Inc. used by permission*)

for the device serving as a display. (In most applications for displays of this type, the polarization of the light source and the polarizer through which the image are observed are generally physically separate from the microdisplay component itself, and in practice both functions are commonly provided by a single component.)

It should be noted at this point that at least one company has successfully produced a liquid-crystal microdisplay which does not operate in the reflective mode. Kopin Corporation bases their displays on specially designed silicon circuits, as in the above types, but through a proprietary process transfers the circuitry to a glass substrate, and the microdisplay then constructed on that substrate is basically a miniaturized *transmissive* active-matrix LCD.

The other major class of microdisplay are the electromechanical types, which are characterized by their use of physically deforming or altering the position of structures within the device in order to control light. The most successful example of this class to date has been the "digital micromirror device", or DMD, introduced by Texas Instruments in 1987, and which forms the heart of a technology which T.I. refers to as "digital light processing", or DLP. In these devices, each pixel is actually a movable metal mirror, which tilts back and forth under the control of electrostatic forces driven by the integrated circuit below (as shown in Figure 4-16a). The mirrors' tilt determines whether incoming light will be directed either through an optical system to the viewer (the "white" state for the pixel) or off to a "light trap" and so not seen by the viewer (the "black" state). The DMD has seen considerable commercial success in the conference-room and larger-screen projection markets, and is now one of the most serious challengers to conventional film projection in cinematic entertainment applications. Other examples of electromechanical microdisplays include Silicon Light Machines "grating light valve" device, in which strips are deformed electrostatically to control light via diffraction.

In any of these, however, full-color operation presents a unique challenge for the microdisplay, at least in the case of near-eye applications. These devices are in most cases too small to achieve color through individual color filters for each pixel, as is normally done

MICRODISPLAYS 77

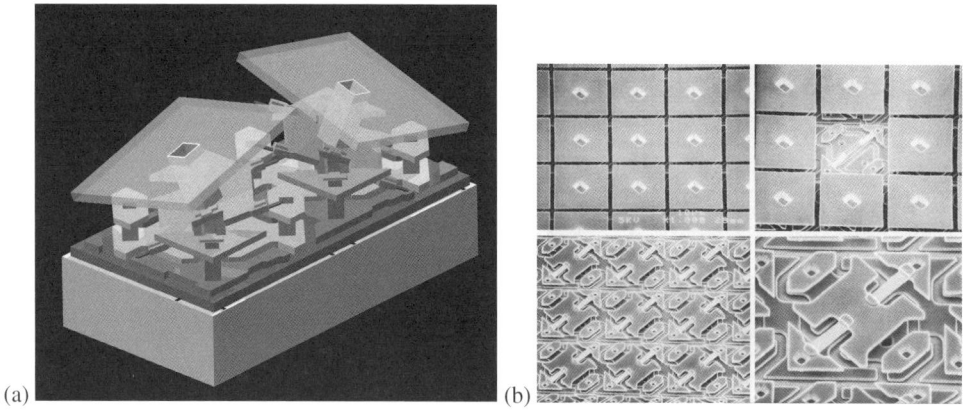

Figure 4-16 The Texas Instruments Digital Micromirror Device, or DMD. In (a), two mirrors are shown in schematic form, illustrating how they may be tilted to direct light in different directions (this occurs due to electrostatic forces from electrodes on the underlying IC. (b) shows a series of photographs of actual mirrors and their underlying support structures. (c) shows the complete device, in its packaging. (*All images courtesy of Texas Instruments, Inc; used by permission.*)

with direct-view LCDs. (And in some cases, such as the electromechanical types, the color-filter method is simply not possible.) Instead, a field-sequential color drive scheme is more commonly employed. In this method, the color image is separated into three fields, one for each primary, and displayed in rapid succession. The light source is similarly switched between the three primary colors, in synchronization with the displayed fields. This results in the appearance of full-color image, and each pixel appears as the proper color over its full area; there are no separate color sub-pixels. In near-eye applications, the light source is most often implemented as a set of light-emitting diodes (LEDs), one in each of the three primaries. Figure 4-17 shows a typical near-eye display employing a reflective LCoS microdisplay with LED illumination.

Microdisplays may also be used as the basis for display systems providing normal "desktop"-sized images, and even beyond to large, group-presentation displays, by projecting the image of the display on a screen of the desired size. Projection displays in general are described in more detail in the next section. The basic mode of operation of the microdisplay, however, is unchanged; it is simply a case of providing a considerably higher level of illumination, and then employing projection optics to image the display at the desired location.

78 DISPLAY TECHNOLOGIES AND APPLICATIONS

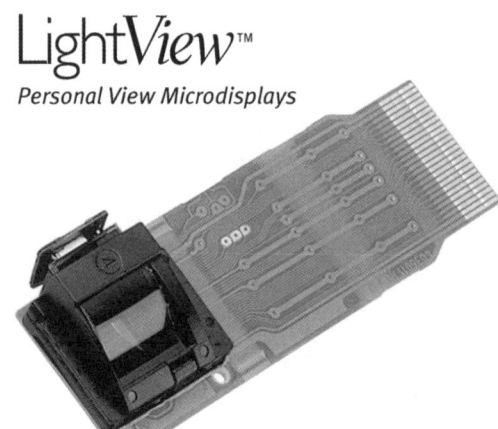

Figure 4-17 An LCoS microdisplay product with an integral light source, for near-eye use. The LCoS device itself is mounted on the flexible substrate, underneath the black plastic structure that carries LEDs (on the upper left of the housing, as seen here) and a curved film which acts as a polarizing beamsplitter. (*Picture courtesy of Displaytech, Inc.; used by permission.*)

Simple in theory, at least; as we will see, projection displays have their own unique set of challenges.

4.12 Projection Displays

Projection displays do not represent a separate class of fundamental display technology, as was the case with the above types, but rather are most often a different application of one of these technologies. This class of display is distinguished by a single common feature: rather than being directly viewed, the display device (or at least the image it creates) is imaged onto or through a passive surface (the projection screen) which is then the location of the image as seen by the user.

Again, this display class generally divides into two – those projections in which the display device itself produces the light (i.e., the display is of one of the emissive technologies), and those in which the device modulates (either in a transmissive or reflective mode) light produced by a separate source. The former type is almost without exception based on an rather extreme modification of the CRT, while the latter has in the past been dominated by liquid-crystal devices and is now a mix of LCD-based systems and those which employ one of the microdisplay technologies. Projection displays are also commonly distinguished by the optical path used; there are *front-projection* types (Figure 4-18a), in which the image source and the viewer are on the same side of the projection screen, and the image is viewed by reflection from the screen surface; and *rear-projection* types (Figure 4-18b), in which the image is projected to the rear of the screen (as seen by the viewer), and is observed through the translucent screen material.

As the length of the path from the projection optics to the screen is often not fixed, projection displays often are not specified by screen size, but rather solely by the light output by the projection optics. This figure, along with the screen size and characteristics, will deter-

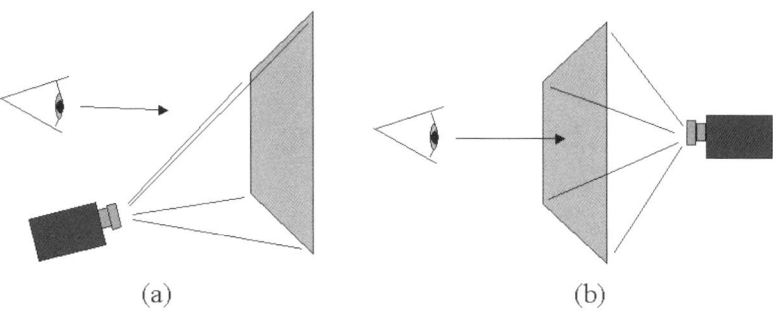

Figure 4-18 Projection display systems are commonly classified as either *front-projection* (a), in which the viewer and image source are on the same side of a reflective screen, and *rear-projection* (b), in which the viewer and image source are on opposite sides of a translucent screen.

mine the "brightness" of the image as ultimately seen by the user. This is where the person familiar with the direct-view display types will often be shocked by an analysis of a projection system; the numerous sources of loss in the path from light source to screen will result in an overall light efficiency which may appear extremely low. It is not uncommon, in some very practical projection display systems, for less than 5% of the original light produced by the source to actually make it in usable form to the screen. It is rare for this figure to get much above 10% or so in all but the CRT types. Still, projection displays are often the only viable option for many applications, such as very-large-screen display (as in cinematic presentation) or as a portable, conference-room type of display. Projection is also beginning to compete well in some traditional direct-view markets, such as the larger desktop monitors and television displays. In these applications, it is not so much the overall efficiency of the optical path which matters, but how much total electrical power will be required by the display. And from this perspective, projection systems are becoming very competitive.

4.12.1 CRT projection

CRT-based projection systems operate by projecting the image generated by the CRT(s) onto the desired projection screen surface. While this could be done with a conventional direct-view CRT type, these do not produce sufficient light to be viable as the basis for a practical projection system. This is particularly true of the common tricolor CRT, which is already suffering from an efficiency problem due to the interception of significant beam current by the shadow-mask structure. For this reason – among others – CRT-based projection displays use very specialized tubes designed specifically for this purpose. These are relatively small-screen (round screens of 17.75 cm (7 inch) and 22.75 cm (9 inch) diameter are currently the most popular), long tubes, producing extremely bright, well-focused images which are then projected by the optical system. The larger-sized tubes not only provide (usually) more light output, but also can provide better focus, geometry, etc., and so are usually the preferred choice for high-resolution displays.

Since projection cannot afford the inefficiencies of the shadow-mask color CRT, full-color operation in such systems is generally achieved in one of two ways: either a single monochrome (white) tube is used in conjunction with a color-filter wheel, in a field-sequential color system; or three separate projection tubes (one with each of the three primary-color phosphors) are employed, with their separate images combined in the optical system. Currently, the three-tube system is by far the most common. Regardless of the number of tubes used, projection CRTs also differ from their direct-view cousins in one other significant regard. Due to the very high light output required, projection CRTs operate with very high beam currents directed to a relatively small screen. This would cause excessive heating and ultimately the rapid destruction of the screen surface, were some form of dedicated cooling not employed. Projection CRT screens may be cooled via circulating fluid (water, oil, or other optically suitable liquids) around and/or over the faceplate, or (if the heat to be dissipated is not excessive) through a passive fluid system which simply couples the heat to an external heat sink.

4.13 Display Applications

We have at this point looked at the basic operation of a very wide range of display technologies, both those in current use and several which hold promise for the future. Clearly, each brings its own unique set of advantages and handicaps to the market. These qualities, which can be compared in terms of cost, size, weight, image quality, environmental suitability, reliability, and the unique operating requirements for each type, result in each being well- or poorly suited to a given application. A comparison of the relative attributes of each technology is shown as Table 4-1.

Table 4-1 Relative attributes of various display technologies[a]

Technology	Bright	Color	Contrast	Viewing angle	Power	Weight	Cost
CRT	Good to excellent	Very good	Good to excellent	Excellent	High	High	Low
Passive LCD	Poor to fair	Poor to fair	Poor to good	Poor to fair	Very low	Low	Low to medium
Active LCD	Good to excellent	Good to very good	Fair to very good	Fair to very good	Low to medium	Low to medium	High
OLED	Very good to excellent	Excellent	Good to excellent	Excellent	Low to medium	Low	High
FED	Very good to excellent	Very good	Good to excellent	Excellent	Medium	Medium	High
Plasma	Good to very good	Very good	Good to very good	Very good	High	High	V. High

[a] These are, of course, somewhat subjective ratings, but should give a good idea of the relative strengths and weaknesses of each type.

As mentioned at the start of this chapter, the cathode-ray tube, or CRT, display has been far and away the most successful single technology in history, at least to this time. The CRT offers very good image quality, great flexibility, and is available in a very wide range of sizes and types, each geared to specific applications. In almost every market in which the CRT competes, it has been the best answer in terms of its cost vs. performance – but this has come through either having no viable competition, as was the case in the early history of electronic displays, or later through retaining a significant cost advantage over any other viable type. If cost is ignored, the CRT becomes far less attractive. It is large, in terms of the overall package required for a given screen size, it is heavy, and it is relatively fragile. Today, a number of the alternative types, most of which fall into the "flat-panel display" (FPD) category, can provide equivalent or superior image quality performance, reliability, and flexibility, in a much lighter and physically smaller overall package. And as the cost penalty associated with these types continues to decline, they are taking more and more market share away from the CRT. It is far too soon to consider the CRT display as obsolete; hundreds of millions are still produced each year, and as of this writing (in 2002), that annual volume is still expected to *increase* through most if not all of the foreseeable future. Still, there is no doubt that eventually the CRT will be completely replaced by a combination of several of the "flat-panel" types.

Distinguishing among these types by the applications to which they are best suited is primarily, at this point, a matter of screen size (Figure 4-19). The two most popular broad categories of FPDs at the present time are the liquid-crystal display (LCD) and the various plasma types. LCDs for now remain dominant in applications requiring small-to-medium size screens; those needing diagonal sizes of perhaps 1 cm to 0.75 m. So far, LCDs over approximately 50–55 cm (20–21 inches) in diagonal size are considered "very large," and are very low-volume, high-cost products aimed as some very specific applications. Conversely, plasma screens require physical structures and operating voltages which do not lend them-

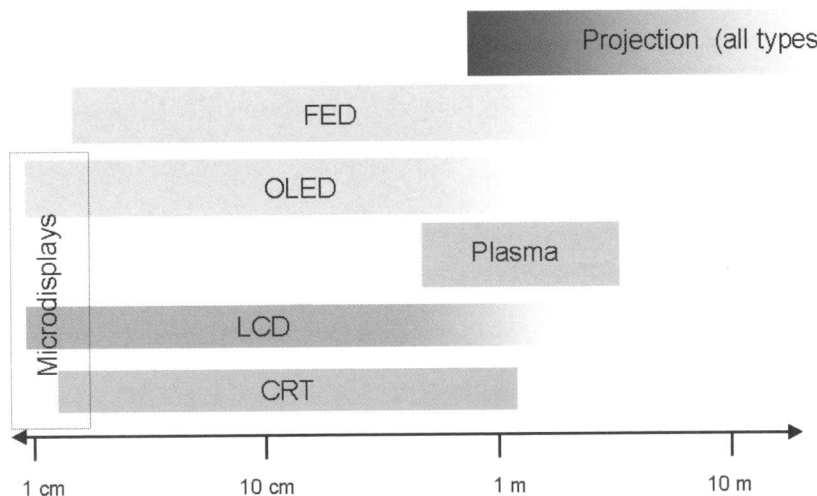

Figure 4-19 Practical size ranges (diagonal) of various display technologies. The upper size limit of many of the flat-panel technologies is not yet clear, as new advances continue to push the potential of the LCD, OLED and FED types. "Microdisplays" are, of course, not a separate technology per se, but rather an adaptation of any of a number of technologies to very small (3 cm or smaller) displays.

selves readily to being scaled down to small displays. Plasma is today the dominant FPD type in roughly the 0.5 to 1.5–2.0 m (diagonal) range. Displays above this size are almost always of the projection type, based on any of a number of basic technologies.

The newer FPDs, at least in terms of their commercialization and acceptance into the market, are the organic-LED (OLED) and field-emission (FED) types. These will be rapidly gaining in market share over the next few years, but are currently expected to enter the market at the low end (in small, portable display applications), and then grow upwards in size and capability. As such, they are direct competitors to the LCD only, and specifically the smaller, low-power types such as the STN-LCDs commonly seen today in portable equipment. These newer technologies are distinguishing themselves on image quality issues, such as viewing angle and color, with the LCD's main advantage over them coming in the area of power (at least for the purely reflective types).

5

Practical and Performance Requirements of the Display Interface

5.1 Introduction

Every engineering decision is, by necessity, a compromise. We are not given infinite resources, either in time, money, or any physical constraints (size, weight, etc.) with which to fulfill any given requirement, and of course we must work within the bounds imposed by the basic laws of physics. Display interfaces are certainly no exception. While to this point we have been concerned primarily with examining the requirements the interface design must meet – the specifications for "getting the job done" – we now must also look at these constraints, the limitations within which the display system designer must work.

Of these, some are a bit too basic or application-specific for much attention here. The constraints of cost, physical space, and so forth will vary with each particular design, and cannot be discussed other than to note that there is generally not an unlimited amount of such resources available. Other factors entering into the interface selection include at least the following, in addition to the basic task of conveying the desired image information.

- The requirement for compatibility with existing standards or previous designs.
- Constraints imposed by regulation or law; these include such things as safety or ergonomic requirements, radiated and conducted interference limits, and similar restrictions. Note that these may be either actual legal requirements for sale of a product into a given country, region, etc. (an example might be the emissions restrictions set by the US Federal Communications Commission), or de-facto requirements set by the market itself (for example, compliance with the standards set by groups such as Underwriter's Laboratories).

84 THE DISPLAY INTERFACE

- The need to carry additional information, power, etc., over the same physical connection, and the effect these will have on the video data transmission and vice versa. An example might be a "display" connection that also carries analog audio channels, power, or supplemental digital information.
- The limits of the physical connection and media which are available and which meet the other requirements imposed on the design.

These are addressed following a review of the requirements determined by the basic image-transmission task itself.

5.2 Practical Channel Capacity Requirements

Fundamentally, the display interface's job is the transmission of information. As such, its basic requirements can be analyzed in terms of the information capacity required of the interface, and from there to the bandwidth and noise restrictions on the physical channel or channels used to carry this information. As noted in the first chapter, the amount of information required is determined by the number of samples or pixels comprising each individual image, multiplied by the rate at which these images are to be transmitted, and the number of fundamental units of information (generally expressed in terms of "bits"). For example, if we assume the usual arrays of pixels in columns and rows, with each such array considered as one "frame" of the transmission, the information rate can be no lower than:

$$\text{Info. rate (bits/s)} = (\text{bits/pixel}) \times (\text{pixels/line}) \times (\text{lines/frame}) \times (\text{frames/s})$$

This represents the minimum rate; in practice, the peak transmission rate required of the interface will be somewhat higher (generally by 5–50%), as there will be unavoidable periods of "dead time" (times during which no image information can be transmitted), resulting from the limitations of the display itself, the image source, or the overhead imposed by the transmission protocol.

It is very important to note that, while the above analysis seems to apply only to "digital" systems (what with the use of bits and discrete samples), information theory tells us that any transmission system may be analyzed in the same or similar manner. Transmissions which are generally labelled as "purely analog" may still be analyzed in terms of their information content and rate in "bits" and "bits/second," through the relationships between sample rate or spatial resolution and bandwidth, and the information content of each sample (bits/sample) with such analog" concepts as dynamic range. Television provides a good example of this. The specifics of the development of the television standards of today are examined in detail later, but for now we note that the effective resolution provided by the US television standard, in the vertical direction, is roughly equivalent to 330 lines/frame. If the system is to provide equal resolving ability along both axes, the horizontal resolution should equal, in "digital" terms, approximately 440 pixels/line (given the 4:3 image aspect ratio of television). At a line rate of approximately 15.75 kHz, with roughly 80% of each line available for the actual image, this would equate to a sample rate of about 8.66 Msamples/s. As the highest fundamental frequency in a video transmission is half the sample rate (since the fastest change that can be made is from one pixel to the next and back again), this would suggest that the bandwidth required be at least 4.33 MHz, very close to the actual value used under

PRACTICAL CHANNEL CAPACITY REQUIREMENTS 85

the US standard. The number of bits required for each of these effective sample or pixels may be directly obtained from the dynamic range, which in imaging terms is equivalent to contrast.

Ignoring for the moment the problems introduced by the non-linearities of vision, the display, or the image capture system, we could simply approximate the number of bits of information per sample as the base 2 log of the dynamic range. (As there are, however, such non-linearities in the system, any value obtained in this manner should be viewed as the absolute minimum information required per pixel, at least if linear encoding is assumed.) Through this sort of a rough calculation, we might expect to adequately convey a monochrome ("black and white") television signal in 7–8 bits/pixel, and so we would expect to be able to convey a "TV-grade" black-and-white transmission through a channel capable of a data rate of approximately 60–70 Mbits/s.

The capacity of any real-world channel is, of course, limited. This limitation results fundamentally from two factors: the bandwidth of the channel, in the proper sense of the total range of frequencies over which signals may be transmitted without unacceptable loss, and the amount of noise which may be expected in the channel. Simply put, the bandwidth of the channel limits the rate at which the state of the signal can change, while noise limits our ability to discriminate between these states. If we are conveying information through a change of signal amplitude, for example, it does little good to define states separated by a microvolt if the noise level far exceeds this. The theoretical limit on the information capacity of any channel, regardless of its nature or the transmission protocol being used, was first expressed by Claude Shannon in his classic theorem for information capacity in a noisy, band-limited channel, as:

$$\text{Capacity (bits/s)} = BW \times \log_2(1+S/N)$$

where BW is the bandwidth of the channel in Hz and S/N is the signal-to-noise ratio. For example, if we have a television transmission occupying 4.5 MHz of bandwidth, and with a signal strength and noise level such that the signal-to-noise ratio is 40 dB (10,000:1), we can under no circumstances expect to receive more than about 60 Mbits/s of equivalent information, roughly what we believed was required for this transmission. (The analysis of these last few paragraphs, while far from rigorous, does tell us something about the approximate channel requirements for video transmission, but also provides significant clues as to how television and similar "analog" systems behave in the presence of noise. Consider the effect of a reduced signal-to-noise ratio in this system, and the actual experience of observing a television broadcast under high-noise conditions.)

Standard broadcast television, however, is actually near the low end of the information rate scale, when compared to other systems and applications. Computer displays, medical imaging systems, etc., typically use far higher pixel counts and frame rates. Fortunately, at least from the perspective of ease of analysis, these are commonly considered as having discrete, well-defined pixels, with fixed spatial formats. A graph showing the pixel clock rate required for a range of standard image formats, at various frame rates, is shown as Figure 5-1. This chart assumes an overhead of about 25% in all cases, a typical value for systems based on the requirements of CRT displays. Note that the pixel rates shown here range from a low of a few tens of MHz to several hundreds of MHz. In their simplest forms, most of the systems using these formats will employ straightforward RGB color encoding, and generally assume at

86 THE DISPLAY INTERFACE

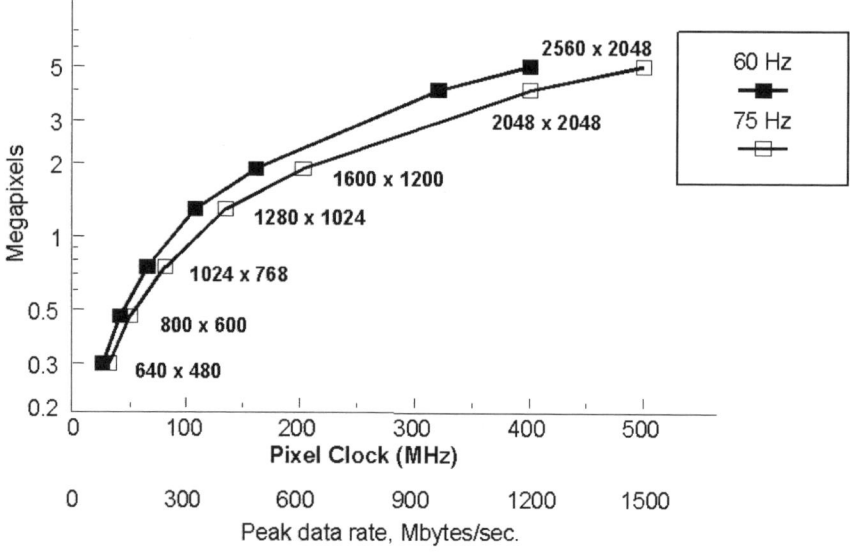

Figure 5-1 Pixel rates for common image formats. These are based on an assumption of 25% overhead (blanking) in each case. If the typical 24 bits/pixel (3 bytes) is assumed, the transmission of such video signals represents peak data rates, in megabytes per second, of simply three times the pixel rate.

least 8 bits/color (24 bits/pixel) will be required for realistic images. This results in the peak data rates listed as the second line of labels for the X-axis.

We again are reminded that image transmission, especially at the high frame rates required for good motion portrayal and the elimination of "flicker", is an extremely demanding task. In fact, motion video transmissions are among the most demanding of any information-transmission problem, with the task complicated by the fact that the data must continue to flow in a steady stream to produce an acceptable, convincing representation at the display.

5.3 Compression

This tells us why, for instance, video transmission is for the most part not practical with wireless interfaces, unless some sophisticated techniques are used. With the exception of broadcast television, which is a unique combination of some ingenious compromises, "wireless", over-the-air transmission of imagery has been restricted to either very low resolution, low frame rates, or both. The "PicturePhones" shown at the 1964 World's Fair may have been intriguing, but they were woefully impractical at the time; the telephone network could not provide the capacity needed to support a nation of video-enabled telephones. Today, such devices are finally appearing on the market – although not yet with the image quality that some may expect, and we are also seeing the advent of high definition broadcast television, or HDTV. Dr. Shannon's theorem has not been invalidated, however; instead, more efficient means of conveying image information have been developed. These all fall under the general term *compression*, meaning a reduction in the actual amount of information which must be transmitted in order to convey an acceptable image. Compression methods fall into two general categories – *lossless* compression techniques, which take advantage of the high level of redundant information present in most transmissions (enabling the removal of information with no impact on the end result), and *lossy* compression, which in any form literally deletes some of the information content of the transmission, in the expectation that it will not significantly impact the usability of the end result.

A simple example of lossless compression might be given as follows: Suppose I am sending you a video signal which comes from a camera aimed at a blank white wall. This represents a situation in which the signal contains an extreme amount of redundant information; rather than repeatedly sending the same image over and over again, it would be far more efficient to simply send it once, and then to send a command which tells the display to continue to show that same image until I send a different one. This assumes certain capabilities in the receiving display, but it does significantly reduce the load on the transmission channel. However, there is a price to pay for this; if the signal were corrupted during the transmission of that one image, say by a "spike" of noise on the line which affected several lines, that corrupted image will continue to be displayed for a long time. Removing redundant information always increases the vulnerability of the system to noise, since the remaining information now carries greater importance – if it is not received correctly, there is no additional information coming in through which it may be corrected.

Lossy compression, in one of its simplest forms, is seen in standard broadcast television. In the above examples, you may have noted that some of the numbers did not seem to add up; if television gives us the equivalent of 330 lines of resolution per frame, and yet is oper-

Figure 5-2 Interlaced scanning, a simple form of lossy compression used in analog television. Rather than transmitting all lines of the image in a single frame, the frame is divided into even and odd fields (one containing only the even-numbered scan lines (solid in the diagram above), and the other the odd). These are interleaved as shown above at the receiver, to create the illusion of the full vertical resolution being refreshed at the field rate. (Note that interlaced systems normally will require an odd number of lines per frame, such that one may consider each field as including a "half line." This accounts for the offset needed to properly interleave the two fields.)

88   THE DISPLAY INTERFACE

ating at a 60 Hz rate, the line rate given (15.75 kHz) was too low. In fact, you may recall the standard number of 525 total lines per US standard television frame, and so we would expect an even higher rate – something in excess of 30 kHz! Such a rate, however, would have required a video bandwidth far in excess of what was allowable, and so a simple form of lossy compression was employed. Instead of sending the full 525 lines of the frame every 1/60 of a second, the standard television system sends the odd lines during the first 1/60 second period, and then the even lines during the next. Exactly half of each frame has been removed, but we still see an acceptable image at the receiver as the viewer's visual system merges the two resulting "fields" (Figure 5-2). There clearly has been a loss of information, however, and this is where the "330 lines" number comes from. The original 525 line frame actually has around 480 active lines (those containing image content), but this *interlaced transmission* technique forces a reduction the effective resolution delivered at the final display. "Interlacing" is, fundamentally, a simple form of lossy compression.

Specific details of both the analog broadcast television systems, and the more sophisticated compression techniques employed in digital high-definition television, are covered in later chapters. At this point, it should be noted that state-of-the-art compression techniques have been demonstrated which reduce the required data transmission rate by factors of over well over 50:1, while still providing very high quality images at the final display.

5.4 Error Correction and Encryption

Not all of the processing performed on signals, especially in modern digital systems, results in a reduction in the amount of data to be transmitted. Two processes which can impose significant additional requirements on the channel capacity are the use of error detection and/or correction techniques, and the use of various types of data encryption. Additional data encoding methods may also be encountered which add to the burden of the interface by increasing the total amount of data to be conveyed.

Robust error detection/correction methods are only rarely employed in the case of image data transmission, especially in a motion-video system. Typically, error rates are sufficiently low such that the errors that do occur are not noticeable in the rapidly changing stream of images. However, critical applications may require images to be known to be error-free to a high-degree. This might occur, for instance, in the case of high-resolution still images in the medical field. Error detection and correction may also be incorporated in some systems employing high levels of compression, due to the increased importance of receiving the remaining data correctly, as noted above. In any event, the techniques employed may in general be viewed as the functional opposite of lossless compression; to protect against errors, some degree of redundancy must be added back into the transmitted information.

Encryption is more commonly seen in video or display interfaces than error correction, at least when dealing with uncompressed data. The term is used here to refer to any technique that modifies or encodes the data for the purpose of security; essentially to render it useless unless you are able to decrypt it (and therefore are presumably an authorized user of the information). In the case of video information, the more common application of data encryption comes in the form of various copy-protection schemes. These are intended to make it impossible (or at least, impractically difficult) to make an unauthorized copy of the material, most often "entertainment" imagery such as movies or television programming. While some forms of copy protection have been used with analog video transmission, this became significantly more important with the advent of digital interfaces and recording. As these technologies potentially allow for "perfect" reproductions in practically unlimited generations, ensuring the security of copyrighted material was a major concern to the developers of digital display interface standards. These are examined in more detail in Chapter 11, but for now it is important to note that such techniques may carry a penalty in terms of the total data required to transmit the image.

Other forms of data encoding may be required to optimize the characteristics of the transmission itself. One example common in current practice is the encoding used in the "Transition Minimized Differential Signalling" (TMDS) interface standard, now the most widely used digital interface for PC monitors. This is also examined in greater detail in Chapter 11, but for now we note that the encoding required under TMDS is of the "8 to 10" variety – meaning that every 8 bits of the original information is encoded as 10 bits in the transmitted stream. This is done not for error correction or encryption purposes, but rather to both minimize the number of transitions on the serial data stream and to "DC balance" the transmission (such that the transmitted signal spends half the time in the "high" state, and the other half "low.") Note that this represents a 25% increase in the required bit rate of the transmission over what would be expected from the original data.

5.5 Physical Channel Bandwidth

Having looked at the data rates required for typical display interfaces, we must now turn to the characteristics of the available physical channels, both wired and wireless. In the case of wireless, "over-the-air" transmission, we can quickly see that full-motion video is typically going to be restricted to the higher frequency ranges, where there is spectrum available to meet the requirements of such high data rates. Again, broadcast television is an excellent example; even with the relatively low resolution of standard TV, the minimum channel width in use today is the 6 MHz standard channel of the "NTSC" system used in North America; other countries use "channelizations" as wide as 8 MHz. This restricts television transmission to the VHF range and above; in the US, for example, the lowest allocated television channel occupies the 54–60 MHz range. This single channel is roughly six times the width of the entire "AM" or "medium wave" broadcast band; fewer than five such channels would occupy the entire "short wave" spectrum from 1.5 to 30 MHz.

The UHF (above 300 MHz), microwave (above 1 GHz) and even higher frequencies are best suited for wide-bandwidth video transmissions, but these frequencies are problematic in terms of being restricted to line-of-sight transmission, and requiring significantly higher transmitter power, for acceptable terrestrial transmission, than the short wave and VHF bands. Recently, the advent of direct broadcast by satellite (DBS) systems, using compressed digital transmission, has opened a new paradigm for wireless television transmission, using extremely high frequencies but requiring only small, relatively simple receiving systems. Terrestrial broadcasting is also benefiting from digital compression and transmission techniques, which permit multiple standard definition signals, or a single transmission of greatly enhanced definition, to be transmitted in the standard 6–8 MHz terrestrial broadcast channel, and at lower power levels than required for conventional analog television. These systems are examined in more detail in Chapter 12.

The high-resolution, high-frame-rate, progressively scanned images common in computer graphics, medical imaging, and similar applications still require higher data rates than is currently practical for standard wireless transmission systems. Thus, the "display interface" in these fields almost always requires some form of physical medium for the transmission channel. This would almost always be some form of wired connection, or, in the one significant example which spans the gap between "wired" and "wireless," a connection employing optical fiber.

Wired interconnects, employing twisted-pair, coaxial, or triaxial cabling, are capable of carrying signals well into the gigahertz range, and so (assuming adequate transmitter and receiver devices), are certainly capable of dealing with practically any video transmission which is likely to be encountered. However, such connections are not without their own set of problems. First, due to the high frequencies involved, wired video connections almost always require consideration as a transmission-line system, meaning that the characteristic impedance of the line and the source and load terminations must be carefully matched for optimum results. Impedance mismatches in any such system result not only in the inefficient transfer of signal power, but also distortion of the signal through reflections travelling back and forth along the line.

A wired connection also results in the potential for exposing the signal to outside noise sources, through capacitive or inductive coupling or straight EM interference, and conversely makes it possible for the signal itself to radiate as unwanted (and potentially illegal, per various regulatory limits) electromagnetic interference, or EMI. This adds the requirement that the line and its terminations not only be impedance-controlled, but generally that some form of "shielding" be incorporated. External signals are not the only concern here; in systems using multiple parallel video paths, as in the case of separate RGB analog signals or even parallel digital lines, the signals must be protected from interfering with each other (a problem generally referred to as *crosstalk*).

Systems employing multiple physical paths, such as the standard RGB analog video interconnect mentioned above, or digital systems with separate paths for various channels of data and their clock, must also present the data to the receiver without excessive misalignment, or *skew*, in time. This requires that the effective path length of all channels be carefully matched, meaning that not only must the physical lengths be held to close tolerances, but also that the velocity of propagation along each line be similarly matched. Such rigorous requirements on the physical medium – the cable – itself can add considerable cost to the system, and so an important distinction between various interfaces is often the relative skew of the source outputs, and the skew tolerance of the receiver circuit design. More tolerant designs at each end of the line permit more generous tolerances on the cabling, and so lower costs.

All practical physical conductors exhibit non-zero resistance, and no practical insulating material offers infinite resistance. In addition, any practical cable design will also exhibit a characteristic impedance which is not flat across the spectrum, and will present significant capacitive and/or inductive loads on the source. In simpler terms, this means that all cables are lossy to a certain degree, and further that this loss and other effects on the signal will not be independent of frequency. This will limit the length of cable of a given type which may be used in the system, at least in a single length. Quite often, active circuitry – a *buffer amplifier* in analog systems, or a *repeater* in digital connections – must be inserted between the source and receiver (display) to achieve acceptable performance over a long distance.

Finally, placing a conductive link between two physically separate products always introduces concerns of a signal quality and regulatory nature, in addition to some non-obvious potential problems from the standpoint of the DC operation of the system. Due to the common requirement for the signal conductors to be shielded, connecting video cables of any type most often means a direct connection between the grounded chassis of the two pieces of equipment Noise potentials, or in fact any potential difference between the two separate products, will result in unwanted currents flowing via this connection (Figure 5-3a). These can, of course, result in noise appearing on the signal reference connection, and so interfere with the signal itself, but also can be radiated by the cable or by the equipment at the opposite end of the cable from the noise source. Ensuring that the high-frequency currents on the connection remain balanced both reduces cable emissions and improves the signal quality; this can be achieved, or at least approached, through the use of true differential signal drivers and receivers, and/or by adding impedance into the path of possible common-mode currents (for example, by placing ferrite toroids around the complete cable bundle, Figure 5-3b).

A DC connection for the signals themselves can also be problematic. Given the relatively low signal voltages typical of display interfaces, even providing a substantial ground connection through the cable will not ensure that the source and receiving devices are at the same DC reference potential. Besides the problem with unwanted currents, as mentioned above, offsets in the reference at either end of the cable can degrade or disable the operation of the driver or receiver circuits. This will very often result in a requirement for AC coupling (either capacitive or inductive) at least at one end of the cable.

Before moving to specific concerns for analog and digital interfaces, we should also

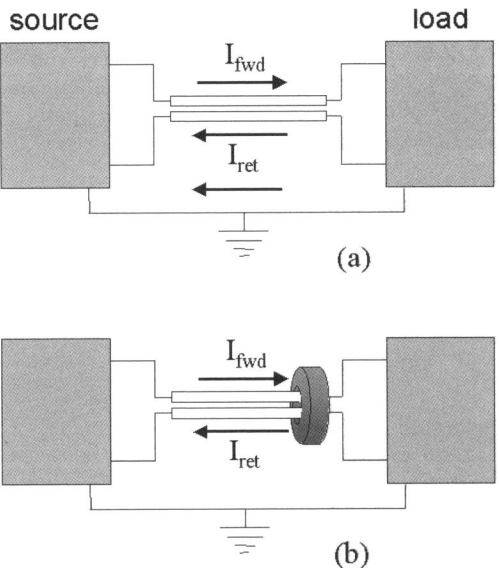

Figure 5-3 Return currents and noise in wired signal transmission systems. In any such system, it is desirable that both the "forward" signal current and the return travel along the intended path, as defined by the cable. However, any additional connection (such as the safety grounds) between the source and load devices represents a potential return path, and signal return current will flow along this path in proportion to its impedance vs. the desired return (a). This is both a source of undesirable radiated emissions and a means whereby noise may be induced into the transmitted signal. Increasing the impedance of these unwanted paths may be achieved by adding a ferrite toroid around the conductor pair (b); as both the intended current paths pass through this, it represents no added inductance for these.

briefly consider an alternative mentioned earlier. Optical connections, using light guided by optical fibers as the transmission medium, solve many of the problems, described above, of electrical, wired interfaces. They are effectively immune to crosstalk and external noise sources, cannot radiate EMI, and can typically span longer distances than a direct-wired connection. Further, since optical-fiber interfaces do not involve an electrical connection between the display unit and its host system, no safety current or noise concerns exist. We might also expect optical connections to provide practically unlimited capacity; the visible spectrum alone spans a range of nearly 400 terahertz (4×10^{10} Hz), which one would think would be enough to accommodate any data transmission. Unfortunately, it has proven difficult to manufacture electro-optical devices capable of operation at very high data rates. Only recently have practical systems capable of handling gigahertz-rate data become available. The optical cabling itself has also been historically relatively expensive, and difficult to terminate properly. But optical connections are rapidly moving into the mainstream, and promise to become significant in the display market in the near future.

5.6 Performance Concerns for Analog Connections

In the case of an analog interface, the job of the physical connection is to convey the signal from source to receiver with a minimum of loss and as little added noise and distortion as possible. (Distortion can, in fact, be considered as a form of noise in the broadest sense of the word – anything that is not a part of the intended information is noise.) In short, we are attempting to maximize the ratio of signal to noise at the receiver. Given the range of frequencies used in video, and the typical lengths of the physical interconnect, this first of all requires that the impedance of the transmission path be maintained at a constant value throughout the path. Impedance discontinuities result in reflections of part of the signal power, which means both a loss of power and potentially a distortion of the signal. At the same time, the signal must be protected from external noise sources (including other signals which may be carried by adjacent conductors), which generally means some form of shielding and/or filtering. Finally, the materials used and the design of the cabling and connectors must be chosen such that signal losses be kept to a minimum, consistent with the other requirements.

5.6.1 Cable impedance

Any physical conductor (or more properly, conductor pair, since there must always be some return path) may be modelled as a series of elements as shown in Figure 5-4. In this model, the L and C elements represent the distributed inductance and capacitance of the cable, respectively, while the R and G elements are distributed losses. These losses are the result of the resistance of the conductor itself (modelled as the distributed R), and the fact that the insulating material between conductors can never have infinite resistance (modelled in this case as a distributed conductance, G). If we assume an infinite length of the cable being modelled – which is then an infinite number of such sections – the impedance seen looking "into" the cable at the source end is given by

PERFORMANCE CONCERNS FOR ANALOG CONNECTIONS 93

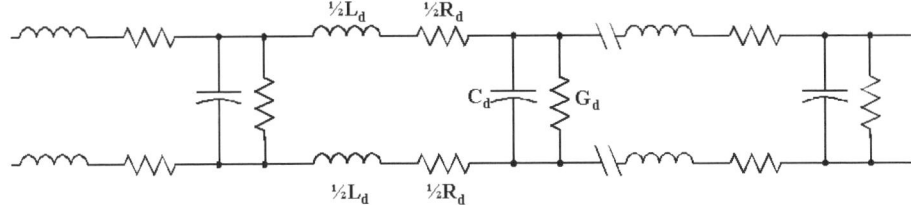

Figure 5-4 Distributed-parameter model of a conductor pair. Any cable may be modelled as an infinite series of distributed elements as shown above; these represent the distributed series inductance and resistance of the conductors (the series elements L_d and R_d), plus the capacitance and conductive losses between the conductors (C_d and G_d). These are considered as being distributed uniformly along the length of the cable. The values of each is given in terms of their value per unit distance; for example, the distributed capacitance is typically given in terms of picofarads per meter (pF/m).

$$Z = \sqrt{\frac{L + j\omega R}{C + j\omega G}}$$

This is referred to as the *characteristic impedance* of the cable, or as it would more commonly be referred to at frequencies where this is of concern, the *transmission line*. Per this model, this impedance varies with frequency. However, note that the frequency dependence in this case is associated with the resistive and conductive loss factors; if it can be assumed that the product of these factors and the frequency in question are small compared to the distributed L and C elements, this equation reduces to

$$Z = \sqrt{\frac{L}{C}}$$

Under these circumstances, then, the characteristic impedance of the cable is independent of frequency, and depends solely on the distributed inductance and capacitance of the cable. These are determined by the size and configuration of the conductors, plus the characteristics of the insulating or *dielectric* material between them. In general, the conductors may be configured so as to minimize the distributed inductance or capacitance, but not both simultaneously. Minimizing the distributed inductance generally means minimizing the "loop area" defined by the conductors in question, which implies placing the conductors in close proximity and ideally causing the currents in both directions to follow the same average path in space. Coaxial construction (Figure 5-5), in which one conductor is completely surrounded by the other, is an example of a configuration which does this. However, minimizing the distributed capacitance is a matter of keeping as little of the conductors are possible in close proximity; the farther the conductors are separated, or the lower the area placed in proximity, the lower the capacitance. The best that can usually be done in this direction, in terms of a practical conductor configuration, is the case of parallel conductors held a fixed distance apart by an insulating support, the "twinlead" type of cable (Figure 5-6a). As this construction results in much greater "loop area" than a coaxial design, the capacitance reduction comes at the expense of greater inductance. In the coaxial cable, exactly the opposite happens – the distributed inductance is minimized at a cost of greater capacitance. As a result,

94 THE DISPLAY INTERFACE

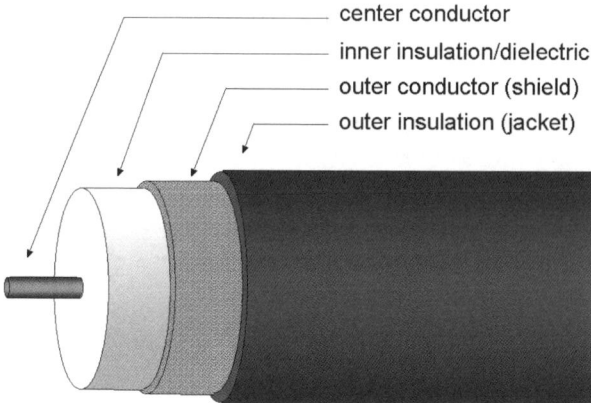

Figure 5-5 Coaxial cable construction. This form is ideal for ensuring that the forward and return currents follow the same average path in space, but at a cost of increased capacitance and losses, and therefore a lower characteristic impedance, than other cable designs.

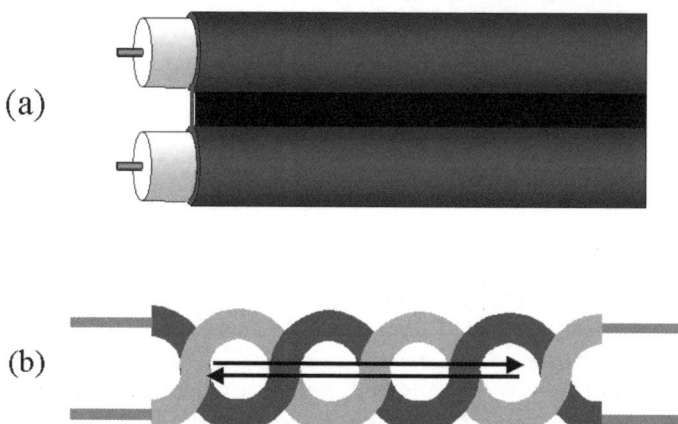

Figure 5-6 Twinlead and twisted-pair construction. In the twinlead type (a), the conductors are held parallel and at a fixed separation distance by the outside jacket; the portion between the conductors is often made as thin as practical, to minimize losses and inter-conductor capacitance. This form of construction provides significantly lower losses and capacitance than the coaxial design, and so a higher characteristic impedance, but does not provide coaxial cable's "self-shielding" property. Twisting a pair of conductors together (b) holds the two in close physical proximity, and causes the forward and return currents to follow the same average path. This minimizes both the possibility of induced noise on the lines, and the degree of unwanted radiation from the cable.

coaxial cable types tend to have lower characteristic impedances (most commonly in the 50–100 Ω range) as compared with twinlead types (commonly 100–300 Ω, and in some cases as high as 600–800 Ω). The most common standard characteristic impedance for analog video interconnects is 75 Ω, so these are almost always constructed of coaxial cable.

A straight "twinlead" cable design (Figure 5-6a), while fairly common in some applications (such as low-loss antenna cabling for television), is not commonly used for video signal cabling, as it suffers from being very vulnerable to external noise sources. Both capacitively and inductively coupled noise can be reduced significantly, however, by simply twisting the pair of conductors as shown in Figure 5-6b. Such a twisted-pair cable provides approximately the same characteristic impedance as the twinlead type, other factors being equal, but noise is reduced as external sources will couple to the cable in the opposite sense each "loop" in the pair, resulting ideally in cancellation of the noise. Twisted-pair construction is very common in communications and computer interfaces, such as telephone wiring and many computer-networking standards. Some very inexpensive analog video cables have been produced and sold using shielded twisted-pair construction, but these are adequate only for very low-frequency applications and very short distances. They are to be avoided for any serious analog video use.

5.6.2 Shielding and filtering

While both the coaxial and twisted-pair configurations can be considered "self-shielding" to some degree, an additional conductor is often added to cabling to provide further protection from external noise sources, and/or to reduce radiated emissions from the cable. This is more commonly required in the case of the twisted-pair type, whose performance in practice tends to be further from the theoretical ideal than is commonly the case for a coaxial cable. An added shield will affect the characteristics of the cable to some extent; the degree to which these may change will depend on the precise configuration of the shield and conductors, as well as whether or not the "shield" is truly used simply as a shield, or if there is the possibility of signal or return current being carried via this path.

Other factors affecting the effectiveness of such shielding include the type of material and construction of the shield itself, and the quality of its connection at either end. Ideally, the shield would be perfectly conductive and completely cover the inner conductors of the cable, neither of which is, of course, achievable in practice. Low-resistance shielding with the required flexibility is most often provided through the use of a braided copper layer, but such braids cannot generally provide 100% coverage. Conductive foils, often a Mylar or similar plastic-film layer with a conductive layer on one side, can provide better coverage than a braided shield, but may also represent a significant inductance (depending on the construction of the foil shield) and will typically exhibit a higher series resistance than the heavier braided conductors. The most effective shielding which still retains sufficient flexibility is a combination of the two – a film/foil layer providing 100% coverage, in intimate contact with a braided shield to create a low-resistance/inductance path. As a lower-cost (and somewhat less effective) alternative, a "foil" shield may also be provided with a "drain wire", which is simply an exposed conductor running the length of the cable assembly, in contact with the conductive portion of the film or foil. (Note also that, per the earlier discussion regarding safety standards, any conductor which has the possibility of carrying fault currents will likely be required to demonstrate that a specified minimum current can be carried for at least a certain minimum time. This requirement can in many applications constrain the type and design of the shield layer and it connections at either end.)

Protection from external noise sources may also come in the form of filtering, which is often designed into the cable assembly and/or connectors. Many connector types intended for video applications are offered in "filtered" version, generally meaning that the signal connections pass through a ferrite material that adds inductance into the signal path. This of course limits the bandwidth of the connection by increasing the impedance at higher frequencies. While this can often be effective in removing unwanted high frequencies – those which do not affect the image quality, but which may cause problems at the receiver or be radiated as unwanted EMI – some care must be taken to ensure that the use of such measures truly does not impact the quality of the displayed image at all desired timings.

Another form of "filtering" in the cable involves the addition of a ferrite "core" (a toroid) over the entire cable bundle, or at least over individual signal/return pairs. In theory, since both the "outbound" and "inbound" currents for any given signal pass through the ferrite, but in opposite directions, this adds no inductance to the signal path itself. However, noise coupling in to the lines from an external source is presumed to be induced on both conductors in the same sense; i.e., it is "common mode" noise. In that case, the added ferrite material places significantly greater impedance in the path of the noise, and so preferentially reduces its magnitude vs. that of the signal. Ferrites are also often used in this manner as a countermeasure against electromagnetic interference, or EMI. The theoretical basis here is the same as for the reduction of common-mode noise; since both the "outbound" and "inbound" currents pass through the ferrite, the signal path sees a low impedance only if both the currents are matched. This effectively places a higher impedance in any other possible return path, and so helps to maintain the match between these two currents. With equal and opposite currents on the cable, radiated emissions are minimized.

5.6.3 Cable losses

It is not possible, of course, to produce any practical physical cable with zero losses. Even if the conductors themselves were lossless, there would be the unavoidable effects of the capacitive coupling between the conductors, etc. Not surprisingly, cable loss generally increases with frequency (owing both to these capacitive effects, plus the "skin effect" increase in the series resistance of the conductors), and are lower for low-capacitance types. A "twinlead" or twisted-pair configuration typically provides lower loss than coaxial; larger coax cables will provide lower loss than smaller types of the same characteristic impedance for the same reasons. The material and construction of the insulation also is a major factor in determining both the cable loss characteristics and the bulk capacitance of the cable. Low-loss designs often use a foamed-plastic dielectric, which of course replaces much of what would have been solid plastic with air. The extreme case of this approach, in coaxial cables, is *hardline*, in which the space between the center conductor and a solid metal shield is mostly air (or a dry, relatively inert gas such as nitrogen); plastic spacers are used only to support the center conductor within the outer, pipe-like shield. (This is used only in very critical applications, and never to my knowledge in any common display connection; it is mentioned here only in passing.)

Cable loss characteristics are most commonly stated in terms of decibels (dB) of loss per a given distance, most often as dB/100 m or db/100 feet. The resistance and capacitance of the

Table 5-1 Comparison of characteristics for various typical video cable types.

Type/description	Construction	OD (mm)	Velocity factor[a]	Capacity (pF/m)	Loss MHz	dB/100 m
75 Ω minimum coaxial cable (Belden 9221)	30 AWG stranded center, foamed HDP[b] dielectric, tinned copper braid shield (89% coverage), black PVC jacket	2.46	0.78	56.8	1	2.3
					5	5.2
					10	7.2
					50	16.7
					100	23.9
					200	34.4
					400	50.9
					1000	87.3
Standard 75 Ω coaxial cable (RG-59/U type; Belden 8241)	23 AWG solid center, polyethylene dielectric, bare copper braid shield (95% coverage), PVC jacket	6.15	0.66	67.3	1	2.0
					10	3.8
					50	7.9
					100	11.2
					200	16.1
					400	23.0
					1000	39.4
Precision 75 Ω video cable (RG-59/U type; Belden 1505A)	20 AWG solid center, gas-injected foam polyethylene dielectric, 100% coverage foil plus 95% coverage tinned copper braid, PVC jacket	5.97	0.83	53.1	1	0.95
					10	2.85
					71.5	6.89
					135	8.86
					270	12.5
					360	14.5
					720	21.3
					1000	25.6
Standard 75 Ω coaxial cable (RG-6/U type; Belden 8215)	21 AWG solid center, polyethylene dielectric, 2 bare copper braids (total 97% coverage), polyethylene jacket	8.43	0.66	67.2	1	1.3
					10	2.6
					50	6.2
					100	8.9
					200	13.4
					400	19.4
					1000	32.1
Standard 75 Ω coaxial cable (RG-11/U type; Belden 9292)	14 AWG solid center, foamed polyethylene dielectric, 100% coverage foil plus 61% coverage tinned copper braid shield, PVC jacket	10.29	0.84	52.8	1	0.6
					10	1.6
					50	3.0
					100	4.3
					200	5.3
					400	7.6
					1000	14.1
300 Ω twinlead cable (Belden 8230)	2 bare copper-covered steel conductors, parallel; brown polyethylene insulation	1.83× 10.16	0.80	11.8	100	3.6
					200	5.6
					300	7.2
					500	10.2
					900	14.8

[a] Velocity of signal propagation along cable as a fraction of c.
[b] HDP, high-density polyethylene.
Source: Belden, Inc. Master Catalog.

cable are also stated in similar units, such as Ω/km or pF/100 feet. However, the loss numbers will generally be stated for multiple frequencies covering the range of typical uses expected for that cable; the specifications of coaxial cables for PC video interconnect use, for example, will typically provide figures for loss within at least the 1–1000 MHz range.

In the case of analog video interconnects, losses in the cable represent a loss of dynamic range of the signal; the image will generally remain usable, and amplification within the display can compensate to some degree for this loss, at the expense of increasing the noise as well. The insertion of buffer or distribution amplifiers into the path will often be required to maintain a usable signal over long distances (generally, greater than a few tens of meters). A more serious problem may result from the loss of amplitude in the synchronization signals, as these may be degraded to the point at which they are unusable by the display or are confused with noise. In either case, the display will not be able to provide a usable image at all. The sync signals are in many cases particularly vulnerable, as they are often produced by relatively simple output circuits with limited drive capability. Often, the sync outputs are simply standard TTL drivers, not really intended to drive lengthy cable runs.

A sampling of cable specifications for coaxial types intended for video use is given in Table 5-1. Note the differences in capacitance, loss, and velocity factors for the various dielectrics and cable diameters.

5.6.4 Cable termination

As in any transmission-line situation, it is important in analog video applications not only to use cable of the proper impedance, but also to ensure that the cable is properly terminated. Again, the norm for analog video systems has almost always been 75 Ω, and so the display inputs ideally provide a purely resistive 75 Ω load across the full range of video frequencies of interest. If the display inputs were, in fact, to provide this ideal termination, the impedance of the source (the video output) would be irrelevant (as long as the source could drive the line with the proper amplitude signal), as no signal would be reflected by the load. However, as we will see, real-world inputs are rarely even close to the ideal, and so source termination is also an important consideration in preventing reflections on the cable and the resultant "ghosting" in the displayed image.

The easiest method of terminating the cable at the display input is to simply place a resistance of the proper value across the input (Figure 5-7a). The display takes its input across this resistance, through a buffer amplifier stage (which is presumed to have a very high input impedance compared to the value of the terminating resistor). This method, though, will not provide a constant impedance over any but the lowest frequencies. It is can be used successfully for standard baseband television, as such signals do not exceed a few MHz, but may not be satisfactory for higher-frequency use (such as PC monitors). The problem is that the next stage – the amplifier – typically presents a load which is significantly capacitive, and in addition the terminating resistance itself can present a significant parasitic capacitance. As a result, the impedance seen at the display input shows the characteristics of a parallel R-C combination, at least up to a certain frequency. At some point, parasitic inductances – including those in the terminating resistor, as well as the typical coupling capacitor between that termination and the first stage of the video amplifier – will begin to dominate, and the input impedance will again increase, often well beyond the intended nominal value.

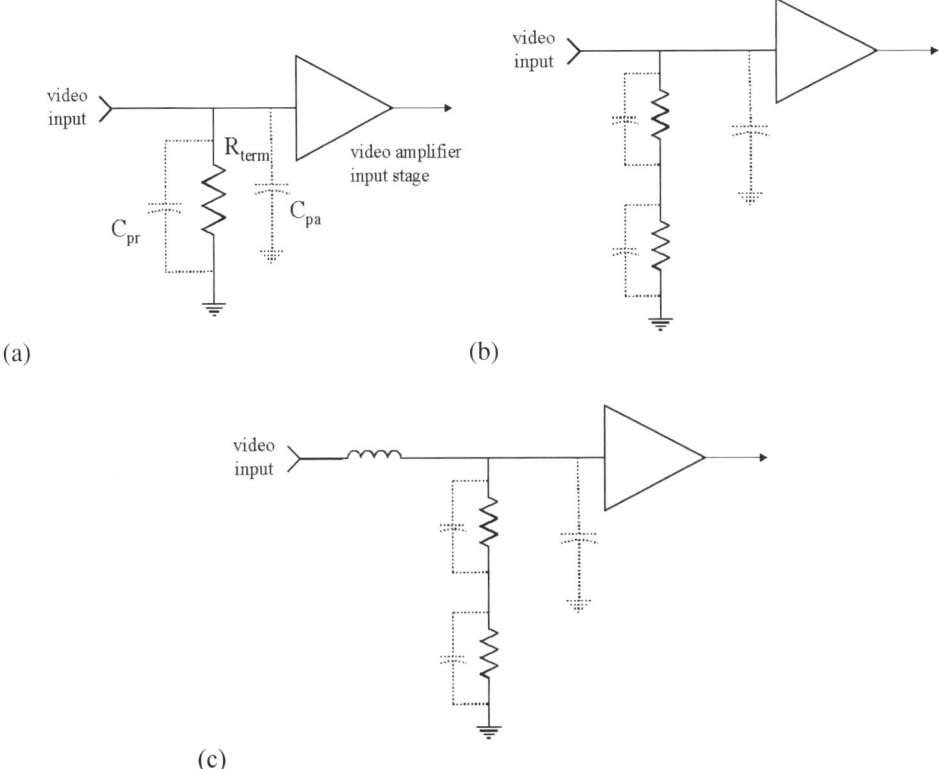

Figure 5-7 (a) Simple resistive shunt termination. This will terminate the video transmission line in the proper impedance at low frequencies. However, parasitic capacitances, including those across the termination resistor itself (C_{pr}) as well as the expected capacitive portion of the video amplifier's input impedance, will significantly reduce the total effective impedance seen by the input signal at high frequencies. This form of termination can often result in significant reflections and "ghosting" of the video signal. (b) Splitting the shunt termination. Dividing the single terminating resistor into two in series (e.g., if the original R_{term} was 75 Ω, using two 39-Ω resistors in series) improves the situation by breaking up the parasitic capacitance C_{pr}. (c) The addition of a series impedance to the terminating network, especially an inductive reactance, will serve to compensate for the increasing capacitive effects at high frequencies and maintain the proper termination of the signal, but at the cost of reduced signal amplitude at the input to the video amplifier itself.

Some slightly more elaborate termination schemes can provide significantly better results. First, a common method of dealing with the parasitic capacitance of the termination resistor is to break this into two resistors of half (or slightly higher) the desired total value (Figure 5-7b). This places the parasitic capacitances of the two in series, reducing the total capacitance of the termination. It may also be desirable to introduce a small series resistance, or even a small inductance, between the input connector and the shunt termination, to help maintain the impedance as seen by the cable over those frequencies at which it would normally decline (Figure 5-7c). This does result in a slight increase in the impedance at low frequencies, and a loss of signal across the termination. However, these effects may be negligible – and in terms of minimizing reflections on the cable, it is always preferable to be somewhat over the desired terminating impedance than under it by the same amount. Obviously, much more elaborate termination networks can be designed, which would present an even better (more constant) load impedance to the line, but these are generally beyond the limits of practicality in mass-market designs.

As mentioned, source (output) termination is not as great a concern as that at the load (display input) end of the cable, and in fact is generally not done as well in terms of providing a well-matched output over the entire frequency range. Depending on the characteristics of the signal driver, a simple resistive termination may be all that is provided (if that). However, many computer-graphics cards employ a somewhat more complex output network, which provides both resistive termination and filtering of the output. As noted earlier, some simple filtering may be provided by choosing a "filtered" connector, which generally refers to one in which the signal pins are surrounded by a ferrite material for extra inductance. This can be effective in reducing high-frequency noise on the line, but may attenuate desired high-frequency components in the video signal too much, especially in the case of a "high resolution," high-refresh-rate output.

The need for more sophisticated output filtering comes from requirements to minimize radio-frequency interference (RFI) emissions from the display and/or the cable. Very often, the video outputs of computer graphics cards are capable of significantly faster signal edge rates (rise/fall times) than is actually required by the display. In other words, the limitations of the display device are such that faster signal transitions, beyond a certain point, do not result in any visible improvement in the displayed image. In such cases, very fast edge rates become a liability; they do no contribute to the image quality, but the high frequencies they represent are potential sources of RFI. (Very high frequencies in the signal are particularly troublesome, as these are often the frequencies most effectively radiated by the display or cable assembly.) Therefore, edge rate control becomes an important tool in reducing unwanted emissions.

5.6.5 Connectors

Various connector standards are discussed in later chapters, but a few words are appropriate at this point regarding the effect of the connector choice on the video signal path. While practically any connector can be made to work in a display interconnect design (and often it seems that practically all types *have* been used!), the connector choice can have a significant impact on the performance of the interface.

The role of the connector is basically the same as that of the cable itself: to convey the signal with minimum loss and distortion, while protecting it from outside influences. To this basic task is added the requirement that connectors provide the ability to break the connection – otherwise, there would be no need for a connector, or more properly a connector pair, in the first place. This function brings with it the need for mechanical ruggedness – the ability to withstand repeated connection and disconnection while maintaining the electrical performance characteristics – and generally a need for some degree of mechanical security while in the connected state (i.e., you do not want to the connector to be *too* easy to disconnect, to prevent unwanted failure of the interface).

Electrically, the connector system is governed by the same theory as the cable, in terms of needing to provide a stable and constant impedance, its behavior in terms of shielding performance, and so forth. The importance of the connector in this regard is, however, admittedly far less than that of the cable in all but the most critical applications, due to the much shorter electrical length of the signal path through the connector. Still, any discontinuity in the characteristic impedance of the path or a break in the shielding or return paths can have a very significant impact on the signal quality. Quite often, a given connector type will work quite well *if* everything is "just right"; one hallmark of a good connector choice is that its design is robust enough to ensure proper performance without undue effort on the part of the manufacturer or user.

A good example of this is the ubiquitous "VGA" connector of the PC industry, which is covered in more detail in Chapter 9. At first glance, one would not expect this connector to be a good choice for high-frequency video signal applications, and in fact it is not by any objective measure. However, it has benefited from a truly enormous installed base, and so has been pressed far beyond its original performance requirements by the need for "backward" compatibility. But this connector – basically, a higher-density, 15-pin version of the standard 9-pin D-subminiature type used often in other computer applications (such as the relatively low-speed "serial port" common on PCs) – has several factors working against it. First, none of the contacts are well suited to the connection of coaxial cables; when using the "VGA" connector with video cables, the most common termination method is to attach short lengths of wire to the two conductors of the coax, and then solder those (usually by hand)

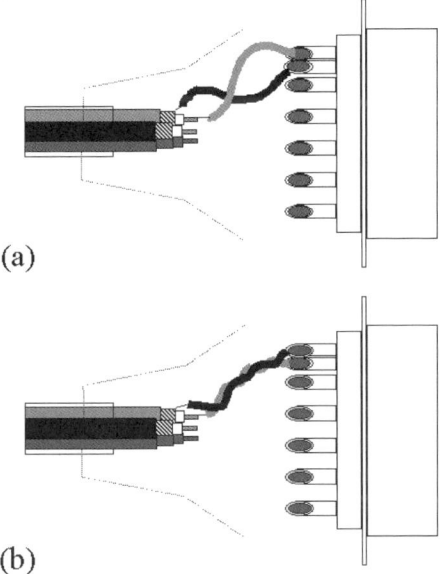

Figure 5-8 Termination problems using the "VGA" 15-pin connector. As this connector family was not original intended for use with coaxial cables, there can be significant problems resulting from the method of connecting such cables. In a typical VGA cable assembly, three miniature coaxial cables (supplied in a single cable bundle) are connected to the pins of the connector using short lengths of wire tack-soldered to the shield and center conductor. However, if care if not taken in routing these wires prior to the addition of the overmolded shell (dotted outside line), this may result in an impedance discontinuity due to the large loop area thus formed (a). Simply twisting the wires together (b), to ensure that they remain in close proximity, will result in a significantly improved connection from the standpoint of maintaining the characteristic impedance of the line, although still not as good as a true coaxial connection. (Additional wiring and pins not shown for clarity.)

onto the connector's contacts. This can result in a significant impedance discontinuity within the connector, unless care is taken to make sure that the wires remain in close physical proximity (Figure 5-8). (Numerous examples can be found where this is not the case.) Next, the contacts themselves – including the contact between the outer shells of the plug and receptacle, commonly used as a connection for the overall cable shield – are not particularly robust, and often fail to make a solid, low-impedance connection for high frequencies. Finally, the connector design itself was not intended to provide a constant impedance, and (depending on the lead configuration used) can result in a fairly large impedance discontinuity even if all else works correctly. Contrasting this connector with one specifically designed for high-frequency video, such as the "13W3" or "DVI" types (also detailed in Chapter 9) is an interesting exercise.

5.7 Performance Concerns for Digital Connections

Attempting to draw a clear distinction between "analog" and "digital" signalling is often based on numerous unstated (and often unrealized) assumptions regarding these terms (see Chapter 6). It is much more difficult than is commonly assumed to point out meaningful inherent distinctions between "analog" and "digital" *signals*, and truly many of the concerns discussed above for the "analog" connection also apply to what are commonly said to be "digital" types. The signal must still be conveyed by the physical interconnect with minimum loss and distortion, and these applications generally require a very high bandwidth over which this must be achieved. However, "digital" in the context of a display interface generally means any of several possible binary-encoded systems, using a clock to latch discrete packets of information at the receiver. These may be of either the *serial* (single bits of information transmitted in sequence over a single physical connection) or *parallel* (multiple bits received simultaneously for each clock pulse, over multiple physical connections) types, but the basics are the same in either case. (Serial transmission will, however, obviously require higher rates on a per-line basis, if the same total amount of information is to be conveyed.)

The fact that these systems employ binary encoding (only two possible valid states for the signal, the simplest but least efficient form of "digital" transmission) implies that the sensitivity of the signal to noise is reduced. The receiver must only be capable of distinguishing between these states, rather than resolving the much smaller changes required in an analog transmission. (This does not, however, translate to a noise-immunity advantage for digital interfaces in general, as discussed in Chapter 6.) However, the relative importance of the timing of signal increases dramatically. Consider the problem of transmitting a 1280 × 1024 image at a 60 Hz refresh rate, at 24 bits per pixel. If this were to be done using a serial transmission of binary data over a single line, the minimum bit rate on that line would be

$$1280 \text{ (pixels/line)} \times 1024 \text{ (lines/frame)} \times 24 \text{ (bits/pixel)} \times 60 \text{ (frames/s)} = 1.89 \text{ Gbits/s}$$

This means that each bit transmitted is only a little over 500 ps in duration; during that time, the signal must unambiguously reach the desired state for that bit and be clocked into the receiver. This requires both an extremely fast rise and fall time for the signal, and also that this signal and the clock used to latch it into the receiver be properly aligned, to within a tolerance of absolutely no worse than ±250 ps! Given that a 250 ps change the signal position relative to the clock can be caused by a few centimeters' difference in the effective cable lengths for each, it is very apparent that this is an extremely challenging task. (In fact, no digital interface system with the capability of achieving this rate on a single line has yet been brought to the display market.)

To achieve the levels of performance required for a digital display interface, much the same concerns apply to the cable and connector choices as in the analog video case previously discussed. Impedance control remains important, as does protection both from outside noise sources and from possible radiated emissions by the cable assembly itself. In addition, the cable material and design must be chosen so as to minimize differential delays, or *skew*, between the various signals carried and their clocks, and also to enable the fast transitions required. Further, to obtain the necessary data rates required in video applications, digital interfaces of these types will commonly require more physical conductors than comparable analog systems. System impedances are generally higher (commonly in the 90–150 Ω range). To meet all of these requirements within the constraints of a reasonably sized cable assembly, shielded-twisted-pair or shielded ribbon cable construction is typical.

A very useful method for quickly evaluating the performance of a digital transmission system is the *eye diagram*, as shown in Figure 5-9. This is created by observing the digital signal line in question on an oscilloscope, using the appropriate clock signal from the digital interface as the trigger, and transmitting a data signal which ideally alternates between the

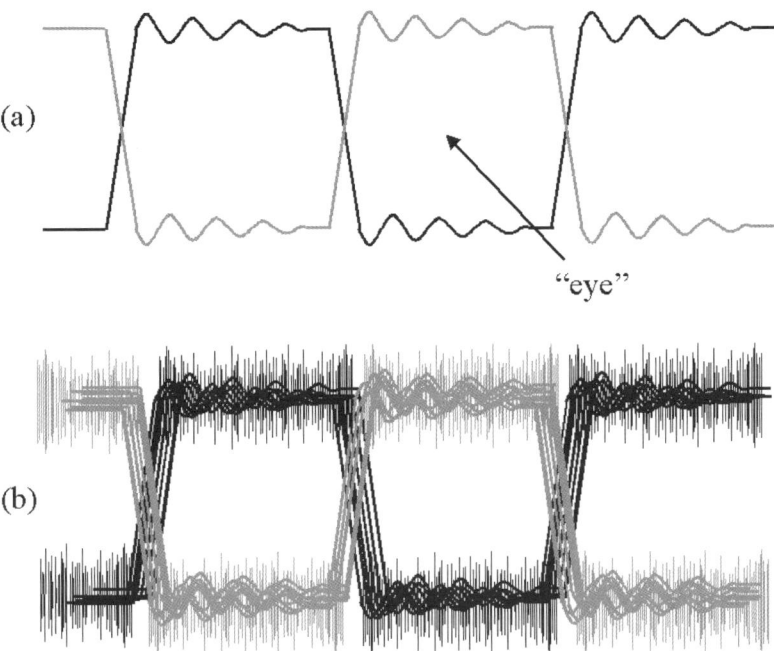

Figure 5-9 An "eye diagram". Using an oscilloscope set up to display overlapping traces of a digital transmission, an "eye" is formed between the two traces (a). Noise, jitter, and amplitude instabilities will all have the effect of reducing the area of the "eye," making it a quick visual check of the quality of a digital transmission. There must be sufficient margin between the high and low states, and a sufficiently long period during which the signal is unambiguously in one state or the other, to permit each transmitted state or "symbol" to be reliably identified in the system.

high and low states. (However, an acceptable "eye" display can often be obtained simply by monitoring an actual data transmission.) Care must obviously be taken in the selection and use of the oscilloscope and its probes, to minimize the influence of these on the measurement. The goal is that the "eye" appear to be as open as possible; vertical separation of the traces corresponds to noise margin, while lateral separation and the position (in time) of the data relative to the clock gives a visual indication of the effects of skew and *jitter* (short-term variations in the position of the data signal edges relative to each other or to the clock reference).

Details of many of the digital interface and transmission systems in current use, in both the computer and television industries, are given in Chapters 10 and 12.

6

Basics of Analog and Digital Display Interfaces

6.1 Introduction

It seems that in any discussion of electrical interfaces, the question of "analog vs. digital" types always comes up, with considerable discussion following as to which is better for a given application. Unfortunately, much of the ensuing discussion is often based on some serious misunderstandings about exactly what these terms mean and what advantages and disadvantages each can provide. This has, at least in part, come from the tendency we have of trying to divide the entire world of electronics into "analog" and "digital". But while these two do have significant differences, we will also find that there are more similarities than are often admitted, and some of the supposed advantages and disadvantages to each approach turn out to have very little to do with whether the system in question is "analog" or "digital" in nature.

In truth, it must be acknowledged that the terms "analog" and "digital" do not refer to any significant physical differences; both are implemented within the limitations of the same electrical laws, and one cannot in many cases distinguish with certainty the physical media or circuitry intended for each. Nor can one distinguish, with certainty, "analog" from "digital" signals simply by observing the waveform. Consider Figure 6-1; it is not possible to say for certain that this signal is carrying binary, digital information, or is simply an analog signal which happens at the time to be varying between two values. Is there really any such thing as an "analog signal" or a "digital signal", or should we look more carefully into what these terms really mean?

There has been an unfortunate tendency to confuse the word "analog" with such other terms as "continuous" or "linear," when in fact neither of these is necessary for a system to be considered "analog". Similarly, "digital" is often taken to mean "discrete" or "sampled" when in fact there are also sampled and/or discrete analog systems. Much of what we know –

Figure 6-1 An electrical signal carrying information. Many would be tempted to assume that this is a "digital" signal, but such an assumption is unwarranted; it is possible this could be an analog signal that happens to be varying between two levels, or perhaps it only has two permissible levels. The terms "analog" and "digital" properly refer only to two different means of encoding information onto a signal – they say nothing about the signal itself, only about how it is to be interpreted.

or at least think we know – about the advantages and disadvantages of the two systems is actually based on such misunderstandings, as we will soon see.

Fundamentally, at least as we consider these terms from an interfacing perspective, we must recognize that the terms "analog" and "digital" actually refer to two broad classes of methods of encoding information as an electrical signal. The words themselves are the best clues to their true meanings; in an *analog* system, information about a given parameter – sound, say, or luminance – is encoded by causing *analogous* variations in a different parameter, such as voltage. This is really all there is to analog encoding. Simply by calling something "analog," you do *not* with certainty identify that encoding as linear or even continuous. In a similar manner, the word "digital" really indicates a system in which information is encoded in the form of "digits" – that is to say, the information is given directly in numeric form. We are most familiar with "digital" systems in which the information is carried in the form of binary values, as this is the simplest means of expressing numbers electronically. But this is certainly not the only means of "digital" encoding possible.

Even with this distinction clear, there is still the question of which of these systems – if either of them – is necessarily the "best" for a given display interfacing requirement. There remain many misconceptions regarding the supposed advantages and disadvantages of each. So with these basic definitions in mind, let us now consider these two systems from the perspective of the suitability of each to the problem of display interfacing.

6.2 "Bandwidth" vs. Channel Capacity

One advantage commonly claimed for digital interface systems is in the area of "bandwidth" – that they can convey more information, accurately, than competing analog systems. This must be recognized, however, as rather sloppy usage of the term bandwidth, which properly refers only to the frequency range over which a given channel is able to carry information without significant loss (where "significant" is commonly defined in terms of decibels of loss, with the most common points defining channel bandwidth being those limits at which a loss of 3 dB relative to the mid-range value is seen – the "half-power" points.) With the bandwidth of a given channel established, the theoretical limits of that channel in terms of its information capacity is given by Shannon's theorem for the capacity of a noisy channel, as noted in the previous chapter:

$$C \text{ (bits/s)} = BW[\log_2 (S/N +1)]$$

where C is the channel capacity, BW is the 3 dB bandwidth of the channel in hertz, and S/N is the ratio of signal to noise power in this channel. Note that while this value is expressed in units of bits per second, there is no assumption as to whether analog or digital encoding is employed. In fact, this is the theoretical limit on the amount of information that can be carried by the channel in question, one that may only be approached by the proper encoding method. Simple binary encoding, as is most often assumed in the case of a "digital" interface, is actually not a particularly efficient means of encoding information, and in general will not make the best use from this perspective of a given channel. So the supposed inherent advantage of digital interfaces in this area is, in reality, non-existent, and very often an analog interface will make better use of a given channel. We must look further at how the two systems behave in the presence of noise for a complete understanding.

6.3 Digital and Analog Interfaces with Noisy Channels

A significant difference between the two methods of encoding information is their behavior in the presence of noise. Some make the erroneous assumption that "digital" systems are inherently more immune to noise; this is not the case, if a valid comparison between systems of similar information capacity is to be made. This is essentially what Shannon's theorem tells us; ultimately, the capacity of any channel is limited by the effect of noise, which, as its level increases, makes it increasingly difficult to distinguish the states or symbols being conveyed via that channel. Consider the signals shown in Figure 6-2. The first is an example of what most people think of as a "digital" signal, a system utilizing binary encoding, or possessing only two possible states for each "symbol" conveyed. (In communications theory, the term symbol is generally used to refer to the smallest packet of information which is transmitted in a given encoding system; in this case, it is the information delimited by the clock pulses.) The second represents what many would view as an "analog" system – in the case of a video interface, this might be the output of a D/A converter being fed eight-bit luminance information on each clock. The output of an 8-bit D/A has 256 possible states – but if each of these states can be unambiguously distinguished, meaning that the difference between adjacent levels is greater than the noise level – eight bits of information are thereby transmitted on each clock pulse. If the clock rate is the same in each case, eight times as much information is being transmitted over any given period of time by the "analog" signal as by the "digital". The third example in the figure is an intermediate case – three bits of information being transmitted per clock, represented as eight possible levels or states per clock in the resulting signal.

This may seem to be a very elementary discussion, but it should serve to remind us of an important point – that which we call "analog" and "digital" signals are, again, simply two means of encoding information as an electrical signal. The "inherent" advantages and disadvantages of each may not be so clear-cut as we originally might have thought.

This comparison of the three encoding methods gives us a clearer understanding of how the noise performance differs amongst them; the simple binary form of encoding is more

108 BASICS OF ANALOG AND DIGITAL DISPLAY INTERFACES

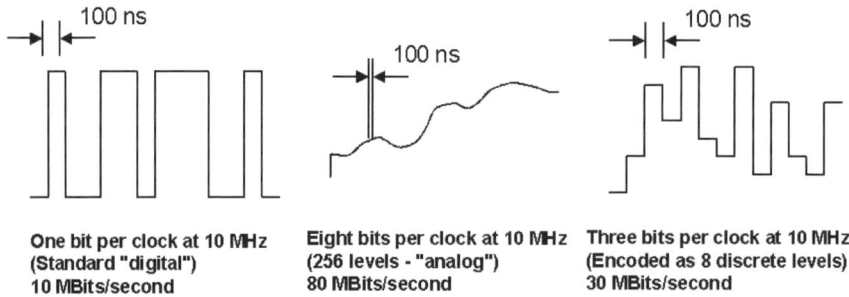

One bit per clock at 10 MHz
(Standard "digital")
10 MBits/second

Eight bits per clock at 10 MHz
(256 levels - "analog")
80 MBits/second

Three bits per clock at 10 MHz
(Encoded as 8 discrete levels)
30 MBits/second

Figure 6-2 Three signals using different encoding methods. The first, which is a "digital" signal per the common assumptions, conveys one bit of information per clock, or 10 MBit/s at a 10 MHz clock rate. Many people would assume the second signal to be "analog" – however, it could also be viewed as encoding eight bits per clock (as any of 256 possible levels), and so providing eight times the information capacity as the first signal. The third is an intermediate example: three bits per clock, encoded as eight discrete levels. As long as the difference between least-significant-bit transitions may be unambiguously and reliably detected, the data rates indicated may be maintained.

immune to noise, if only because the states are readily distinguishable (assuming the same overall signal amplitudes in each case). But it cannot carry as much information as the "analog" example in our comparison, unless the rate at which the symbols or states are transmitted (more precisely, the rate at which the states may change) is increased dramatically. This is a case of trading off sensitivity to random variations in amplitude – the "noise" we have been assuming so far – with sensitivity to possible temporal problems such as skew and the rise/fall time limitations of the system.

There is one statement which can be made in comparing the two which is generally valid: in an analog system, by the very fact that the signal is varying in a manner analogous to the variations of the transmitted information, the less-important information is the first to fall prey to noise. In more common terms, analog systems degrade more gracefully in the presence of noise, losing that which corresponds to the least-significant bits, in digital terms, first. With a digital system, in which the differences between possible states generally does not follow the "analog" method, it is quite common to have all bits equally immune – and also equally vulnerable – to noise. Digital transmissions tend to exhibit a "cliff" effect; the information is received essentially "perfectly" until the noise level increases to the point at which the states can no longer be distinguished, and at that point all is lost (see Figure 6-3). Intermediate system are possible, of course, in which the more significant bits of information are better protected against the effects of noise than the less-significant bits. This is, again, one of the basic points of this entire discussion – that "analog" and "digital" do not refer to two things which are utterly different in kind, but which rather may be viewed in many cases as points along a continuum of possibilities.

PRACTICAL ASPECTS OF DIGITAL AND ANALOG INTERFACES 109

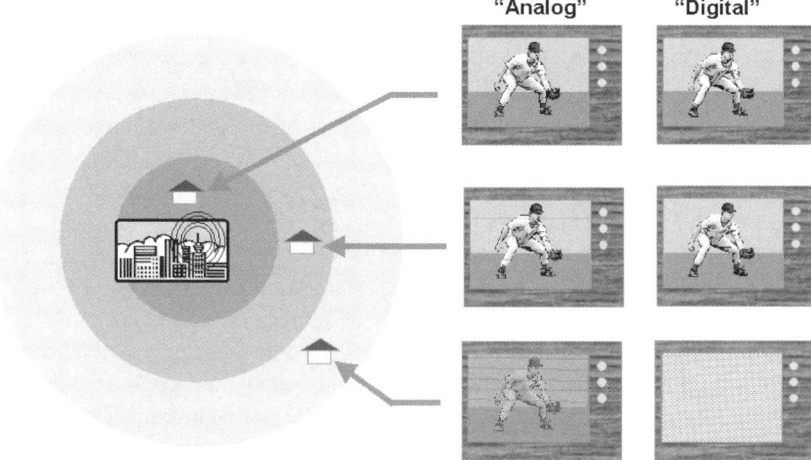

Figure 6-3 Analog and digital systems (in this case, television) in the presence of noise. Analog encoding automatically results in the least-significant bits of information receiving less "weighting" than the more significant information, resulting in "graceful degradation" in the presence of noise. The image is degraded, but remains usable to quite high noise levels. A "digital" transmission, in the traditional sense of term, would normally weight all bits equally in terms of their susceptibility to noise; each bit is as likely as another to be lost at any given noise level. This results in the transmission remaining essentially "perfect" until the state of the bits can no longer be reliably determined, at which point the transmission is completely lost – a "cliff effect."

6.4 Practical Aspects of Digital and Analog Interfaces

There are some areas in which digital and analog video interfaces, at least in their most common examples, differ in terms of the demands placed on physical and electrical connections and in the performance impact of these. As commonly implemented, a "digital" interface typically means one which employs straight binary encoding of values, transmitted in either a parallel (all the bits of a given sample sent simultaneously) or serial fashion. "Analog" interfaces, in contrast, are generally taken to mean those which employ a varying voltage or current to represent the intended luminance (or other quantity) at that instant. If we are comparing systems of similar capacity, and/or those which are carrying image data of similar format and timing, then some valid comparisons can be made between these two.

Let us consider both digital and analog interfaces carrying a 1280 × 1024 image at 75 frames/s, and in the form of separate red, green, and blue (RGB) channels. It will further be assumed that each of these channels is providing an effective eight bits of luminance information for each sample (i.e., "24-bit color", typical of current computer graphics hardware). If we assume that 25% of the total time is spent on "blanking" or similar overheard, then the peak pixel or sample rate (regardless of the interface used) is

$$\frac{75 \text{ Hz} \times 1280 \times 1024}{0.75} = 131.07 \text{ MHz}$$

We will consider three possible implementations; first, a typical "analog video" connection, with three physical conductors each carrying one color channel. Second, a straightforward parallel-digital interface, with 24 separate conductors each carrying one bit of information per sample, and finally a serial/parallel digital interface as is commonly used in PC systems today – three separate channels, one each for red, green, and blue, but each carrying serialized digital data.

If the "analog" system uses standard video levels, typically providing approximately 0.7 V between the highest ("white") and lowest ("black" or "blank") states, then in an 8 bit per color system, each state will be distinguished by about 2.7 mV. Further, at a 131 MHz sample rate, each pixel's worth of information has a duration of only about 7.6 ns. Clearly, this imposes some strict requirements on the physical connection. Over typical display interface distances (several meters, at least), signals of such a high frequency will demand a controlled-impedance path, in order to avoid reflections which would degrade the waveform. Further, the cable losses must be minimized, and kept relatively constant over the frequency range of the signal in order to avoid attenuation of high-frequency content relative to the lower frequencies in the signal. The sensitivity of such a system to millivolt-level changes also argues for a design which will protect each signal from crosstalk and from external noise, if the intended accuracy is to be preserved. The end result of all this is that analog video interfaces typically require a fairly high-quality, low-loss transmission medium, generally a coaxial cable or similar, with appropriate shielding and the use of impedance-controlled connector system.

The simple parallel digital system might be expected to fare a bit better in terms of its demands on the physical interface, but this is generally not the case. While the noise margin would be considerably better (we might assume a digital transmission method using the same 0.7 V p-p swing of the analog video, just for a fair comparison), this advantage is generally more than negated by the need for eight times the number of physical conductors and connections. And even with this approach, the problems of impedance control,. noise, and crosstalk cannot be completely ignored.

But what of the serial digital approach? Surely this provides the best of both worlds, combining a low physical connection count with the advantages – in noise immunity, etc. – of a "digital" approach. However, there is again a tradeoff. In serializing the data, the transmission rate on each conductor increases by a factor of eight – the bit rate on each line will be over 1 GHz, further complicating the problems of noise (both in terms of susceptibility and emissions) and overall signal integrity. Such systems generally have to employ fairly sophisticated techniques to operate reliably in light of expected signal-to-signal skew, etc.. Operation at such frequencies again requires some care in the selection of the physical medium and connectors.

It is also often claimed that one of the two systems provides some inherent advantages in terms of radiated noise, and the ability of the system using the interface to comply with the various standards in this area. However, the level of EM noise radiated by a given interface, all else being equal, is primarily determined by the overall signal swing and the rate at which the signal transitions. There is no clear advantage in this regard to an interface using either "analog" or "digital" encoding of the information it carries.

6.5 Digital vs. Analog Interfacing for Fixed-Format Displays

An often-heard misconception is that the fixed-format display types, such as the LCD, PDP, etc., are "inherently digital"; this belief seems to come from the confusion between "digital" and "discrete" which was mentioned earlier. In truth, the fact that these types employ regular arrays of discrete pixels says little about which form of interface is best suited for use with them. Several of these types actually employ analog drive at the pixel level – the LCD being the most popular example at present – regardless of the interface used. It should be noted that analog-input LCD panels are actually fairly common, especially in cases where these are to be used as dedicated "television" monitors (camcorder viewfinder panels being a prime example). The factor which provides the functional benefit in current "digital interface" monitors is not that the video information is digitally encoded, but rather that these interfaces provide a pixel clock, through which this information can be unambiguously assigned to the discrete pixels. The difficulty in using traditional "CRT-style" analog video systems with these displays comes from the lack of such timing information, forcing the display to generate its own clock with which to sample the incoming video.

Most often, analog-input monitors based on one of the fixed-format display types, such as an LCD monitor, will derive a sampling clock from the horizontal synchronization pulses normally provided with analog video (Figure 6-4). This is the highest-frequency "clock" typically available, although it still must be multiplied by a factor in the hundreds or thousands to obtain a suitable video sampling clock. This relatively high multiplication factor, coupled with the fact that the skew between the horizontal sync signal and the analog video is very poorly controlled in many systems employing separate syncs, is what makes the job of properly sampling the video difficult. The results often show only poor to fair stability at the pixel level, which sometimes leads to the erroneous conclusion that the analog video source is "noisy." But it should be clear at this point that the difficulties of properly implementing this sort of connection have little to do with whether the video information is presented in analog or digital form; "digital" interfaces of the most common types would have even worse problems were they to be deprived of the pixel clock signal.

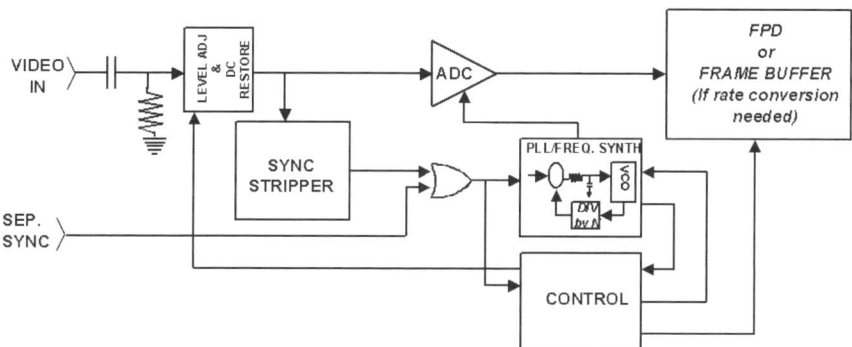

Figure 6-4 Regeneration of a sampling clock with standard analog video. In almost all current analog video interface standards, no sampling clock is provided – making it difficult to use such interfaces with fixed-format displays such as LCDs. To permit these types to operate with a conventional analog input, the sampling clock is typically regenerated from the horizontal synchronization signal using PLL-based frequency synthesis techniques. This can work quite well, but in some cases may be difficult due to poor control of the sync skew and jitter relative to the video signal(s).

112   BASICS OF ANALOG AND DIGITAL DISPLAY INTERFACES

Still, digital interfaces have proven to be worthwhile for many fixed-format, non-CRT display types. This results in part from the lack of pixel-level timing information in the conventional analog standards, as reviewed above, and also in part from the fact that these display technologies have traditionally been designed with "digital" inputs, to facilitate their use as embedded display components in various electronic systems (calculators, computers, etc.). There is also one other significant advantage when fixed-format displays are to be used as replacements for the CRT in most computer systems. Due to the inherent flexibility of the CRT display, "multi-frequency" monitors (those which can display images over a very wide range of formats and update rates) have become the norm in the computer industry. The only way to duplicate this capability, when using display devices providing only one fixed physical image format, is to include some fairly sophisticated image scaling capabilities, achieved in the digital domain. With the need for digital processing within the display unit itself, a "digital" form of interface most often makes the most sense for these types.

6.6 Digital Interfaces for CRT Displays

With the rise of non-CRT technologies and the accompanying development of digital interface standards aimed specifically at these displays, there has quite naturally been considerable discussion regarding the transition of the CRT displays to these new interfaces. Several advantages have been claimed in order to justify the adoption of a digital interface model by this older, traditionally analog-input, technology. Some of these have been discussed already, but there are several new operational models for the CRT that have been proposed in this discussion. These deserve some attention here.

Several of the claimed advantages of digital-input CRT monitors are essentially identical to those previously mentioned for the LCD and other types; that such interfaces are "more immune" to the effects of noise, provide "higher bandwidth", improved overall image quality as a result of these, etc.. The arguments made in the previous section still apply, of course; they are not dependent on the display technology in question. The CRT-based monitor is even less concerned with the one advantage "digital" interfaces provide for the non-CRT types: the presence of pixel-level timing information (a "pixel clock" or "video sampling clock"). The CRT, by its basic nature and inherent flexibility, is not especially hampered by an inability to distinguish discrete "pixels" in the incoming data stream. The important concern in CRT displays is that the horizontal and vertical deflection of the beam be performed in a stable, consistent manner, and this is adequately provided for by the synchronization information present in the existing analog interface standards. To justify the use of a digital interface in these products, then, we must look to either new features which would be enabled by such an interface (and not possible via the traditional analog systems), or the enabling of a significant cost reduction, or both.

The argument for using digital interfaces with CRT displays as a cost-reduction measure is based on the assumption that this will permit functions currently performed in the analog domain to be replaced with presumably cheaper digital processing. Most often, the center of attention in this regard is the complexity required of "fully analog" CRT monitors in providing the format and timing flexibility of current multi-frequency CRT displays. Admittedly, this does involve significant cost and complexity in the monitor design, with results which sometimes are only adequate. However, there is little question that the final output of the deflection circuits must be "analog" waveforms, to drive the coils of the deflection yoke. So simplification of the monitor design through digital methods is generally assumed to mean that the display will be operated at a single fixed timing (only one set of sweep rates), thereby turning the CRT itself into a "fixed-format" display! In this model, as in the LCD and other monitor types, the adaptation to various input formats would be performed by digital image scaling devices, producing as their output the one format for which (presumably) the display has been optimized. (The other possible alternative would be to simply operate the entire system at a single "standard" format/timing – however, this is not dependent on the adoption of a digital-interface model, and in fact is the norm for some current CRT-dominated applications such as television or the computer workstation market.)

6.7 The True Advantage of Digital

From the preceding discussion, the reader may now be under the impression that there is no real advantage to digital interfaces over analog, or that the author is somehow "anti-digital".

This is not the case; the purpose so far has been to show that the terms "analog" and "digital" do not, as is commonly assumed, describe two vastly different and unrelated fields of study, and in fact are governed by the same basic constraints as any other electrical system. Many of the supposed differences between the two do not have anything to do with how the information to be transmitted is encoded, but rather on completely unrelated factors.

Given this, why should we expect a continued transition away from the established analog standards and toward the much-hyped "all-digital" future? It should now be apparent that, as long as "digital" interfaces are being used in a manner that simply duplicates the functions of their analog predecessors, there is little practical impetus for such a change.

However, the advantage of digital interfaces, for display applications as well as many others, lies not in the advantages at the interface itself but rather what providing information in this form will enable. The true advantage of digital encoding of video information lies in the ease with which such information may be processed and stored. In the analog domain, storing even a single frame – let alone a significant portion of a video stream – requires complex and relatively expensive equipment, and the quality of the signal will inevitably be degraded by the process. And there are also no practical analog means through which the image information can be significantly altered: you cannot change the image format, the frame rate, etc., readily. But all of these and more are easily achieved in the digital domain through inexpensive processors. Digital memory sufficient to store hours of video has become relatively cheap – and this storage will be essentially "perfect", with no degradation of the image quality at all. Digital signal processing and storage is also the primary enabling factor in practically all of the recent advances in display systems – from the compression techniques used to make HDTV a practical reality, to the encryption techniques used to protect copyrighted imagery, and on to the scaling and processing methods that make fixed-format displays a viable alternative to the traditional CRT. These are what truly makes a digital representation worthwhile, and each is covered in some depth in later chapters.

6.8 Performance Measurement of Digital and Analog Interfaces

Before proceeding on to review current analog and digital interface standards, it is appropriate to look at some of the techniques and terminology used to measure and describe signal quality in each type. In both cases, observation of the signal itself, at relevant points in the

114 BASICS OF ANALOG AND DIGITAL DISPLAY INTERFACES

system, via an oscilloscope (and often more sophisticated equipment) remains the primary means of determining signal "goodness." However, there are different areas of concern for analog and digital signalling, and somewhat different terminology used to describe the relevant points of each. There are also some additional, more specialized techniques which should be covered, having to do with the characterization of other parts of the system.

6.8.1 Analog signal parameters and measurement

Many of the parameters and measurements discussed in this section have been defined, at least for computer video purposes, by the VESA Video Signal Standard (VSIS) and the related VESA test procedure document, "Evaluation of Analog Display Graphics Subsystems." Similar standards and procedures have been published by such bodies as the EIA and SMPTE. The reader is directed to those standards for further information on the specifics for each.

A typical analog signal, as might be seen on an appropriate wideband oscilloscope, is shown in Figure 6-5. This might be one small portion of an analog video transmission, but in showing several transitions it demonstrates many of the important parameters for video signal characterization. In this example, the signal in question includes synchronization pulses (as is common in television practice, but rarely seen in computer display systems), but the concepts will apply to any analog video system.

First, assuming that we have ensured that the source of this signal is delivering a full "black to white" swing, observation of the signal provides a quick check of the absolute and relative amplitudes of the signal and its various components. Note that most analog video systems are intended for use with AC-coupled inputs at the display or receiver, and so a

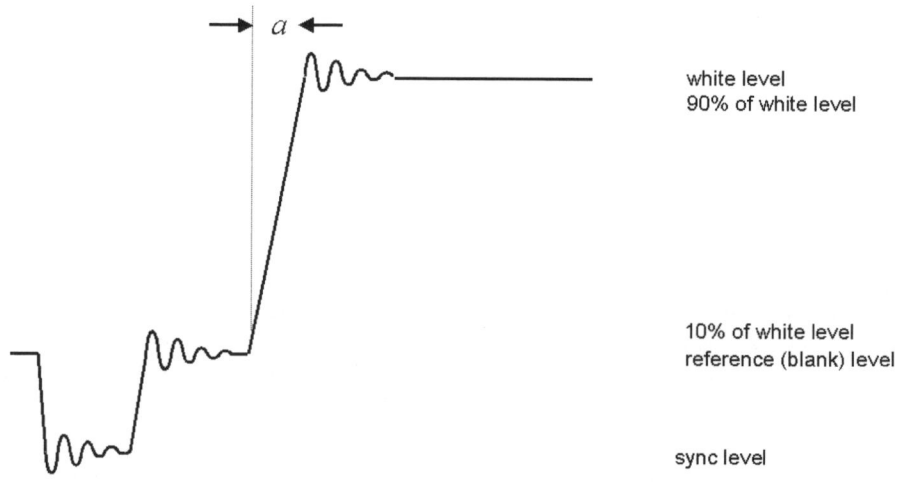

Figure 6-5 A portion of a typical analog video signal as it might appear on an oscilloscope, showing a horizontal sync pulse and a transition from the blanking level to the reference white level. Rise and fall times for such signals are normally measured between the 10% and 90% points on each transition, as shown. Note the "ringing" (damped oscillation) which follows each transition.

PERFORMANCE MEASUREMENT OF DIGITAL AND ANALOG INTERFACES

measurement of the signal offset from the local reference (ground) may or may not be meaningful. Besides the amplitudes themselves, though, the signal behavior during the transitions is among the most important observations that can be obtained from such a view. The parameters which describe this behavior include the following.

6.8.1.1 Rise/fall time

Among the most basic of indicators of the "sharpness" of the video signal – and related to the bandwidth of the channel required for its successful transmission – are the rise and fall times. These are as indicated by item "*a*" in Figure 6-5, and are simply measurements of the duration of the transition, between specified points. Typically, standard rise and fall time measurements are made between the 10% and 90% of full-scale points, on a full-amplitude (i.e., from "black" to "white") transition. It is possible, of course, to use other defined points – and one might ask why the 0% and 100% values are not used instead. The reason for this is the difficulty of locating these points at a time which is unambiguously part of the transition itself, and not during some other transient event such as the overshoot/ringing portions of the transition.

6.8.1.2 Overshoot/undershoot; settling time

As in any practical system, the signals observed in analog video interfaces will never show "perfect," instantaneous transitions between any two states. Owing to the ever-present reactive elements in any real circuit, there will always be some degree of "overshoot" and "ringing" associated with a change of signal level. These are typically as shown in Figure 6-6. While they are unavoidable, careful attention to the design and construction of the interface, and especially the termination components, can greatly minimize their impact. If small enough in both amplitude and duration, these phenomena will have essentially no visible

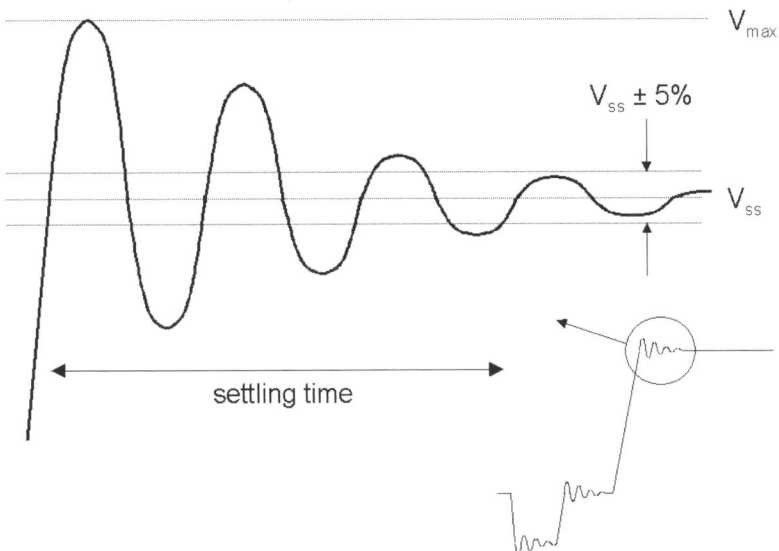

Figure 6-6 Overshoot, ringing, and settling time. The ringing following the full-range transition of Figure 6-5 is magnified here, showing details of how overshoot and settling time are defined.

116 BASICS OF ANALOG AND DIGITAL DISPLAY INTERFACES

effect on the displayed image. ("Undershoot" is commonly used to refer to the same phenomena as "overshoot", but on a negative-going transition. We will assume that the two can be treated identically here.)

The amplitude of the overshoot may be given simply by noting the amount to which the final steady-state amplitude of the signal is exceeded. In other words, referring again to Figure 6-6, the overshoot amplitude is $V_{max} - V_{ss}$ (or in the case of undershoot, $V_{ss} - V_{min}$). However, unless the over/undershoot is truly excessive, an even more important concern may be the settling time. This may be characterized by noting the time between a given point of the transition – commonly, either the 90%-of-full-scale point, or the first point *after* the initial overshoot when the signal passed the final, steady-state value, are used – to the time at which the signal settles to within a certain margin (typically, ±5–10%) of the final value.

Other parameters of potential concern are not readily observable from a simple observation such as this, and require either specific signals or reference to other signals for their determination. These include jitter, skew, integral and differential linearity, and monotonicity. Signals representative of those used to measure these factors are shown in Figures 6-7 and 6-8. Note that jitter and skew are measures of the signal's accuracy and stability *in time*, relative to a given reference. In typical analog video systems, this reference may be a sampling clock (if one is provided or accessible), or more commonly the synchronization signals used by the system. Skew, jitter, and other mismatches (including those of amplitude) may also be characterized between multiple analog channels, as in an RGB three-channel video interface.

6.8.1.3 Skew
Skew is the difference in temporal location of supposedly simultaneous events between two (or more) channels. For example, Figure 6-7 shows a situation in which the three channels of

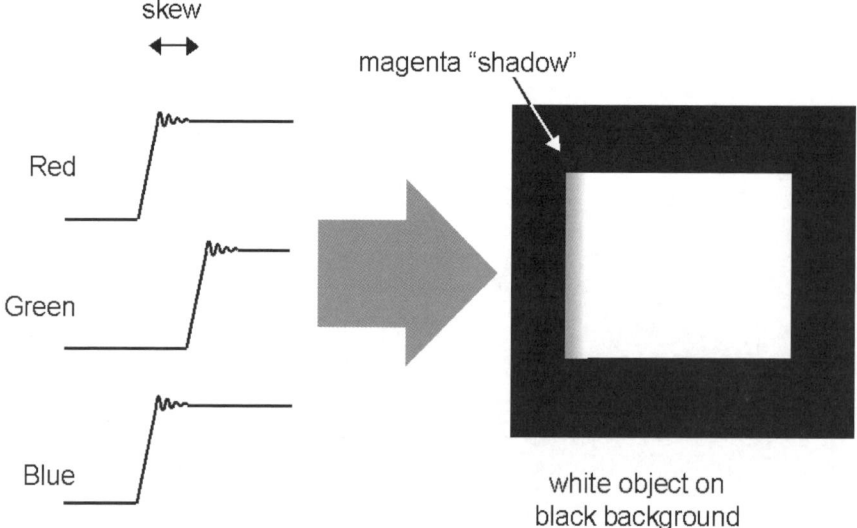

Figure 6-7 Skew. Here, the green signal transition has been delayed relative to the red and blue; this might result in the appearance shown at right, as what was supposed to have been a white square has a magenta edge. Should the transition back to the black level be similarly delayed, there would also of course be a green edge on the right side of the square.

PERFORMANCE MEASUREMENT OF DIGITAL AND ANALOG INTERFACES 117

a video interface are all supposed to be making identical transitions (this might represent the start of a white area on the display), but the green is noticeably later than the other two. The time difference, as measured between identical points on the waveform, would then be stated as the *skew* between the two signals (or between a given signal and an agreed-to reference). In terms of the appearance of the resulting image, this particular case would result in what should have been a clean black-to-white transition on the screen appearing to have a magenta "shadow" to the left (Figure 6-7b). Skew results from mismatches in the path delay of one channel relative to another, either from physical length differences in the channel (cabling, PC traces, etc.) or different cumulative propagation delays within discrete components.

6.8.1.4 Jitter

Jitter is similar to skew, in that is relates to the location of a particular portion of the waveform in time, but differs in that skew refers to a steady-state error while jitter is the amplitude of expected or observed transient temporal errors. Jitter may, in other words, be viewed as the "noise" in the location of the waveform in time, relative to another signal or established reference. Most often, jitter is observed by using the system clock (the clock intended for sampling or latching the signal in question) as the trigger or reference for the oscilloscope observing the signal, and noting the amplitude of the variation in the signal edge position. This is shown in the example of Figure 6-8.

Jitter may also occur, and be measured, between a given reference point within a signal and a later point within that same signal; in this case, it represents the cumulative timing in-

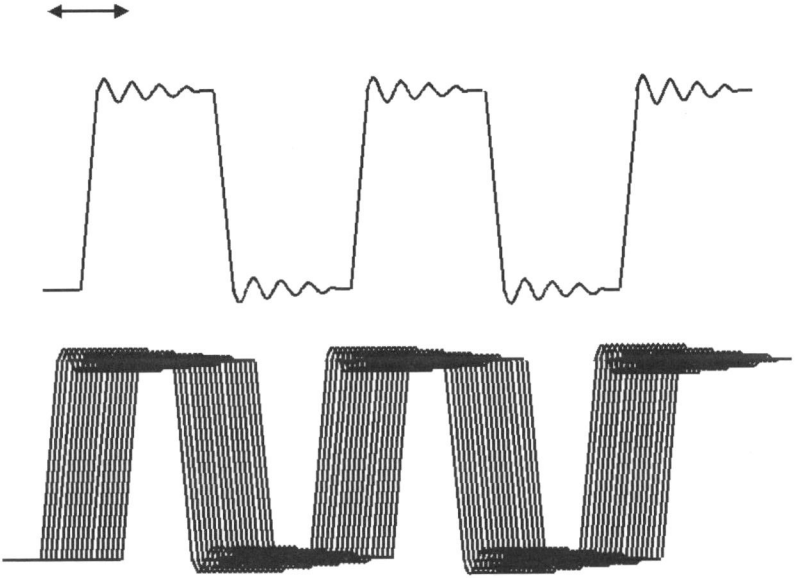

Figure 6-8 Jitter. Here, the upper trace is used as the reference clock, and we are observing the jitter (temporal instability relative to the reference) of a similar signal shown below. The excursion in time of the edge of the signal being measured, relative to the reference, is the jitter; it may be expressed in terms of the absolute maximum jitter, average, etc..

118 BASICS OF ANALOG AND DIGITAL DISPLAY INTERFACES

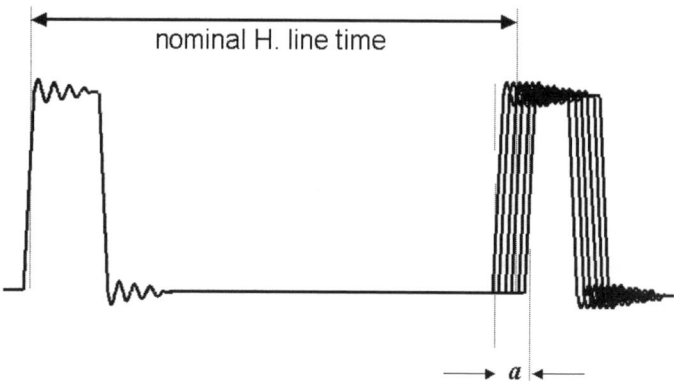

Figure 6-9 Jitter within a single signal. In this case, the position of the leading edge of a horizontal sync pulse is observed to be unstable relative to the preceding pulse; i.e., the duration of a horizontal time is not perfectly fixed.

stability or error between the two points. An example of this is shown in Figure 6-9, in which a horizontal sync pulse edge has been defined as the reference point, and the jitter is being measured as the maximum peak-to-peak variation in the location of the subsequent edge.

6.8.1.5 Linearity

Linearity error can refer to either non-linearities introduced into a signal as it has been passed by a supposedly linear operation (i.e., an amplifier) or in the generation of that signal (as in the digital-to-analog converter (DAC) common to most computer-graphics video outputs). Linearity may be specified and measured in a number of different ways. The VESA video standard defines two measurements for linearity, *differential* and *integral*. This assumes, as is often the case in computer graphics systems, that the analog signal is being created under the control of digital signals (via a digital-to-analog converter), and so may be expected to have discrete and identifiable "steps" from one value to the next. With reference to Figure 6-10, differential linearity is stated as the difference between adjacent values, normalized to the expected value corresponding to one LSB change (and therefore is ideally 1.000). Expressed mathematically, this is

$$\text{DNL}(n) = \frac{V(n) - V(n-1)}{V_{\text{LSB(ideal)}}}$$

The integral linearity error, on the other hand, is simply the difference between the actual signal level at any given point, compared to its ideal value, again normalized to the ideal LSB value, or

$$\text{INL}(n) = \frac{V(n) - n \times V_{\text{LSB(ideal)}}}{V_{\text{LSB(ideal)}}}$$

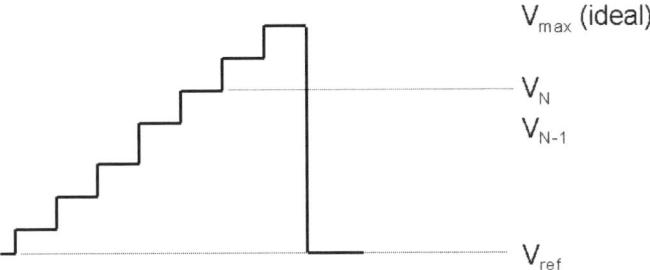

Figure 6-10 Linearity. In this simple example, the output of a 3-bit (eight level) digital-to-analog converter is being checked. There are obvious linearity problems, as not all transitions are of equal amplitude. The *differential linearity error* is the difference between the amplitude of a given transition and the theoretical value for that transition; in this case, a transition of one least-significant-bit (LSB) should be equal to the ideal $V_{max}/7$. (Note that the LSB value for an N-bit system is $V_{max}/(2^N - 1)$.) The integral linearity error is the total error between the measured amplitude at a given level, and the theoretical value for that level (for level N, this is $N \times [V_{max}/(2^N - 1)]$).

As noted, these direct measurements of signal linearity are based on the assumption that the signal is, in fact, created as a series of discrete steps. It is also possible to measure and express linearity error by arbitrarily selecting reference points on the input signal, and then noting the error between the actual and expected value (assuming a linear process) of the output signal.

Historically, other measurements have been used to indirectly measure the linearity of a given analog system. These generally are based on the fact that non-linearities in a system will result in the generation of spurious spectral components, not present in the original signal but related to the original components. An example of one such is the statement of the *total harmonic distortion* (THD) of a system, which is a measurement of the energy present in such spurious harmonics as a percentage of the total output signal.

6.8.2 Transmission-line effects and measurements

As noted above, analog signals are susceptible to distortion and error arising from improper terminations and similar "transmission-line" concerns. While the effect these have on the signal can certainly be characterized through many of the measurements already discussed, it is important to be able to recognize such situations when they are seen.

The only real difference between what are classed as "transmission-line" phenomena and other non-ideal behaviors of a system is the introduction of a time factor; the delay introduced by the physical length of the transmission channel. This has the effect of separating, or spreading out in time, the effects of distortion-causing elements. For example, the most common "transmission-line" problems generally result from impedance mismatches at various point in the signal path, and the reflections which result from these. In simple circuit analysis, an incorrect impedance results only in less efficient power transfer. But if a significant time delay is introduced between a signal source and the load, or receiver, of that signal, the analysis becomes significantly more complex. The lumped impedance of the source, the distributed or characteristic impedance of the signal path (the physical conductors), and the

120   BASICS OF ANALOG AND DIGITAL DISPLAY INTERFACES

termination of that path provided by the receiver, plus the path length between all points of concern, must all be considered.

Consider Figure 6-11a, which shows a simplified version of an idealized analog video system. Analog video specifications have generally been written assuming a system impedance standard of 75 Ω, as is shown here. And should all of the indicated impedances exactly meet this requirement, no problems arise. However, in Figure 6-11b, an impedance mismatch exists; in this case, the characteristic impedance of the video cable is incorrect. This results in reflections being generated at the cable endpoints, as shown, and the resulting distortion of the signal. In video systems, depending on the length of the path between signal source and receiver (and so the delay between the two) such reflections can result in the visible effect known as "ghosting" of images in the display. However, the distortion of the signal "edges" or transitions will often result in more subtle effects. Deviations from the ideal here, as in Figure 6-12, can result in changes in the system timing should they occur on an edge used as a clock or timing reference, such as would be the case with a synchronization signal or sampling clock edge. Sufficient distortion may even result in additional, spurious triggering of the circuit the signal is to clock.

Detection and characterization of such problems in practical systems can be done through observation of these effects in the signal as viewed at the source or receiver, but more spe-

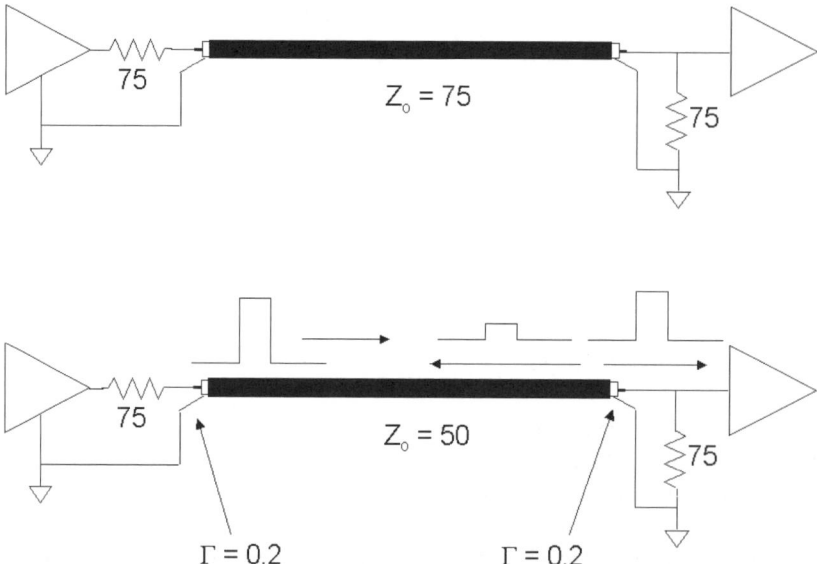

Figure 6-11 Transmission-line effects in video cabling. In a properly matched system (top), source, load, and cable impedances are all identical (the standard for video is a 75-Ω system impedance, as shown here), and the video signal may be transmitted with no reflections at any point. If any of the impedances do not match, as in the lower figure, the impedance discontinuity at that point generates a reflection. In this example, a cable of the incorrect impedance has been used, making for a mismatch condition at both the source and load end. Inserting a 50-Ω cable into a 75-Ω system results in a reflection coefficient ($\Gamma = (Z_L - Z_O)/(Z_L + Z_O)$) of 0.2 at both; any pulse arriving at either termination will produce a reflection of 20% the amplitude of that pulse. With reflections produced at both ends of the cable, such conditions can result in visible "ghost" images, in addition to causing the signal integrity problems discussed in the text.

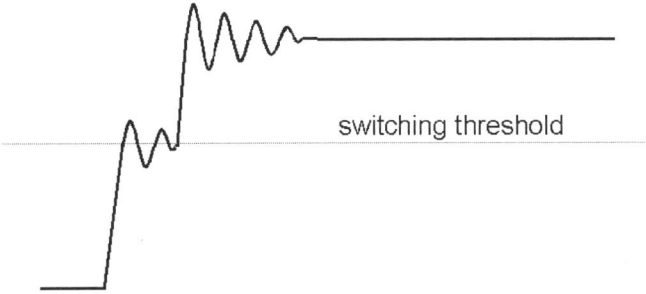

Figure 6-12 Impedance mismatches may also result in problems at signal transitions, as shown here. In this example, a mismatch results in an initial transition to a lower amplitude than would otherwise be the case; the next transition to the final value does not occur until the reflection has returned from the other end of the cable. Should the first transition – and possible subsequent ringing - occur around the switching threshold of an input connected to this line, in the case of a logic signal, such distortions of the signal edge can result in unwanted triggering.

cialized equipment is also available to analyze transmission-line systems. The *time-domain reflectometer*, or TDR, is a common example. This device operates by launching pulses into the signal path, and analyzing the magnitude, polarity, and timing of the reflections it then sees. Such instruments are capable of providing considerably more detailed looks at the system's impedance characteristics, including detection of mismatches.

6.8.3 Digital systems

From the earlier discussion, it should be apparent that "digital" and "analog" signals are not really different in kind; the two classifications are, after all, simply different means of encoding information onto what remains just an electrical signal. Therefore, it should not be surprising that many of the same problems can impact "digital" systems, and many of the same measurements are still relevant. Obviously, some will not be; in a simple binary-encoded system, for instance, the linearity of the channel conveying the signal is not of particular concern. However, rise and fall time, jitter, skew, and so forth are still very much a concern in many digital applications. Even greater importance is often assigned to those factors that may affect the "edges" or transitions of the signal, due to the often-greater importance of the timing of these transitions in digital systems.

Due to the nature of most digital systems, however, a single, relatively simple examination of the signal can often serve to determine the acceptability of the system's performance. In the majority of digital systems, the primary concerns with the signal's integrity center on being able to unambiguously identify one of two possible states at certain discrete points in time, as determined by a clock signal. Therefore, if the signal can be determined to provide sufficient noise margin, and to be settled at the correct level at the time of the active clock edge, the system may be expected to perform correctly. This may be quickly checked through the observation of the *"eye pattern"* using an oscilloscope. An eye pattern is produced by generating a signal with approximately a 50% average duty cycle (i.e., it spends as much time in the high state as in the low); alternating between the two states is ideal. Using the appropriate clock signal as the trigger source (the same clock which the intended receiver

122 BASICS OF ANALOG AND DIGITAL DISPLAY INTERFACES

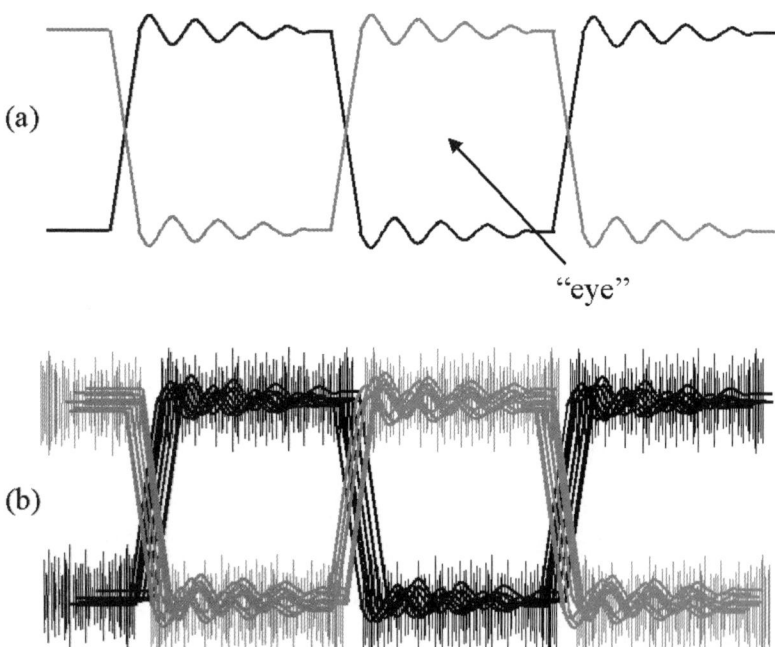

Figure 6-13 Eye patterns. Commonly used as a quick visual check of digital signal quality, an "eye pattern" is produced by causing transitions of the signal in question to overlay on an oscilloscope display as shown in (a). Noise on the signal, amplitude deviations, jitter, etc., will all tend to close the "eye," as shown in (b). The open area of the eye shows the period during which the states of the signal can be properly distinguished in both amplitude and time; should the "eye" close entirely, the signal is unusable.

would normally use to sample the data), the oscilloscope timebase is adjusted such that multiple states of the signal are displayed within the same portion of the screen. This results in the pattern shown in Figure 6-13a.

Most digital oscilloscopes, and many analog instruments with storage or "variable persistence" features, may be set to show multiple samples of the signal, overwritten upon each other. Unless each has exactly the same amplitude and timing characteristics, differences here will show up as a "smearing" of the signal traces around the nominal amplitude values and edge locations (Figure 6-13b). The effects of jitter smear the location of the signal edges in time; noise and amplitude errors smear the trace in the vertical axis. The worse either or both of these become, the more "closed" the "eye" (the inner portion of each period) will appear. Should the eye close entirely, or at least to the point at which the state of the signal cannot be unambiguously determined during the expected sampling time, the transmission will experience errors.

7

Format and Timing Standards

7.1 Introduction

A large part of the many standards having to do with displays and display interfaces concern themselves not with the nature of the physical or electrical connection, but rather solely with the format of the images to be communicated, and the timing which will be used in the transmission. In fact, these are arguably among the most fundamental and important of all display interface standards, since often these will determine if a given image can be successfully displayed, regardless of the specific interface used. Established format and timing standards often ensure the usability of video information over a wide range of different displays and physical interfaces. In some cases, such as the current analog television broadcast standards, the specifications for the scanning format and timing are fundamental to the proper operation of the complete system.

But how are these standards determined, and what factors contribute to these selections? This chapter examines the needs for image format and timing standards in various industries and the concerns driving the specific choices made for each. In addition, we cover some of the problems which arise when, as so often is the case, electronic imagery created under one standard must be converted for use in a different system

7.2 The Need for Image Format Standards

As was noted in Chapter 1, the vast majority of current electronic imaging and display systems operate by sampling the original image for representation as an electronic signal. In most "fully digital" systems, the image is represented via an array of discrete samples, or "pixels" (picture elements), but even in analog systems such as the conventional broadcast televisions standards, there is a defined structure in the form of separate scan lines. Clearly, the greater the number of pixels and/or scan lines used to represent the image, the greater the amount of information (in terms of spatial detail) that will be conveyed. However, there will also obviously be a limit to the number of samples per image which can be supported by any practical system; such limits may be imposed by the amount of storage available, the physical limitations of the image sensor or display device, the capacity of the channel to be used to convey the image information, or a combination of these factors.

Even if such limitations are not a concern – a rare situation, but at least imaginable – there remains a definite limit as to the number of samples (the amount of image detail which can be represented) in any given application. It may be that the image to be transmitted simply does not contain meaningful detail beyond a given level; it may also be the case that the viewer, under the expected conditions of the display environment, could not visually resolve additional detail. So for almost any practical situation, we can determine both the resolution (in the proper sense of the term, meaning the amount of detail resolvable per a given distance in the visual field) required to adequately represent the image, as well as the limits on this which will be imposed by the practical restriction of the imaging and display system itself.

But this sort of analysis will only serve to set an approximate range of image formats which might be used to satisfy the needs of the system. A viewer, for instance, will generally not notice or care if the displayed image provides 200 dpi resolution or 205 dpi – but the specifics of the image format standard chosen will have implications beyond these concerns. Specific standards for image formats ensure not only adequate performance, but also the interoperability of systems and components from various manufacturers and markets. Many imaging and display devices, as previously noted, provide only a single fixed array of physical pixels, and the establishment of format standards allows for optimum performance of such devices when used together in a given system. Further, establishment of a set of related standards, when appropriate, allows for relatively easy scalability of image between standard devices, at least compared to the situation that would exist with completely arbitrary formats.

The selection of the specific image and scanning formats to be used will come from a number of factors, including the desired resolution in the final, as-displayed image; the amount of storage, if any, required for the image, and the optimum organization (in terms of system efficiency) of that storage; and the performance available within the overall system, including the maximum rate at which images or pixels may be generated, processed, or transmitted. As will be seen shortly, an additional factor may be the availability of the necessary timing references ("clocks") for the generation or transmission of the image data. The storage concern is of particular importance in digital systems; for optimum performance, it is often desired that the image format be chosen so as to fit in a particular memory space,[1] or permit organization into segments of a convenient size. For example, many "digital" formats have been established with the number of pixels per line a particular power of two, or at least divisible in to a convenient number of "power of 2" blocks. (Examples being the 1024 × 768 format, or the 1600 × 1200 format – the latter having a number of pixels per line which is readily divisible by 8, 16, 32, or 64.)

One last issue which repeatedly has come up in format standard development, especially between the television and computer graphics industries, has been the need for "square

---

[1] Examples of such include the 1152 × 864 pixel format used by Apple Computer; this is a 4:3 aspect ratio, square-pixel format, which comes very close to exactly filling a "1 meg" (1024 × 1024, or 1,048,576 locations) address space. Each line is also 9 × 128 pixels in length, another fairly convenient number.

pixel" image formats – meaning those in which a given number of picture elements or samples represents the same distance in both the horizontal and vertical directions. (This is also referred to as a "1:1 pixel aspect ratio," although since pixels are, strictly speaking, dimensionless samples, the notion of an aspect ratio here is somewhat meaningless.) Television practice, both in its broadcast form and the various dedicated "closed-circuit" applications, comes from a strictly analog-signal background; while the notion of discrete samples in the vertical direction is imposed by the scanning-line structure, there is really no concept of sampling horizontally (or along the direction of the scan line). Therefore, television had no need for any consideration of "pixels" in the standard computer-graphics sense, and certainly no concern about the image samples being "square" when digital techniques finally brought sampling to this industry. (Instead, as will be seen in the later discussion of digital television, sampling standards were generally chosen to use convenient sampling clock values, and the non-square pixel formats which resulted were of little concern.) In contrast, computer-graphics practice was very much concerned with this issue, since it would be very inconvenient to have to deal with what would effectively be two different scaling factors – one horizontal and one vertical – in the generation of graphic imagery. As might be expected, this has resulted in some degree of a running battle between the two industries whenever common format standards have been discussed.

7.3 The Need for Timing Standards

Image format standards are only half of what is needed in the case of systems intended for the transmission and display of motion video. Again, as noted in Chapter 1, the rendition of motion is generally through the display of still images in rapid succession; establishing a standard format for the two-dimensional aspect of these images still leaves the problem of timing the overall sequence. And even when motion is not a major concern of the display system, most display technologies will still require continuous refresh of the image being displayed. The timing problem is further not limited to the rate at which the images themselves are presented; it is generally not the case that the entire image is produced instantaneously, or all the samples which make up that image are transmitted or presented at the same time. Rather, the image is built up piece-by-piece, in a regular, steady sequence of pixels and lines. The timing of this process (which typically is of the type known as raster scanning) is also commonly the subject of display timing standardization.

At the frame level, meaning the rate at which the complete images are generated or displayed, the usual constraints on the rate are the need for both convincing motion portrayal, and the characteristics of the display device (in terms of the rate at which it must be refreshed in order to present the appearance of a stable image). Both of these factors generally demand frame rates at least in the low tens of frames per second, and sometimes as high as the low hundreds. Restricting the maximum rate possible in any given system are again the limitations of the image generation and display equipment itself, along with the capacity limitations of the channel(s) carrying the image information.

But within this desirable range of frame rates, why is it desirable to establish specific standards, and to further standardize to the line and pixel timing level? Standardized frame rates are desirable again to maximize interoperability between different systems and applications, and further to reduce the cost and complexity of the equipment itself. It is generally simpler and less expensive to design around a single rate, or at worst a relatively restricted range, as opposed to a design capable of handling any arbitrary rate. This same reasoning generally applies to a similar degree, if not more so, to the line rate in raster-scanned displays. A further argument for standardized frame rates comes from the need to preserve convincing motion portrayal as the imagery is shared between multiple applications. While frame-rate conversion is possible, it generally involves a resampling of the image stream and can result in aliasing and other artifacts. (An example of this is discussed later, in the methods used to handle the mismatch in rates between film and video systems.)

7.4 Practical Requirements of Format and Timing Standards

Several additional practical concerns must also be addressed in the development of format and timing standards for modern electronic display systems. Among these are those standards which may already have been established in other, related industries, along with the constraints of the typical system design for a given market or application.

For the largest electronic display markets today, which are television (and related entertainment display systems) and the personal computer market, clearly the biggest single "outside" influence has come from the standards and practices of the motion-picture industry. Film obviously predates these all-electronic markets by several decades, and remains both a significant source of the images used by each and, in turn, is itself more and more dependent on television and computer graphics as a source of imagery. Film has had an impact on display format and timing standards in a number of areas, some less obvious than others.

Probably the clearest example of the influence of the motion-picture industry is in the aspect ratios chosen for the standard image formats. Film, in its early days, was not really constrained to use a particular aspect ratio, but it soon became clear that economics dictated the establishment of camera and projector standards, and especially standards for the screen shape to be used for cinematic presentation. These early motion-picture standards efforts included the establishment of the so-called "Academy standard" (named for the Academy of Motion Picture Arts & Sciences), and its 4:3 image aspect ratio.[2] During the development of television broadcast standards in the 1940s, it was natural to adopt this same format for television – as motion pictures were naturally expected to become a significant source of material for this new entertainment medium. This ultimately led to cathode-ray tube (CRT) production standardizing on 4:3 tubes, and 4:3 formats later became the norm in the PC industry for this reason. With the 4:3 aspect ratio so firmly established, the vast majority of non-CRT graphic displays (such as LCD or plasma display panels) are today also produced with a 4:3 aspect ratio! Ironically, the motion-picture industry itself abandoned the 4:3 format standard years ago, in large part due to the perceived threat of television. Fearing loss of their audience to TV, the film industry of the 1950s introduced "widescreen" formats[3] such as "Panavision" or "Cinemascope," to offer a viewing experience that home television could not match. Today, though, this has led to television now following film into the "widescreen"

---

[2] The standard "Academy format" is often stated as 1.37:1, rather than 1.33:1 (or 4:3). However, this slightly wider specification includes that portion of the film devoted to the optical soundtrack; the actual image aperture is 4:3, a ratio first seen in silent films.

[3] It should be noted that the formats of cinematic presentation often do not match those of the image as it actually appears on the film. Many film standards employ "anamorphic" photography and projection, in which the image is both captured and presented through special optics which intentionally distort it (and then restore the intended geometry during projection) for more efficient storage on the film stock.

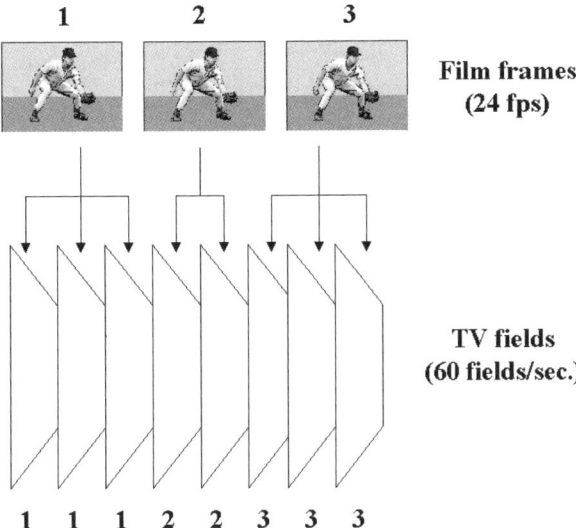

Figure 7-1 "3-2 pulldown." In this technique for displaying motion-picture film within an interlaced television system, successive frames of the 24 frames/s film are shown for three and then two fields of the television transmission. While this is a simple means of adapting to the approximately 60 Hz TV field rate, it can introduce visible motion artifacts into the scene.

arena – with a wider image being one of the features promoted as part of the HDTV systems now coming to market.

The motion-picture industry has also had some relationship, although less strongly, with the selection of frame rates used in television and computer display standards. Television obviously came first. In establishing the standard frame and field rates for broadcast TV, the authorities in both North America and Europe looked first to choose a rate which would experience minimum interference from local sources of power-line-frequency magnetic fields and noise. In North America, this dictated a field rate of 60 fields/s, matching the standard AC line frequency. Similarly, European standards were written around an assumption of a 50 fields/s rate. But both of these also – fortunately – harmonized reasonably well with standard film rates. In North America, the standard motion-picture frame rate is 24 frames/s. This means that film can be shown via standard broadcast television relatively easily, using a method known as "3:2 pulldown". As 24 and 60 are related rates (both being multiples of 12), 24 fields/s film is shown on 60 fields/s television by showing one frame of the film for three television fields, then "pulling down" the next frame and showing it for only two fields – and then repeating the cycle (Figure 7-1). In Europe, the solution is even simpler. The standard European film practice was (and is) to use 25 frames/s, not 24[4] – and so each frame of film requires exactly two television fields. (The "3:2 pulldown" technique is not without some degree of artifacts, but these are generally unnoticed by the viewing public.)

---

[4] When North American films are shown in Europe, they are simply shown at 25 fields/s – a 4% speedup from standard. Similarly, European films run 4% slow when shown on North American projectors. This change is generally negligible, except in those situations where it is important to maintain the absolute pitch of tones on the soundtrack, etc..

With the base frame and field rate chosen, the establishment of the remaining timing standards within a given system generally proceeds from the requirements imposed by the desired image format, and the constraints of the equipment to be used to produce, transmit, and display the images. For example, the desired frame rate and the number of lines required for the image, along with any overhead required by the equipment (such as the "retrace" time required by a CRT display) will constrain the standard line rate. If we assume a 480-line image, at 60 frames/s, and expect that the display will require approximately 5% "overhead" time in each frame for retrace, then we should expect the line rate to be approximately:

$$\frac{480 \text{ lines/frame} \times 60 \text{ frames/s}}{1.00 - 0.05} = 30{,}315.8 \text{ Hz}$$

However, the selection of standard line rates and other timing parameters generally does not proceed in this fashion. Of equal importance to the format and frame rate choices is the selection of the fundamental frequency reference(s) which will be used to generate the standard timing. In both television and computer-display practice, there is generally a single "master clock" from which all system timing will be derived. Therefore, the selection of the line rate, pixel rate, and other timing parameters must be done so as to enable all necessary timing to be obtained using this reference, and still produce the desired image format and frame rate.

In the broadcast television industry, for example, it is common practice to derive all studio timing signals and clocks from a single master source, typically a highly accurate cesium or rhubidium oscillator which in turn is itself calibrated per national standard sources, such as the frequency standards provided by radio stations WWV and WWVH in the United States.[5] The key frequencies defined in the television standards have been chosen so that they may be relatively easily derived from this master, and so the broadcaster is able to maintain the television transmission within the extremely tight tolerances required by the broadcast regulations.

Computer graphics hardware has followed a similar course. Initially, graphics boards were designed to produce a single standard display timing, and used an on-board oscillator of the appropriate frequency as the basis for that timing (or else derived their clock from an oscillator elsewhere in the system, such as the CPU clock). As will be seen later, computer display timing standards were initially derived from the established television standards, or even approximated those standards so that conventional television receivers could be used as PC displays. The highly popular "VGA" standard, at the time of its introduction, was essentially just a progressive-scan, square-pixel version of the North American television standard timing, and thus maintained a very high degree of compatibility with "TV" video.

However, the need to display more and more information on computer displays, as well as the desire for faster display refresh rates (to minimize flicker with large CRT displays used close to the viewer) led to a very wide range of "standard" computer formats and timings being created in the market. This situation could very easily have become chaotic, were it not for the establishment of industry standards groups created specifically to address this area (as

---

[5] These stations operate on internationally allocated standard frequencies of 2.500, 5.000, 10.000, 15.000 and 20.000 MHz (WWV only), and provide highly accurate time-of-day information as well as being a primary frequency reference. Other broadcast stations elsewhere in the world provide similar services. For information specifically regarding the WWV/WWVH broadcasts, contact the US National Institute of Standards and Technology, or visit their web site (http://www.boulder.nist.gov/timefreq/index.html).

will be seen in the next section). But along with the growing array of formats and timings in use came the development of both graphics board and displays which would attempt to support all of them (or at least all that were within their range of capabilities). These "multi-frequency" products are not, however, infinitely flexible, and the ways in which they operate have placed new constraints on the choices in "standard" timings.

In general, computer graphics systems base the timing of their video signals on a synthesized pixel or "dot" clock. As the name would imply, this is a clock signal operating at the pixel rate of the desired video timing. In the case of single-frequency graphics systems, such a clock was typically produced by a dedicated quartz-crystal oscillator, but support of multiple timings requires the ability to produce any of a large number of possible clock frequencies. This is most often achieved through the use of phase-locked-loop (PLL) based frequency synthesizers, as shown in Figure 7-2. In this type of clock generator, a single reference oscillator may be used to produce any of a number of discrete frequencies as determined by the two counters or dividers (the "÷M" and "÷N" blocks in the diagram), such that

$$f_{out} = f_{ref} \frac{M}{N}$$

While the set of possible output frequencies can be relatively large (depending on the number of bits available in the M and N dividers), this approach does not provide for the generation of any arbitrary frequency with perfect accuracy. In addition, large values for the M and N divide ratios are usually to be avoided, as these would have a negative impact on the stability of the resulting clock. Therefore, it is desirable that all clocks to be generated in such a system be fairly closely related to the reference clock.

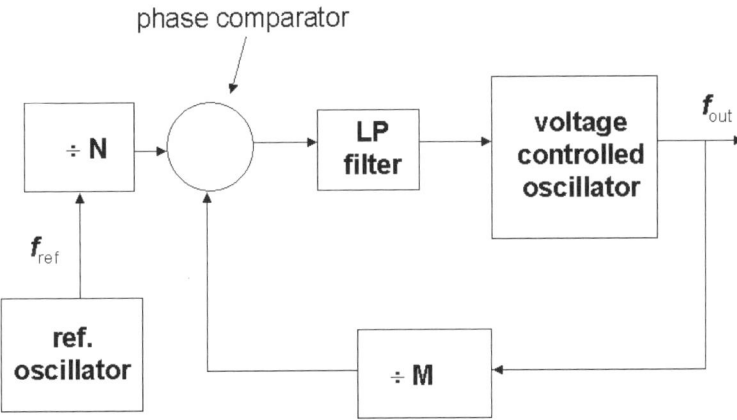

Figure 7-2 PLL-based frequency synthesis. In this variant on the classic phase-locked loop, frequency dividers (which may be simple counters) are inserted into the reference frequency input, and into the feedback loop from the voltage-controlled oscillator's output. These are shown as the "÷M" and "÷N" blocks. The output frequency (from the VCO) is then determined by the reference frequency f_{ref} multiplied by the factor (M/N). If the M and N values are made programmable, the synthesizer can produce a very large number of frequencies, although these are not "infinitely" variable.

Unfortunately, this has not been the case in the development of most timing standards to date. Instead, timing standards for a given format and rate combinations were created more or less independently, and whatever clock value happened to fall out for a given combination was the one published as the standard. This has begun to change, as the various industries and groups affected have recognized the need to support multiple standard timings in one hardware design. Many industry standards organizations, including the EIA/CEA, SMPTE, and VESA, have recently been developing standards with closely related pixel or sampling clock frequencies. Recent VESA computer display timing standards, for instance, have all used clock rates which are multiples of 2.25 MHz, a base rate which has roots in the television industry.[6]

7.5 Format and Timing Standard Development

As noted, the primary markets served today by electronic imaging are, broadly, entertainment video (broadcast television and other consumer-oriented entertainment services) and the computer industry. To be sure, there are other electronic imaging and display applications, but more and more the standards developed for TV and PC displays have come to dominate the field. Several organizations, operating in one or the other fields (or in some cases both) are responsible for these.

Due to the nature of broadcast communications, many of the standards in the television industry derive from regulations established by the government agencies in various countries, such as the Federal Communications Commission in the United States, the Radiocommunications Agency in the UK, the Ministry of Posts & Telecommunications in Japan, etc.. As radio waves do not respect national borders, though, there is also a need for the various national regulations to be coordinated, especially in terms of frequency allocations but also to some degree the specifics of the technical details of various standards. This task falls primarily on the International Telecommunications Union, or ITU, originally established in 1865 to coordinate telegraph (and then telephone) standards, but since 1947 an agency of the United Nations charged with the overall coordination of electronic communications standards. Many ITU standards were originally published under the name of the Comité Consultatif International Teléphonique et Telégraphique (CCITT), a subsidiary group which ceased to exist as a separate body following an ITU reorganization in 1992.

In terms of industry standards organizations, or professional associations which do a significant amount of standards development in this field, the leading players include the Electronic Industries Association (EIA) and Consumer Electronics Association (CEA, formerly the Consumer Electronics Manufacturers Association), along with the Society of Motion Picture and Television Engineers (SMPTE) in the television market. In the computer graphics and display field, the leading organization in the creation of format and timing standards has been the Video Electronics Standards Association (VESA), which was founded in 1991 to address just this need. However, it should also be noted that a considerable number of PC video "standards" were developed prior to the establishment of this group, and were in fact originally simply "de facto" standards which came through the success of various manufac-

---

[6] 2.25 MHz is the lowest common multiple of the line rates used in the standard 525-line, 59.94 Hz and 625-line, 50 Hz television scanning standards, and has also been used as the base rate from which many digital video sampling clock standards have been derived.

turers' products. This has resulted in some obvious incoherence in the standards for display timings first introduced in the 1980s, a problem which is industry is just now addressing.

While the details of the various television and computer-video systems and standards differ considerably, the basics in terms of format and timing standards are directly comparable. All of these systems are based on raster-scanning systems, with the details of the timings generally based on the needs of CRT-based display devices. (Only recently has either industry begun to consider the problem of timing standards oriented specifically to non-CRT displays.) The television industry has traditionally employed 2:1 interlaced scanning (as a method of reducing the bandwidth required for the broadcast transmission of a given image format), but even this is now changing to some degree with the introduction of digital television systems.

7.6 An Overview of Display Format and Timing Standards

While a truly exhaustive list of current format and timing standards used in electronic displays and imaging devices is nearly impossible to produce, at least an overview of those in most common use is of interest for comparative and reference use. Table 7-1 provides a list of image and display formats from a number of industries and applications, along with some brief comments on their origins and usage. Table 7-2 then provides some details of the timing specifications for the more common computer and television formats, where such exist.

Some explanations of the terminology of these standards is in order at this point. First, the pixel formats given in Table 7-1 refer to the *active* or *addressable* image space. The complete video signal, whether analog or digital, will include information corresponding to the image itself, but also will generally have additional "overhead" requirements as previously mentioned. In the simplest situation – that of an analog video signal – the "overhead" is in the *blanking period*, that portion of the signal which is intentionally left free of active content, or of information corresponding to a part of the image itself. The requirement for such periods is imposed by the needs of the various imaging and display hardware technologies, which must have some idle time between each scanned line and frame or field in order to reset and prepare for the next. This can amount to 25% or more of the total signal time, especially when dealing with (or having to accommodate) CRT-based display systems. The blanking period almost always contains the signals which provide synchronization information to the display – those pulses or other signals which identify the start of a new line, field, or frame. It is common, then, to divide the overall blanking period into three sections, as shown in Figure 7-3. The period prior to the synchronization pulse is the "front porch," a name resulting from the appearance of a typical video signal as displayed in standard television practice (in which it is presented at it appears at the cathode of a CRT, with the sync pulses positive-going). The remainder of the blanking period is then divided into the sync pulse itself and the "back porch" (which is any remaining blanking time following the end of the sync pulse). It is common in analog video signal and timing standards to use either the beginning of the blanking period or the beginning of the sync pulse itself as the reference point from which the rest of the line or frame timing is defined.

FORMAT AND TIMING STANDARDS

Table 7-1 An overview of common image and display formats.

Pixel format ($H \times V$)	Name/description	Controlling standard/organization	Comments
176 × 144	Quarter-CIF; video teleconferencing and similar low-res video apps.	CCITT/ITU H.261	
320 × 240	Quarter-VGA ("QVGA")	PC industry	Viewfinders, other low-resolution
352 × 288	"Common Image Format" (CIF); video teleconference standard	CCITT/ITU H.261	
640 × 480	"VGA" (Video Graphics Adaptor) standard; "square pixel NTSC"	PC industry (VESA standards)	
720 × 480	Standard format for digital 525/60 video; 13.5 MHz sampling1	CCIR-601	Non-square pixels
720 × 576	Standard format for digital 625/50 video; 13.5 MHz sampling1	CCIR-601	Non-square pixels
768 × 483	$4f_{sc}$ sampling[a] of 525/60 video	SMPTE 244M	Non-square pixels
768 × 576	"Square-pixel" 625/50 digital video		
800 × 600	"Super VGA" (SVGA) PC standard	PC industry (VESA standards)	
854 × 480	Widescreen (16:9) 480-line format	LCD/PDP TV displays	[b]
948 × 576	$4f_{sc}$ sampling[a] of 625/50 video		Non-square pixels
1024 × 576	Widescreen (16:9) 576-line format		
1024 × 768	"Extended Graphics Adaptor" (XGA) PC standard	PC industry (VESA standards)	
1152 × 864	Apple Computer 1 Mpixel standard	Apple Computers	
1280 × 720	16:9 HDTV standard format	ATSC	
1280 × 960	4:3 alternative to SXGA (below)	PC industry	
1280 × 1024	"Super XGA" (SXGA) PC standard, originally used in workstations	Unix workstations	5:4 aspect ratio
1365 × 768	"Wide XGA"; 16:9, 768-line format	LCD/PDP TV displays	[b]
1440 × 1050	"SXGA+"; first seen in notebook PC LCD panels	PC industry	
1600 × 1200	"Ultra XGA" (UXGA) PC standard	VESA	
1920 × 1080	16:9 HDTV standard format	ATSC	
1920 × 1200	16:10 widescreen PC displays	PC industry	[c]
2048 × 1152	16:9 European HDTV format	DVB-T	
2048 × 1536	"Quad XGA" PC display format	VESA	

[a] For details on digital video sampling standards, see chapter 10.
[b] Both of these formats exist in several versions, differing in the exact pixels/line count. For instance, the 854 × 480 format is also seen as 848 × 480, 852 × 480, and 856 × 480. (An exact 16:9 format would require 853.33 pixels/line.)
[c] The PC industry has begun to standardize on 16:10 displays for "widescreen" applications, in contrast to the standard 16:9 formats of HDTV.

Table 7-2 Comparison of selected TV and computer display timing standards. Used by permission of VESA.

Format	V. rate (Hz)	Pixel rate (MHz)	H total (µs)	H FP (µs)	HS (µs)	H BP (µs)	H rate (kHz)	V total (lines)	V FP (lines)	VS (lines)	V BP (lines)
525/60(I)[a] (orig.)	60.000	n/a	10.7	1.5	4.7	4.5	15.750	262.5	3	3	16
525/60(I)[a] (color)	59.94+	n/a	10.7	1.5	4.7	4.5	15.734	262.5	3	3	16
625/50(I)[a]	50.000	n/a	12.05	1.5	4.7	5.8	15.625	312.5	2.5	2.5	20
				pixels							
640 × 480[b]	59.94+	25.175	800	16	96	48	31.469	525	10	2	43
	75.000	31.500	840	16	64	120	37.500	500	1	3	16
	85.008	36.000	832	56	56	80	43.269	509	1	3	25
720 × 480 (I)[c]	59.94+	13.500	858	16	HS+BP=122		15.734	262	3	3	16
								263	3.5		16.5
720 × 576 (I)[c]	50.000	13.500	864	12	HS+BP=132		15.625	312	2.5	2.5	19
								313			20
800 × 600[b]	60.317	40.000	1056	40	128	88	37.879	628	1	4	23
	75.000	49.500	1056	16	80	160	46.875	625	1	3	21
	85.061	56.250	1048	32	64	152	53.674	631	1	3	27
1024 × 768[b]	60.004	65.000	1344	24	136	160	48.363	806	3	6	29
	75.029	78.750	1312	16	96	176	60.023	800	1	3	28
	84.997	94.500	1376	48	96	208	68.677	808	1	3	36
1280 × 720[d]	60.000	74.250	1650	70	80	220	45.000	750	5	5	25
1280×1024[b]	60.020	108.000	1688	48	112	248	63.981	1066	1	3	38
	75.025	135.000	1688	16	144	248	79.976	1066	1	3	38
	85.024	157.500	1728	64	160	224	91.146	1072	1	3	44
1600×1200[b]	60.000	162.000	2160	64	192	304	75.000	1250	1	3	46
	75.000	202.500	2160	64	192	304	93.750	1250	1	3	46
	85.000	229.500	2160	64	192	304	106.25	1250	1	3	46
1920×1080[d] (I)	60.000	74.250	2200	45	88	148	33.750	562.5	2	5	15.515
									2.5		

H total, total duration of one line, in microsecond or pixels; H FP, horizontal "front porch" duration; HS, horizontal sync pulse duration; HBP, horizontal "back porch" duration; V total, total duration of one field or frame, in lines; V FP, vertical "front porch" duration; VS, vertical sync pulse duration; V BP, vertical "back porch" duration. (I) denotes an interlaced timing. The vertical values given are for one field; some parameters may have two values given, one for each field. *Note*: Due to the nature of standard for interlaced television systems, the above values for the television-related standards should be used for comparison purposes only. The reader should consult the appropriate standards for detailed information in each case.
[a] Analog broadcast television standards; nominal values given, some derived. [b] Per current VESA standards. [c] Per CCIR-601 sampling standard.
[d] HDTV nominal values derived from SMPTE-274M/296M, EIA-770.3-A.

134 FORMAT AND TIMING STANDARDS

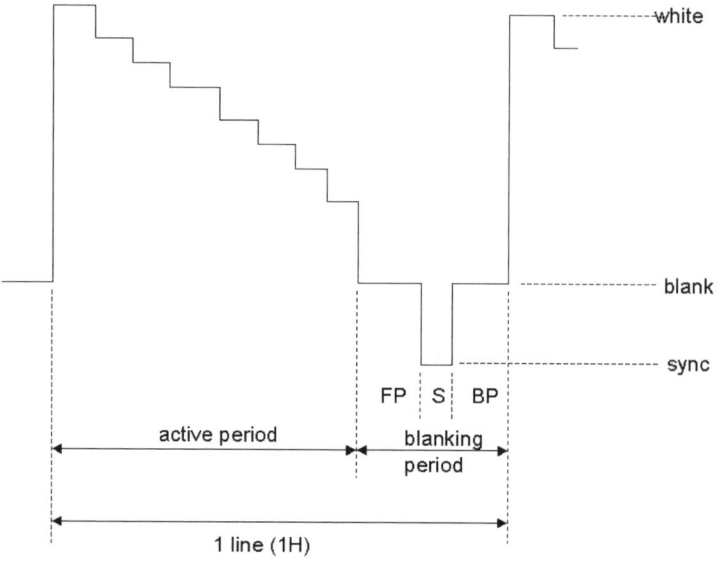

Figure 7-3 Standard nomenclature for the portions of a video signal. In this example, the signal shown corresponds to one complete horizontal line of the image. In terms of defined signal amplitudes, three levels are key: the maximum amplitude the signal may normally attain (which is referred to as the *white* level), the *blank* level (which is typically the reference for all other amplitude definitions), and the *sync* level (the level of the commonly negative-going pulses that provide the synchronization information). (There may also be a fourth defined level, the *black* level – which is generally defined as the lowest signal amplitude that will be encountered during the active image time. If this is omitted, it may be assumed to be identical to the blank level.) In time, the line is divided into active and blanking periods; the latter is further divided into the *front porch*, the *sync pulse* itself, and the *back porch*. The division of the blanking period into these three sub-periods is commonly done, even for those systems which do not provide a sync pulse as part of the video signal (i.e., those with physically separate sync signals). The timing specifications relating to the vertical synchronization (the frame or field timing) uses the same nomenclature (active and blanking periods, front and back porch, etc.), but will normally be specified in terms of line times (e.g., "3H").

In the case of digital video systems, there is often no need for an explicit blanking period, as the information will be placed in digital storage and/or further processed before being delivered to the display itself. However, many "digital" video systems and standards are based on the assumption that an analog signal will be "digitized" in order to create the digital data stream, and so include definitions of the blanking period, sync pulse position, etc., in terms of the sample or pixel period. Doing away with such things entirely, and treating the image transmission as if it were any other digital data communication, is generally not done except in such standards as also support packetization of the video data. These issues are discussed further in Chapter 12, during the review of digital and high-definition television systems.

7.7 Algorithms for Timings – The VESA GTF Standard

To address the need for supporting an ever-growing array of formats and rates in the personal computer industry, the Video Electronics Standards Association introduced a new means of "standard" timing generation in 1996. This was the Generalized Timing Formula, or GTF, standard. GTF does not explicitly define timings, in the traditional sense, at all. Instead, this standard defined an algorithm and established a set of constants which could be used by PC systems – both in host computer firmware and software – and by the display manufacturers to generate timings for any arbitrary combination of image format and frame rate.

At the heart of the GTF system is a definition of a standard "blanking percentage" curve for CRT monitors. You may note, from the timing standards presented in the previous section, that CRT-based displays generally require more blanking time, as a percentage of the total horizontal period, as the horizontal or line rate is increased. At the low end of the standard PC timing range – around 30 kHz horizontal – a typical CRT display might require that 18–20% of the horizontal period be spent in blanking. As the horizontal rate increases, this percentage goes up to 25% and higher. The GTF curve assumes that, within the range of horizontal frequencies likely to be encountered in practice, no more than 30% of the line time will be required for blanking. The curve is therefore defined, using the original, default constants, as:

$$H_{\text{blank}} \text{ (percent)} = \left(30 - \frac{300}{f_h}\right)\%$$

where f_h is the horizontal frequency in kHz. This basic curve is shown as Figure 7-4, with some standard timings plotted on the same axes for comparison.

With a standard blanking-percentage curve defined, there is now a fixed relationship between the pixel clock and the horizontal frequency for any timing. Since the horizontal period is the inverse of the horizontal frequency, and the pixel clock may be defined as the number of pixels in the active (or addressable) video period divided by the duration of that period, the relationship – again using the default parameters as used in the above equation – becomes

$$f_h = \frac{f_{\text{clk}}}{N_h / (1 - H_b)}$$

or

$$f_h = \frac{f_{\text{clk}}}{N_h / (1 - (0.3 - 3.00 / f_h))} = \frac{f_{\text{clk}}}{N_h (0.7 + 3.00 / f_h)}$$

so

$$f_{\text{clk}} = f_h \times N \frac{0.7 + 3.00}{f_h}$$

where f_{clk} is the pixel clock rate, f_h is the horizontal or line rate, and N_h is the number of pixels in the active or addressable line time.

136 FORMAT AND TIMING STANDARDS

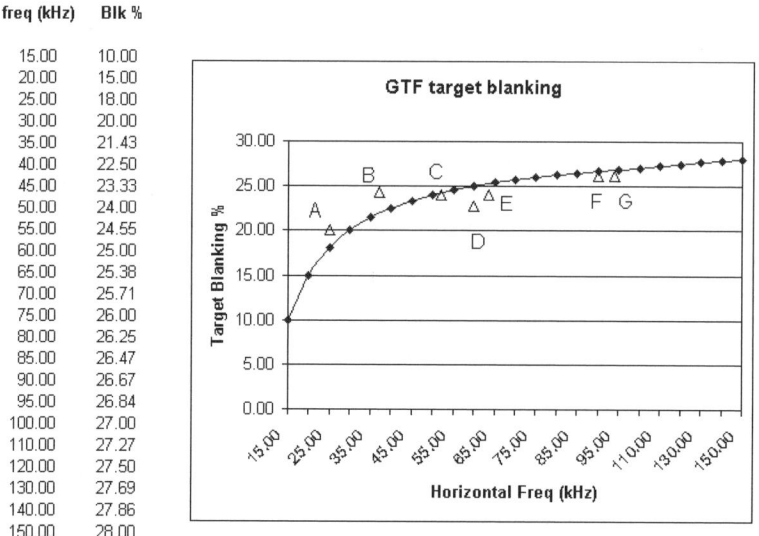

GTF target blanking vs. Horizontal Rate

H freq (kHz)	Blk %
15.00	10.00
20.00	15.00
25.00	18.00
30.00	20.00
35.00	21.43
40.00	22.50
45.00	23.33
50.00	24.00
55.00	24.55
60.00	25.00
65.00	25.38
70.00	25.71
75.00	26.00
80.00	26.25
85.00	26.47
90.00	26.67
95.00	26.84
100.00	27.00
110.00	27.27
120.00	27.50
130.00	27.69
140.00	27.86
150.00	28.00

Figure 7-4 The default GTF blanking curve as discussed in the text. This sets the target blanking percentage for the horizontal timing parameters produced under the standard GTF algorithm. Several VESA-standard timings are plotted along with the curve, for comparison. (A) 640 × 480, 60 Hz; (B) 800 × 600, 60 Hz; (C) 800 × 600, 85 Hz; (D) 1024 × 768, 75 Hz; (E) 1280 × 1024, 60 Hz; (F) 1280 × 1024, 85 Hz; (G) 1600 × 1200, 75 Hz. More recent VESA timing standards have attempted to follow the GTF guidelines as closely as possible.

This fixed relationship is the key to GTF; with either the horizontal frequency or the pixel clock given, the other is also defined. By applying this relationship and a few other simple rules, GTF can then produce a "standard" timing for any format/rate combination.

The GTF algorithm proceeds as follows: first, it must be determined if the timing is to be driven by a requirement that the pixel clock, the horizontal frequency, or the vertical rate. In other words, the standard permits any one of these to be "locked down", at which point the algorithms will set the others as needed to achieve the desired timing as closely as possible. The details of the algorithms are too complex to be covered here; the reader is directed to the VESA Generalized Timing Formula standard for these. However, it is important to note that only one of these three parameters may be selected as the "driving" factor, the one which will be preserved at all costs by the system (assuming that a timing is even possible with the chosen value). This limitation means that the GTF formulas will *not* be suitable as a replacement for explicit timing standards in all applications. If, for example, an exact frame rate must be achieved, *and* this must be done while still using the finite number of clocks which would be available from a frequency-synthesis clock generator (as described earlier), the GTF system may not be able to produce a viable timing.

Assuming that the constraints of GTF are acceptable, and the formulas have successfully been used to determine the pixel clock, horizontal frequency, and horizontal and vertical blanking requirements for the desired timing, the choice of the remaining parameters becomes a simple task. The GTF standard also established some very simple rules for the positioning and duration of the synchronization pulses within the blanking periods. The horizontal sync pulse is always made as close as possible (within the constraints of the hardware) to 8% of the total horizontal period in duration, and positioned such that the end (the deassertion) of the sync pulse be located as closely as possible to the center of the horizontal blanking period. The rules for the vertical sync pulse duration and position are even simpler – GTF vertical syncs are always 3 lines long, and follow a 1 line "front porch." (In an effort to harmonize GTF timings with the more traditional "discrete" timing standards, VESA has recently been applying these same rules to all new timing standards developed by that organization, and further attempting to make the horizontal blanking times used in the those standards fall as closely as possible to the GTF default curve.)

The basic aim of GTF was a simple one. If a monitor declares itself to be "GTF compliant" in the ID information provided to the host system, the host may then generate timings for any given format and rate using this method, with the expectation that the image will still be properly sized and centered by the display. This will occur whenever the monitor detects a "GTF" timing being produced by the host (the sync polarities are used to identify such timings; "GTF" video will have the horizontal sync pulse negative-true, as opposed to the VESA standard for traditional timings of having both syncs positive). The monitor, upon seeing GTF operation flagged by the syncs, "knows" that the timing complies with the GTF defaults and therefore may itself derive the expected blanking times, sync positions, etc., and adjust size and centering accordingly.

The discussion so far has treated the GTF default blanking curve as the only possible choice, and the values used to produce it as mandatory constants. This is not the case, however; in a fully compliant GTF system, these values must be implemented as variables, and are calculated from a set of parameters which may be provided by the display. (If the display does not provide such an alternate, or "Secondary GTF" set, then the defined defaults are to be used. This feature of the GTF standard permits a very wide range of "standard" blanking curves to be defined, through mutual agreement between the host system and the display currently in use. This is intended to free the system from what may be excessive blanking requirements imposed by the default curve (which is admittedly very CRT-oriented, and conservatively so at that), or even to increase the blanking if need be. This "programmable" feature of GTF may only be used if both the host and display support it, and then only if a bidirectional communications channel has been established between the two. (This is needed to ensure that both the host and display understand that the alternate parameters have been communicated and are currently in use by the system.) When GTF was first released, such bidirectional communications or ID capability was relatively rare in PC systems; however, as can be seen in Chapter 11, VESA has also established standards for this, which are now being adopted by the industry.

At this point, we have covered the basic concepts of electronic imaging, the operation of the common display technologies, some of the fundamental practical requirements for display interfaces in general, and now the basic standards of image formats and rates used by these displays. The next three chapters now focus on how the specific standards for several key markets and applications were developed, and the details of each. These include the analog video standards which have been developed for both broadcast television and the personal computer market, followed by the newer and still rapidly growing field of purely digital video interfaces and transmission systems.

8

Standards for Analog Video – Part I: Television

8.1 Introduction

As the CRT may be considered to be the first electronic display technology to enjoy widespread acceptance and success (a success which continues to this day), the analog interface standards developed to support it have been the mainstay of practically all "standalone" display markets and applications – those in which the display device is a physically separate and distinct part of a given system. This chapter and the next examine and describe these, and trace the history of their development from the early days of television to today's high-resolution desktop computer monitors. But since the standards for these modern monitors owe a great deal to the work that has been done, over the past 50 years and more, in television, we begin with an in-depth look at that field.

8.2 Early Television Standards

Broadcast television is a particularly interesting example of a display interface problem, and the solutions that were designed to address this problem are equally intriguing. It was the first case of a consumer-market, electronic display interface, and one which is complicated by the fact that it is a wireless interface – it must deliver image (and sound) information in the form of a broadcast radio transmission, and still provide acceptable results within the constraints that this imposes. The story of the development of this standard is fascinating as a case study, and becomes even more so as we look at how the capabilities and features of the broadcast television system grew while still maintaining compatibility with the original standard. In addition, the histories of television and computer-display standards turn out to be intertwined to a much greater degree than might initially be expected.

The notion that moving images could be captured and transmitted electronically was the subject of much investigation and development in the 1920s and 1930s. Demonstrations of television systems were made throughout this period, initially in very crude form (involving cumbersome mechanical scanning and display systems), but becoming progressively more sophisticated, until something very much like television as it is known today was demonstrated in the late 1930s. Systems were shown both in Europe and in North America during this time; notably, a transmission of the opening ceremonies of the 1936 Olympic Games in Berlin is often cited as the first television broadcast, although it was seen only in a few specially equipped auditoriums. A demonstration of a monochrome, or "black and white" television system by RCA at the 1939 New York City World's Fair – including the first televised Presidential address, by Franklin D. Roosevelt – is generally considered the formal debut of a practical system in the United States. Unfortunately, the outbreak of World War II brought development to a virtual standstill in much of rest of the world. Development continued to some degree in the US, but obviously at a slower pace than might otherwise have been the case.

The goal of television was to deliver a high-resolution moving image within a transmitted bandwidth (and other constraints) that could be accommodated within a practical broadcast system. As it was clear that the display device itself would have to be a CRT, the question of the required image format was in part determined by what could be resolved by the viewer, at expected home viewing distances, and using the largest tubes that could reasonably be expected at the time. It was anticipated that most viewers would be watching television from a distance of approximately 3 m (9–10 feet). Given a picture height of at most 0.5 m, as limited by the available CRTs, it was therefore reasonable (based on the limits of visual acuity) to establish a goal of providing images of at least the low hundreds of lines per frame. And, as it was desirable that television provide "square" resolution (an equal degree of delivered resolution in both the horizontal and vertical directions), this also, when coupled with the target image aspect ratio, becomes the driving force behind the minimum required bandwidth for the television signal.

The original television standards were therefore developed to provide a 4:3 aspect ratio image, with a minimum of several hundred lines per frame, and with a sufficiently high frame rate so as to both deliver acceptable motion and to avoid undesirable flicker as the images were presented on a CRT. In order to minimize the appearance of visible artifacts resulting from interference at the local power-line frequencies, the vertical deflection rate of the CRT display was set to match the power-line rate; 60 Hz in North America and 50 Hz in Europe. However, to transmit complete frames of 400–500 lines or more at this rate would require an unacceptably wide broadcast channel; therefore, all of the original television standards worldwide employed a 2:1 interlaced scanning format. As discussed in Chapter 7, 2:1 interlacing generally requires an odd number of lines per complete frame. Coupling this with the desire for reasonably close horizontal (line) rates in all worldwide standards (to allow for some commonality in receiver design) was a major factor in the final selection of the basic format standards still in use today. These are a 525 lines/frame system, with a 60 Hz field rate, used primarily in North America and Japan; and a 625 lines/frame, 50 Hz standard used in the rest of the world. (For convenience, these are often referred to as the "525/60" and "625/50" standards. It is important to note that these should not properly be referred to as the "NTSC" or "PAL/SECAM" standards. Those more often are used to refer to specific methods of color encoding, which both came later in time and they are discussed in a later section of this chapter.) Earlier experimental standards, notably a 405-line system which had been developed in the UK, and an 819-line system in France, were ultimately abandoned in favor of these.

The basic parameters of these two timing and format standards are shown in Table 8-1.

Table 8-1 Standard television format/timing standards.

	525/60 (original)	525/60 (with color)	625/50
Vertical (field) rate (Hz)	60.00	59.94+	50.00
Scanning format	2:1 interlaced	2:1 interlaced	2:1 interlaced
Lines per frame	525	525	625
Lines per field	262.5	262.5	312.5
Line rate (kHz)	15.750	15.734.26+	15.625
Vert. timing details			
Active lines/field[a]	242.5	242.5	287.5
Blank lines/field	20	20	25
Vert. "front porch"[b] (lines)	3	3	2.5
Vert sync. pulse width[b] (lines)	3	3	2.5
Vert. "back porch"[b,c] (lines)	3 + 11	3 + 11	2.5 + 17.5
Horiz. timing details			
Horiz. line time (μs) (H)	63.492	63.555	64.000
Horiz. active time (μs)[a,d]		52.86	51.95
Horiz. blanking (μs)	10.7[d]	10.7[d]	12.05
Horiz. "front porch"	1.3	1.3	1.5
Horiz. sync pulse width	4.8	4.8	4.8
Horiz. "back porch"[d]	4.5	4.5	5.

Horizontal front porch, sync pulse width, and back porch values are approximations; in the standard definitions, these are given as fractions of the total line time, as: FP = 0.02 H (min); SP = 0.075 H (nom); BP = remainder of specified blanking period.
[a] Not all of the active area is visible on the typical television receiver, as it is common practice to "overscan" the display. The number of lines lost varies with the adjustment of the image size and position, but is typically between 5 and 10% of the total.
[b] Due to the interlaced format used in both systems, the vertical blanking interval is somewhat more complex than this description would indicate. "Equalizing pulses," of twice the normal rate of the horizontal synchronization pulses but roughly half the duration, replace the standard H. sync pulse during the vertical front porch, sync pulse, and the first few lines of the back porch.
[c] The "back porch" timings given here are separated into the time during which equalization pulses are produced (the *postequalization* period) and the remaining "normal" line times.
[d] Derived from other values specified in these standards.

8.3 Broadcast Transmission Standards

Establishing the basic format and timing standards for the television systems was only the first step. The images captured under these standards must then be transmitted as a "radio" broadcast, and again a number of interesting choices were made as to how this would be done.

142 STANDARDS FOR ANALOG VIDEO – PART I: TELEVISION

First, we need to consider the bandwidth required for a transmission of the desired resolution. The 525 lines/frame format actually delivers an image equivalent to approximately 300–350 lines of vertical resolution; this is due to the effects of interlacing, the necessary spot size of display device, etc.. In the parlance of the television industry, this relationship is called the *Kell factor,* defined as the ratio between the actual delivered resolution, under ideal conditions, and the number of lines transmitted per frame. Television systems are generally assumed to operate at a Kell factor of about 0.7; with 485 lines of the original 525 available for "active" video (the other 40 lines constitute the vertical blanking intervals), the "525/60" system is expected to deliver a vertical resolution equal to about 340 lines. If the system is to be "square" – to deliver an equivalent resolution in the horizontal direction – it must horizontally provide 340 lines of resolution per picture height. In terminology more familiar to the computer industry, a 4:3 image with 340 lines of vertical resolution should provide the equivalent of 453 "pixels" for each horizontal line. (It is important to note that these original television standards, defining purely "analog", continuous-scan systems, do not employ the concept of a "pixel" as it is understood in image-sampling terms, or as presented in Chapter 1. The concern here was not for discrete picture elements being transmitted by the system, but only for the visible detail which could be resolved in the final image.)

With a 15,750 Hz line rate, and approximately 20% of the line time required for horizontal blanking/retrace, the horizontal resolution requirement translates to luminance variations (cycles between white and black; using the "pixel" viewpoint, two pixels are required for each cycle) at a fundamental rate of approximately:

$$\frac{227 \text{ (cycles/line)}}{(0.8/15750) \text{ (s/line)}} = 4.47 \text{ MHz}$$

Therefore, we can conclude that to deliver an image of the required resolution, using the formats and frame/field rates previously determined, will require a channel of about 4.5 MHz bandwidth at a minimum, in a simple analog broadcast system. (By a similar calculation, we would expect the 625/50 systems to require somewhat wider channels in order to maintain "square" resolution.)

These numbers are very close to the actual parameters established for the broadcast television standards. In the US, and in most other countries using the 525/60 scanning standard, television channels each occupy 6 MHz the broadcast spectrum. The video portion of the transmitted signal – that which contains the image information, as opposed to the audio – is transmitted using vestigial-sideband amplitude modulation (VSB; see Figure 8-1); this results in the smallest possible signal bandwidth, while not requiring the complexities in the receiver of a true single-sideband (SSB) transmission. The video carrier itself is located 1.25 MHz up from the bottom of each channel, with the upper sideband permitted to extend 4.2 MHz above the carrier, and the vestigial lower sideband allocated 0.75 MHz down from the carrier. The audio subcarrier is placed 4.5 MHz above the video carrier (or 5.75 MHz from the bottom edge of the channel). In all analog television broadcast standards, the audio is transmitted using frequency modulation (FM); in the US, and other countries using the 6 MHz channel standard, a maximum bandwidth of ±200 kHz, centered at the nominal audio subcarrier frequency, is available for the audio information; the US system uses a maximum carrier deviation of ±25 kHz.

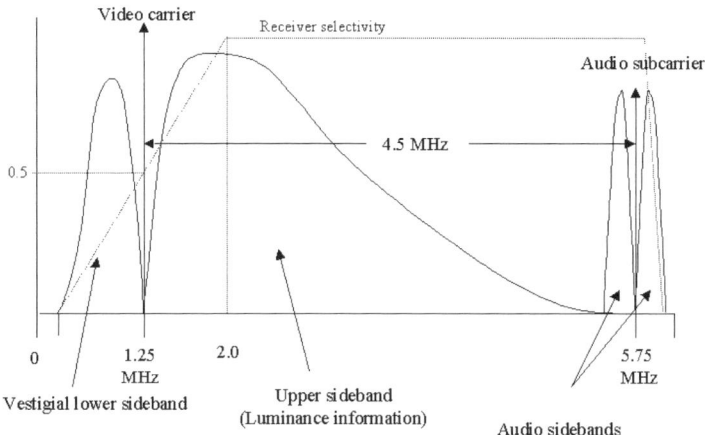

Figure 8-1 The spectrum of a vestigial-sideband (VSB) monochrome television transmission. The specific values shown above are for the North American standard, and most others using a 6 MHz channel (details for other systems are given in Table 8-2). Note that with the vestigial-sideband system, the receiver is required to provide a selectivity curve with an attenuation of the lower frequencies of the video signal, to compensate for the added energy these frequencies would otherwise receive from the vestigial lower sideband. (The shape of the luminance signal spectrum is an example only, and not intended to represent any particular real-world signal.)

The complete set of international standards for channel usage in the above manner was eventually coordinated by the Comité Consultatif International en Radiodiffusion (CCIR), and each standard assigned a letter designator. The 6 MHz system described above, used in the United States, Canada, Mexico, and Japan, is referred to as "CCIR-M". As noted, the 625/50 systems generally required greater video bandwidths, and so other channelizations and channel usage standards were developed for countries using those. A more complete list of the various systems recognized by the CCIR standards is presented later in this chapter, following the discussion of color in broadcast television systems.

One last item to note at this point is that all worldwide television broadcast systems, with the exception of those of France, employ *negative modulation* in the video portion of the transmission. This means that an increase in the luminance in the transmitted image is represented by a *decrease* in the amplitude of the video signal as delivered to the modulator, resulting in less depth of modulation. In other words, the "blacker" portions of the image are transmitted at a higher percentage of modulation than the "whiter" portions. This is due to the system employed for incorporation synchronization pulses into the video stream. During the blanking periods, "sync" pulses are represented as excursions *below* the nominal level established for "black". Therefore, the highest modulation – and therefore the most powerful parts of the transmitted signal – occurs at the sync pulses. This means that the receiver is more likely to deliver a stable picture, even in the presence of relatively high noise levels. The specifications require that the peak white level correspond to a modulation of 12.5%; this is to ensure an adequate safety margin so that the video signal does not overmodulate its carrier. (Overmodulation in such a system results in carrier cancellation, and a problem known as "intercarrier buzz", characterized by audible noise in the transmitted sound. This problem arose again, however, when color was added to the system, as discussed shortly.)

144 STANDARDS FOR ANALOG VIDEO – PART I: TELEVISION

8.4 Closed-Circuit Video; The RS-170 and RS-343 Standards

As an "over-the-air", broadcast transmission, the television signal has little need for (or even the possibility of) absolute standards for signal amplitude; the levels for "white," "black," and intermediate luminance levels could only be distinguished in terms of the depth of modulation for each. Even here, standard levels must be established in a relative sense, and there remains a need for absolute level standards for "in-studio" use – ensuring that equipment which is directly connected (*closed-circuit video*, as opposed to wireless RF transmission) will be compatible.

The video signal is normally viewed as "white-positive," as shown in Figure 8-2. Either as an over-the-air transmission, or in the case of AC-coupled inputs typical of video equipment, this signal has no absolute DC reference available. Therefore, the reference for the signal must be re-established by each device using it, by noting the level of the signal during a specified time. In a video transmission, there are only two times during which the signal can be assumed to be at a stable, consistent reference level – the sync pulses and the blanking periods. Therefore, the blanking level was established as the reference for the definition of all other levels of the signal. This permits video equipment to easily establish a local reference level – a process referred to as *DC restoration* – by clamping the signal a specified time following each horizontal sync pulse, at a point when the signal may safely be assumed to be at the blanking level.

To define the amplitude levels, the concept of defining an arbitrary unit based on the

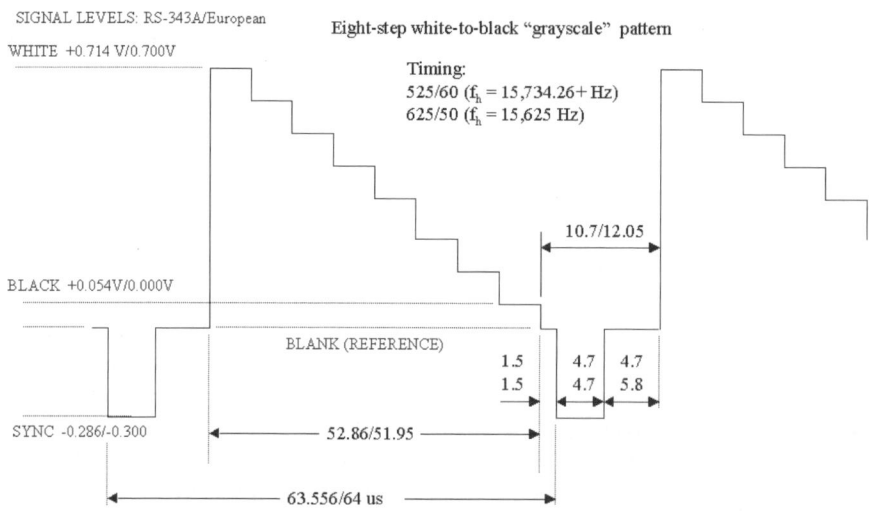

Figure 8-2 One line of a monochrome video signal, showing the standard levels. Where two values are given, as in 0.714V/0.700V, the first is per the appropriate North American standards (EIA RS-343 levels, 525/60 timing) while the second is the standard European (625/50) practice. The original EIA RS-170 standard used the 525/60 timing, but defined the signal amplitudes as follows: Sync tip, – 0.400V; blank (reference level); black, +0.075V (typical); white, +1.000V. Note: All signal amplitudes assume a 75-Ω system impedance. Modern video equipment using the North American system uses a variation of the original 525/60 timing, but the 1.000V$_{p-p}$ signal amplitude originally defined by RS-343.

peak-to-peak amplitude of the signal was introduced. The Institute of Radio Engineers, or IRE, established the standard of considering the amplitude from the blanking reference to the "white" level (the maximum positive excursion normally seen during the active video time) to be "100 IRE units," or more commonly simply "100 IRE." The tips of the synchronization pulses were defined to be negative from the reference level by 40 IRE, for an overall peak-to-peak amplitude of 140 IRE. The US standards for video also established the requirement that the "black" level – the lowest signal level normally seen during the active video time – would be slightly above the reference level. This difference between the blanking and black levels is referred to as the "pedestal" or *setup* of the signal. The nominal setup in the US standards was 7.5 IRE, but with a relatively wide tolerance of ±5 IRE.

The first common standard for wired or closed-circuit video was published in the 1940s by the Electronic Industries Association (EIA) as *RS-170*. In addition to the US system's signal amplitude and timing definitions already discussed, this standard established two other significant requirements applicable to "wired" video. First, the transmission system would be assumed to be of 75 Ω impedance, with AC-coupled inputs. The signal amplitudes were also defined in absolute terms, although still relative to the blanking level. The blank-to-white excursion was set at 1.000V, for an overall peak-to-peak amplitude of the signal of 1.400V (from the sync tips to the nominal white level).

A later "closed-circuit" standard had an even greater impact, in setting the stage for the later analog signal standards used by the personal computer industry. Originally intended as a standard for closed-circuit video systems with higher resolution than broadcast television, *RS-343* closely resembled RS-170, but had two significant differences. First, several new timing standards were established for the "high-res" closed-circuit video. But more importantly, RS-343 defined an overall signal amplitude of exactly 1.000V. The definitions in terms of IRE units were retained, with the blank-to-white and blank-to-sync amplitudes still considered to be 100 IRE and 40 IRE, respectively, but the reduction in peak-to-peak amplitude made for some seemingly arbitrary figures in absolute terms. The white level, for example, is now 0.714V positive with respect to blanking, while the sync tips are 0.286V negative. The "setup" of the black level now becomes a nominal 0.054V. The timing standards of RS-343 are essentially forgotten now, but the basic signal level definitions or later derivatives continue on in practically all modern video systems.

European video standards were developed along very similar lines, but with the significant difference that no setup was used (the blank and black levels are identical). In practice, this results in little if any perceivable difference. Setup was introduced to permit the CRT to be in cutoff during the blanking times, but slightly out of cutoff (and therefore in a more linear range of operation) during the active video time. But this cannot be guaranteed for the long term through the setup difference alone, as the cutoff point of the CRT will drift with time due to the aging of the cathode. So most CRT devices will generate blanking pulses locally, referenced to the timing of the sync pulses, to ensure that beam is cutoff during the retrace times. Therefore, the difference in setup between the European and original North American standards became primarily a complication when mixing different signal sources and processing equipment within a given system.

The European standards also, in doing away with the setup requirement, simplified the signal level definitions within the same 1.000V$_{p-p}$ signal. The white level was defined as simply 0.700V positive with respect to blanking (which remains the reference level), with the sync tips at 0.300V negative. The difference in the sync definition is generally not a problem, but the difference in the white-level definition can lead to some difficulties if equipment designed to the two different standards is used in the same system. At this time, the +0.700/ –0.300 system, without setup, has become the most common for interconnect standards (as documented in the SMPTE 253M specification), although the earlier RS-343 definitions are still often seen in many places.

One last point should be noted with regard to the monochrome video standards. As the target display device – the only display device suitable for television which was known at the time – was the CRT, the non-linearity or gamma of that display had to be accounted for. All of the television standards did this by requiring a non-linear representation of luminance. The image captured by the television camera is expected to be converted to the analog video signal using a response curve that is the inverse of the expected "gamma" curve of the CRT. The various video standards are based on an assumed display gamma of approximately 2.2, which is slightly under the actual expected CRT gamma. This results in a displayed image which is somewhat more perceptually pleasing than correcting to a strict overall linear system response. (Note: a strict "inverse gamma" response curve for cameras, etc., is not followed at the extreme low end of the luminance range. For example, below 0.018 of the normalized peak luminance at reference white, the current ANSI/SMPTE standard (170M) requires a linear response at the camera. This is to correct for deviations from the expected theoretical "gamma curve" in the display and elsewhere in the system, and to provide a somewhat better response to noise in this range.

8.5 Color Television

Of all the stages in the development of today's television system, there is none more interesting, either in terms of the technical decisions made or the political processes of the development, than the addition of color support to the original "black-and-white-only" system. Even at the time of the introduction of the first commercial systems, the possibility of full-color video was recognized, and in fact had been shown in several separate demonstration systems (dating back as far as the late 1920s). However, none of the experimental color systems had shown any real chance of being practical in a commercial broadcast system, and so the original television standards were released without any clear path for future extension for the support of color.

By the time that color was seriously being considered for broadcast use, in the late 1940s and early 1950s, the black-and-white system was already becoming well-established. It was clear that the addition of color to the standard could not be permitted to disrupt the growth of this industry, by rendering the installed base of consumer equipment suddenly obsolete. Color would have to coexist with black-and-white, and the US Federal Communications Commission (FCC) announced that any color system would have to be completely compatible with the existing standards in order to be approved. This significantly complicated the task of adding color to the system. Not only would the additional information required for the display of full-color images have to be provided within the existing 6 MHz channel width, but it would have to be transmitted in such a way so as not to interfere with the operation of existing black-and-white receivers.

Ultimately, two systems were proposed which met these requirements and became the final contenders for the US color standard. The Columbia Broadcasting System (CBS) backed a field-sequential color system which involved rotating color filters in both the camera and receiver. As the sequential red, green, and blue fields were still transmitted using the established black-and-white timing and signal-level standards, they could still be received and properly displayed by existing receivers. These images, displayed on such a receiver, would still appear in total to be a black-and-white image, in the absence of the color filter system (although there were some possible artifacts introduced by the field-sequential system; see Chapter 4). In fact, the filter wheels and synchronization electronics required for color display could conceivably be retrofitted to existing receivers, permitting a less-expensive upgrade path for consumers. (Television receivers were still quite expensive at this time; a typical black-and-white television chassis – with CRT but without cabinet – sold for upwards of $200.00 in 1950, at a time when a new car could be purchased for only about three times this.)

The CBS system, originally conceived by Dr. Peter Goldmark of their television engineering staff, had been demonstrated as early as 1940, but was not seen as sufficiently ready so as to have any impact on the introduction of the black-and-white standards. CBS continued development of the system, though, and it was actually approved by the FCC in October of 1950. However, throughout this period the CBS technique drew heavy opposition from the Radio Corporation of America (RCA), and its broadcast network (the National Broadcasting Company, NBC), under the leadership of RCA's president, David Sarnoff. RCA was committed to development of black-and-white television first, to be followed by their own "all-electronic" compatible color system. RCA's system had the advantage of not requiring the cumbersome mechanical apparatus of the field-sequential CBS method, and provided even better compatibility with the existing black-and-white standards. Under the RCA scheme, the video signal of the black-and-white transmission was kept virtually unchanged, and so could still be used by existing receivers. Additional information required for the display of color images would be added elsewhere in the signal, and would be ignored by black-and-white sets. A key factor enabling the all-electronic system was RCA's concurrent development of the tricolor CRT or "picture tube", essentially the phosphor-triad design of the modern color CRT. This permitted the red, green, and blue components of the color image to be displayed simultaneously, rather than requiring that they be temporally separated as in the CBS method. In 1953, the FCC, acting on the recommendation of the National Television System Committee (NTSC) – a group of industry engineers and scientists, established to advise the Commission on technical matters relating to television – reversed its earlier decision and approved what was essentially the RCA proposal. The new standard would come to be known as "NTSC color."

8.6 NTSC Color Encoding

The NTSC color encoding system is based on the fact that much of the information required for the color image is already contained within the existing black-and-white, or luminance, video information. As seen in the earlier discussion of color (Chapter 4), the luminance, or Y, signal can be derived from the primary (RGB) information; the conversion, using the precise coefficients defined for the NTSC system, is

148 STANDARDS FOR ANALOG VIDEO – PART I: TELEVISION

$$Y = 0.299R + 0.587G + 0.114B$$

If the existing video signal is considered to be the Y information derived in this manner, all that is required for a full-color image is the addition of two more signals to permit the recovery of the original RGB information. What is not clear, however, is how this additional information could be accommodated within the available channel.

Two factors make this possible. First, it was recognized that the additional color information did not have to be provided at the same degree of spatial resolution as the original luminance signal. As discussed in Chapter 2, the human eye does not have the same degree of spatial acuity, in terms of discriminating changes of color (more accurately, *hue*) only, as it does in terms of simple contrast or luminance-only distinctions. Therefore, the additional signals representing the supplemental color information would not require the bandwidth of the original luminance video. Second, the channel is not quite as fully occupied by the monochrome video as it might initially appear. By virtue of the raster-scan structure of the image, the spectral components of the video signal cluster around multiples of the horizontal line rate; as the high-frequency content of the image is in general relatively low, the energy of these components drops off rapidly above and below these frequencies. This results in the spectrum of the video signal having a "picket fence" appearance (Figure 8-3), with room for information to be added "between the pickets."

In the NTSC color system, this was achieved by adding an additional carrier – the color subcarrier – at a frequency which is an odd multiple of one-half the line rate. This ensures that the additional spectral components of the color information, which similarly occur at multiples of the line rate above and below their carrier, are centered at frequencies between

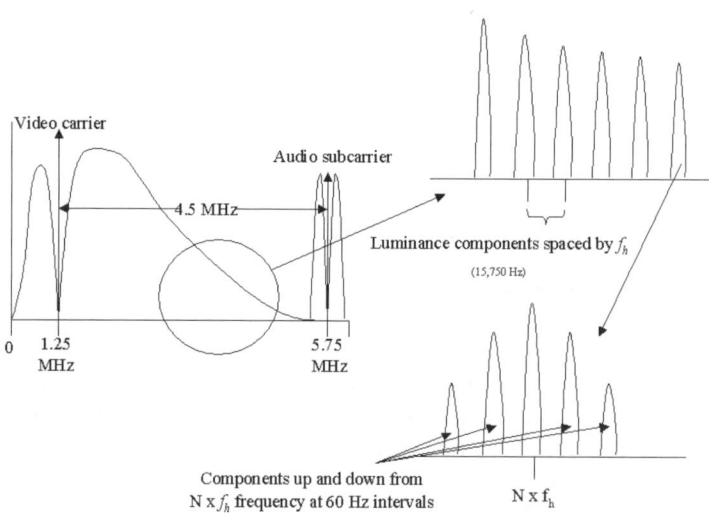

Figure 8-3 Details of the spectral structure of a monochrome video signal. Owing to the raster-scan nature of the transmission, with its regular line and field structure, the spectral components appear clustered around multiples of the line rate, and then around multiples of the field rate. This "picket fence" spectral structure provides space for the color signal components which might not be obvious at first glance – "between the pickets" of the luminance information.

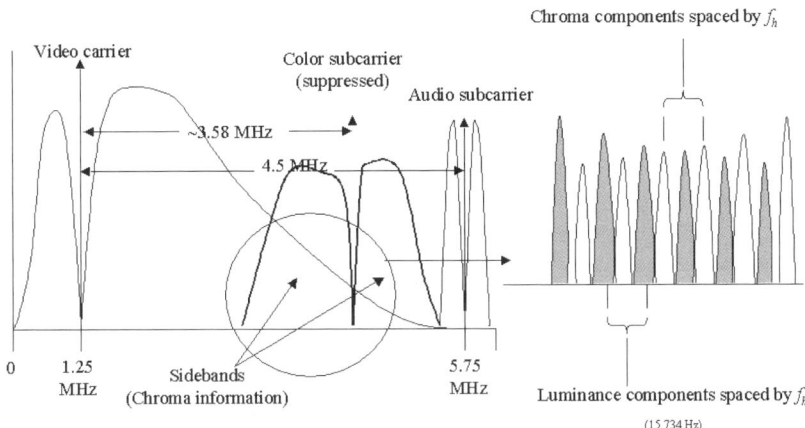

Figure 8-4 Through the selection of the color subcarrier frequency and modulation method, the components of the color information ("chrominance") are placed between the "pickets" of the original monochrome transmission. However, their remained a concern regarding potential interference between the chrominance components and the audio signal, as detailed in the text.

those of the luminance components (Figure 8-4). Two signals carrying the color, or *chrominance*, information are added to the transmission via quadrature modulation of the color subcarrier. Originally, these were named the I and Q components (for "in-phase" and "quadrature"), and were defined as follows:

$$I = 0.596R - 0.275G - 0.321B$$

$$Q = 0.212R - 0.523G - 0.311B$$

These may also be expressed in terms of the simpler *color-difference* signals B–Y and R–Y as follows:

$$I = -0.27(B-Y) + 0.74(R-Y)$$

$$Q = 0.41(B-Y) + 0.48(R-Y)$$

and in fact these expressions correspond to the method actually used to generate these signals from the original RGB source.

(Note: the above definitions of the *I* and *Q* signals are no longer in use, although still permissible under the North American broadcast standards; they have been replaced with appropriately scaled versions of the basic color-difference signals themselves, as noted below in the discussion of PAL encoding.)

At this point, the transformation between the original RGB information and the YIQ representation is theoretically a simple matrix operation, and completely reversible. However, the chrominance components are very bandwidth-limited in comparison to the Y signal, as would be expected from the earlier discussion. The *I* signal, which is sometimes considered roughly equivalent to the "orange-to-cyan" axis of information (as the signal is commonly displayed on a vectorscope), is limited to at most 1.3 MHz; the *Q* signal, which may be considered as encoding the "green-to-purple" information, is even more limited, at 600 kHz bandwidth. It is this bandlimiting which makes the NTSC color encoding process a lossy system. The original RGB information cannot be recovered in its original form, due to this loss.

One other significant modification was made to the original television standard in order to add this color information. As noted earlier, placing the color subcarrier at an odd multiple of half the line rate put the spectral components of the chrominance signals between those of the existing luminance signal. However, the new color information would come very close to the components of the audio information at the upper end of the channel, and a similar relationship between the color and audio signals was required in order to ensure that there would be no mutual interference between them. It was determined that changing the location of either the chroma subcarrier or the audio subcarrier by a factor of 1000/1001 would be sufficient (either increasing the audio subcarrier frequency by this amount, or decreasing the chroma subcarrier to the same degree). Concerns over the impact of a change of the audio frequency on existing receivers led to the decision to move the chroma subcarrier. But to maintain the desired relationship between this carrier and the line rate, all of the video timing had to change by this amount. Thus, in the final color specification, the field rate and line rate both change as follows:

$$\text{New field rate} = 60 \text{ Hz} \times \frac{1000}{1001} = 59.94 + \text{ Hz}$$

$$\text{New line rate} = 262.5 \times 60 \text{ Hz} \times \frac{1000}{1001} = 15,734.26 + \text{ Hz}$$

With these changes, the color subcarrier was placed at 455/2 times the new line rate, or approximately 3.579545 MHz. above the video carrier. These changes were expected to be small enough such that existing receivers would still properly synchronize to the new color transmissions as well as the original black-and-white standard.

With the quadrature modulation technique used to transmit the *I* and *Q* signals, the chroma subcarrier itself is not a part of the final signal. In order to permit the receiver to properly decode the color transmission, then, a short burst (8–10 cycles) of the color subcarrier is inserted into the signal, immediately following the horizontal sync pulse (i.e., in the "back porch" portion of the horizontal blanking time) on each line (Figure 8-5). This *chroma burst* is detected by the receiver and used to synchronize a phase-locked loop, which then provides a local frequency reference used to demodulate the chroma information.

A complete NTSC encoding system is shown in block-diagram form as Figure 8-6.

With the chrominance and luminance (or, as they are commonly referred to, "Y" and "C") components of the signal interleaved as described above, there is still a potential for mutual interference between the two. The accuracy of the decoding in the receiver depends to a great

NTSC "Compatible Color" Signal

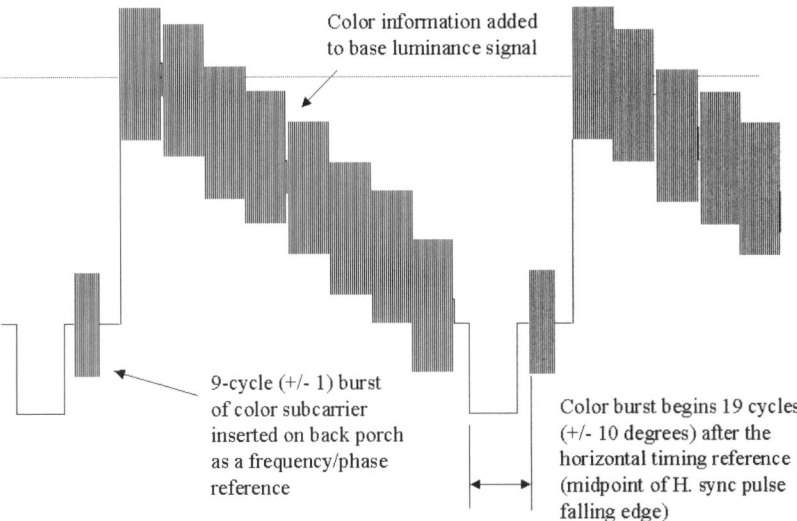

Figure 8-5 The completed color video signal of the NTSC standard, showing the "color burst" reference signal added during the blanking period.

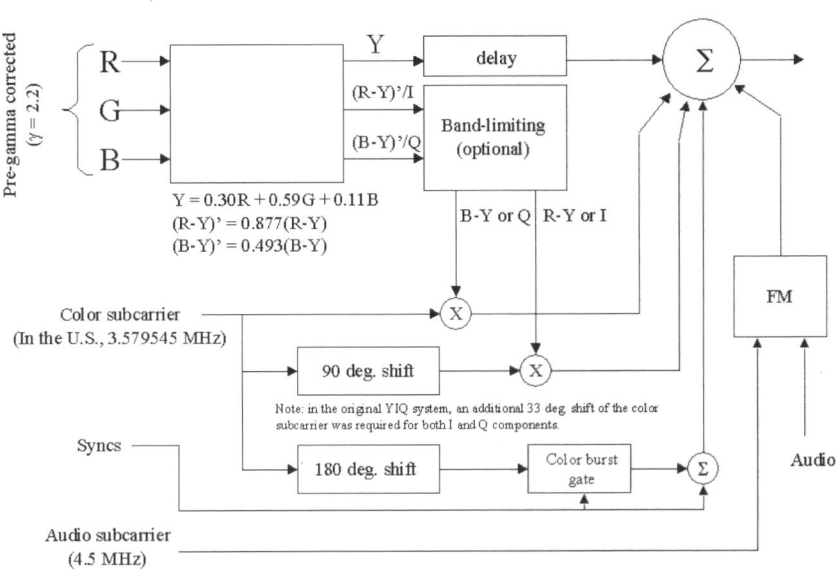

Figure 8-6 Block diagram of NTSC color video encoding system.

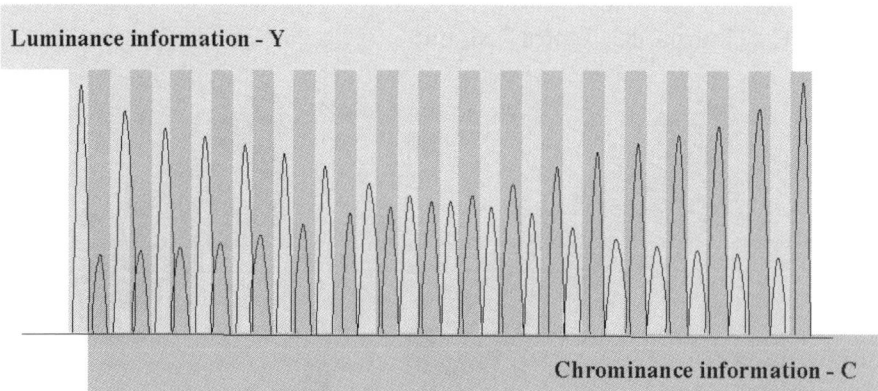

Figure 8-7 The interleaved luminance and chrominance components, showing the filter response required to properly separate them – i.e., a "comb filter".

extent on how well these signals can be separated by the receiver. This is commonly achieved through the use of a *comb filter*, so called due to its response curve; it alternately passes and stops frequencies centered about the same points as the chrominance and luminance components, as shown in Figure 8-7. An example of one possible color decoder implementation for NTSC is shown in Figure 8-8, with a simple comb filter implementation shown in Figure 8-9. This takes advantage of the relative phase reversal of the chroma subcarrier on successive active line times (due to the relationship between the subcarrier frequency and the line rate) to eliminate the chroma components from the luminance channel, and vice-versa. Failure to properly separate the *Y* and *C* components leads to a number of visible artifacts which in practice are quite common in this system. A prime example is

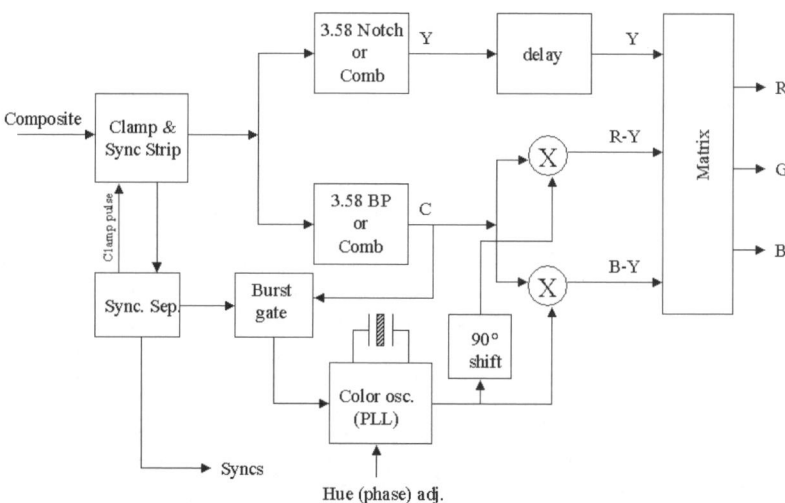

Figure 8-8 Block diagram of a color television decoder for the NTSC color system.

NTSC COLOR ENCODING   153

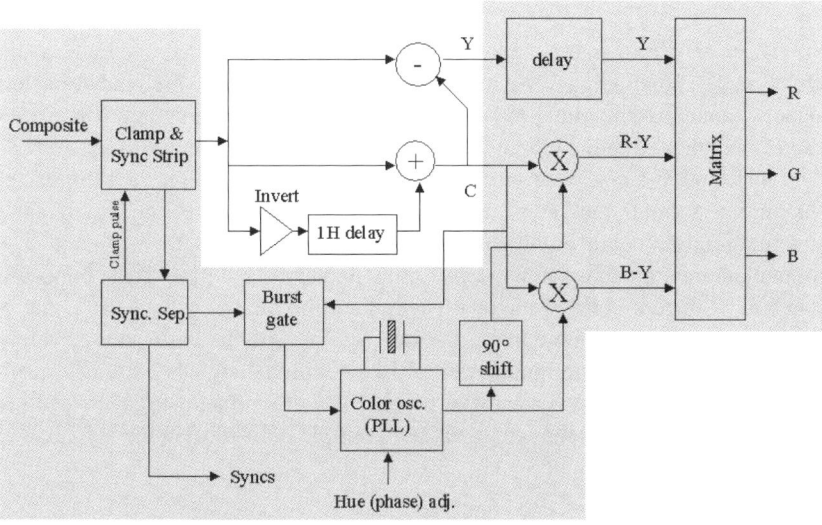

Figure 8-9 The NTSC decoder with a simple comb filter (outside the shaded area). This relies on the relative phase reversal of the chroma subcarrier on alternate transmitted lines to eliminate the chrominance information from the Y signal. This results in a loss of vertical resolution in the chrominance signal, but this can readily be tolerated as noted in the text.

chroma crawl, which refers to the appearance of moving blotches of color in areas of the image which contain high spatial frequencies along the horizontal direction. This results from these high-frequency luminance components being interpreted by the color decoding as chroma information. This artifact is often seen, for instance, in newscasts when the on-camera reporter is wearing a finely striped or checked jacket.

An additional potential problem results from the color components being added to the original luminance-only signal definition. As shown in Figure 8-5, the addition of these signals results in a increase in the peak amplitude of the signal, possibly beyond the 100 IRE positive limit (the dotted line in the figure). This would occur with colors which are both highly saturated and at a high luminance, as with a "pure" yellow or cyan. The scaling of the color-difference signals, as reflected in the above definitions, was set to minimize the possibility of overmodulation of the video carrier, but it is still possible within these definitions for this to occur. For example, a fully saturated yellow (both the R and G signals at 100%) would result in a peak amplitude of almost 131 IRE above the blanking level, or a peak-to-peak signal of almost 171 IRE. This would severely overmodulate the video carrier; with the definition of 100 IRE white as 12.5% modulation discussed earlier, the 0% modulation point and thus the absolute limit on "above-white" excursions is 120 IRE, or 160 IRE for the peak-to-peak signal. Careful monitoring of the signals and adjustments to the video gain levels are required in television production to ensure that overmodulation does not occur. This was considered acceptable at the time of the original NTSC specification, as highly saturated yellows and cyans are rare in "natural" scenes as would be captured by a television camera. However, the modern practice of using various forms of electronically generated imagery, such as computer graphics or electronic titling systems, can cause problems due to the saturated colors these often produce.

8.7 PAL Color Encoding

Due to the effects of World War II, color television development in Europe lagged somewhat behind efforts in North America. By the time the European nations were ready to determine a color broadcast standard (the mid-1960s), the RCA/NTSC encoding system had already been adopted and implemented in the US and Canada. Still, it was clear that this system could not be simply transferred in a completely compatible form; if nothing else, the differences in the standard scanning formats and rates, coupled with the differing European channelization schemes, would require that different color frequency standards be set.

The system adopted by most of Western Europe is very close in its basic concepts to the NTSC standard. It differs in three major aspects. First, no change was made to the original line and frame rates, as had been done in the US. Next, while the basic idea of carrying the color information via quadrature modulation of two additional signals onto a subcarrier was retained, the definition of those signals was simplified. Rather than using the I and Q definitions of NTSC, the new European standards used simple color-difference signals, U and V, defined as

$$U = 0.493(B-Y)$$

$$V = 0.877(R-Y)$$

(It was later realized that these were within the adjustment range of NTSC-standard receivers, and the FCC permitted either these or the original IQ definitions to be used. At this time, the simpler color-difference definitions have essentially displaced the original versions completely.) The final change gave the new standard its common name. In order to minimize one source of color error in the NTSC system, the new European standard reversed the phase of one of the chrominance components (the V, or R–Y, signal) every line. This results in color errors in any given line being more-or-less compensated for by an "inverse" error in the following line, such that the observed result (when the two lines adjacent lines are seen by the viewer) is greater color accuracy. Thus, the new standard was referred to as "*Phase-Alternating-Line*", or *PAL*.

Other than these changes, the PAL system is virtually identical to NTSC, although the European 625/50 scanning formats did result in different frequency and timing definitions.

One other minor change resulted from the phase alternation described above; this results in the spectral components of the two chrominance signals being offset by half the line rate, relative to one another. With this spacing of components, setting the color subcarrier in the same manner as was done for NTSC (at an odd multiple of half the line rate) would have resulted in interference between the luminance signal and one of the chrominance signals. To avoid this, PAL systems had to select a color subcarrier frequency which placed the chroma components at one-quarter the line rate from the luminance components, with an additional offset equal to the frame rate to further minimize interference. The final color subcarrier frequency chosen was

$$\frac{1135}{4} \times 15{,}625 \text{ Hz} + 25 \text{ Hz} = 4.433619 \text{ MHz}$$

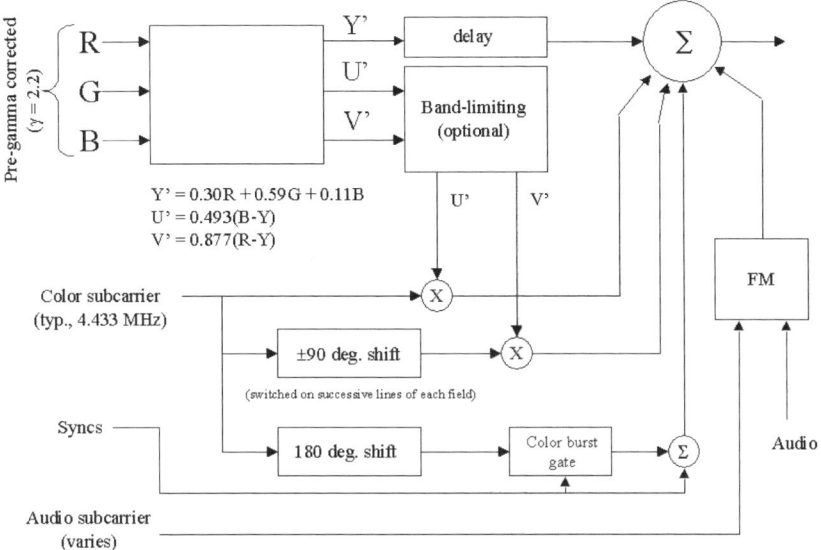

Figure 8-10 Block diagram of PAL encoder.

Note that the PAL standards did *not* involve a change to either the original monochrome line or frame rates of the 625/50 format, or a change to the audio subcarrier frequency. The wider channels used in Europe had permitted a greater spacing between the video and audio carrier frequencies in the first place, so no interference concern arose here with the addition of the chrominance signals. The PAL encoding process is shown in block diagram form in Figure 8-10; note the similarities and differences between this and the NTSC encoder of Figure 8-6.

8.8 SECAM

The observant reader will have noted that the PAL system was adopted by *most* of Western Europe. For a number of reasons, including what must be recognized as some significant political factors, a third – and completely incompatible – system was adopted by France, what was then the Soviet Union, and the former colonies and allies of those two nations.

The SECAM (for "SEquential Colour Avec Memoire") system also utilized two color-difference signals added to the original luminance-only information, but provided these sequentially rather than simultaneously (as had been done in the NTSC and PAL systems). While the luminance information remains continuous, as in the monochrome, the $B-Y$ and $R-Y$ components are transmitted on successive lines. Properly decoding the signal requires the storage of a full line of information in the receiver (and hence the "avec memoire" part of the name), and also results in a reduction of resolution (by a factor of two) of the color information in the vertical direction. This loss of resolution is visually acceptable, however, by the same reasoning which permitted the bandwidth limitations of the chrominance components of the NTSC and PAL systems.

Other incompatibilities of most SECAM systems, relative to NTSC and PAL, include the use of two separate color subcarriers and the use of frequency modulation of these carriers by the color-difference signals. Neither of these carriers is at the same frequency as used in the 625/50 PAL systems. And, as mentioned earlier, transmission of SECAM video generally employs positive modulation of the video carrier by the combined luminance/sync signal, as opposed to the negative modulation standard in PAL and NTSC countries.

Due to the complexities of dealing with SECAM encoding in the production environment, the usage of this system is today almost completely for the actual television transmission. Studio equipment in SECAM countries most often uses PAL encoding or, more recently, is operating on digital composite video.

8.9 Relative Performance of the Three Color Systems

It is very tempting to try to claim performance advantages for one of the three color encoding systems relative to the others, or for either of the basic 625/50 or 525/60 format and timing standards. At the present state of television development, the actual performance differences in terms of as-delivered image quality between any of these is very slight. The phase error problem, which led to the development of PAL from the original NTSC method, has essentially been eliminated in modern NTSC systems; the increased line count of the 625/50 format might deliver somewhat higher potential resolution, but this is often lost to receiver and/or display limitations in the final image. The 60 Hz field rate is claimed to have improved flicker performance over the 50 Hz systems, but many modern "50 Hz" receivers actually deinterlace the transmission and display at 100 Hz. Still, each system still has some very vocal proponents, and the debate continues, albeit on points of ever-decreasing significance. A much more significant change is underway now, with the transition from analog

Table 8-2 CCIR television channelization standards.

CCIR designation	Channel width (MHz)	Video carrier[a] (MHz)	Audio subcarrier offset (MHz)	Chroma subcarrier offset (MHz)	Color encoding
A	obsolete UK 405-line, 50 Hz system				
B	7	1.25	5.5	4.43	PAL
D	8	1.25	6.5	4.43 (PAL)	PAL,SECAM
E,F	obsolete French 819-line, 50 Hz system				
G	8	1.25	5.5	4.43	PAL
I	8	1.25[c]	6.0	4.43	PAL
K,L	8	1.25[c]	6.5	4.25/4.4[b]	SECAM
M	6	1.25	4.5	3.58	NTSC[d]
N	6	1.25	4.5	3.58[e]	PAL,SECAM

[a] Video carrier frequencies are given from the lower channel edge.
[b] The CCIR-K, SECAM system uses two chroma subcarriers and FM modulation of the chroma information, as noted in the text.
[c] The vestigial lower sideband is permitted to extend below the lower channel limit in the I and L standards.
[d] A variant usually referred to as "PAL-M," using the PAL encoding system but in the CCIR-M 6 MHz channel, using the common 3.58 MHz chroma subcarrier frequency and a 525/50 timing, is in use in Brazil.
[e] The chroma subcarrier frequency of PAL-N is close to that of NTSC-M, but not identical.

systems of any variety to full-digital broadcasting. This is examined in more detail in Chapter 12.

8.10 Worldwide Channel Standards

As mentioned earlier, television channel utilization systems are identified using a letter-based system established by the CCIR. With the specifics of the three color encoding systems in common use now understood, Table 8-2 gives the details of the more popular CCIR channelization standards, and the countries or regions in which each is used.

Usage of these standards by country is shown in Table 8-3.

Table 8-3 Usage of channelization standards by country.

CCIR code	Country
B	Australia, Austria, Azores, Bahrain, Belgium, Cyprus, Denmark, Egypt, Finland, Germany, Greece, Hungary, Iceland, India, Indonesia, Israel, Italy, Jordan, Kenya, Luxembourg, Malaysia, Morocco, Netherlands, New Zealand, Norway, Pakistan, Portugal, Saudi Arabia, Singapore, Spain, Sweden, Switzerland, Thailand, Turkey
D	Bulgaria, Czech Republic, Hungary, People's Republic of China, Poland, Russia, Slovakia
G	Australia, Austria, Belgium, Finland, Germany, Greece, Hungary, Israel, Italy, Luxembourg, Netherlands, New Zealand, Norway, Portugal, Romania, Spain, Sweden, Switzerland
I	Hong Kong, Ireland, South Africa, United Kingdom
K	Czech Rep., Hungary, N. Korea?, Poland, Russia
L	France
M	Canada, Japan, Mexico, Peru, Philippines, S. Korea, Taiwan, United States; Brazil (PAL-M)
N	Argentina, Jamaica, Paraguay, Uruguay

8.11 Physical Interface Standards for "Television" Video

With the timing, color-encoding, and signal-level standards reasonably well defined, at least for a given market or region, there are still several possible options for the physical connector standard to be used with these. In the case of analog television interconnects, there is also a separation of applications into the consumer market and the professional/production environment.

8.11.1 Component vs. composite video interfaces

One major distinguishing feature of wired video interfaces is whether they are considered as carrying *component* or *composite* video. While technically a difference in the form of electrical interface, this distinction also has a great impact on the physical connector choice. Simply put, a composite video interface is one which carries the signal in the same form as an over-the-air transmission; the color information is encoded per the appropriate standard and composited into a single electrical signal along with the luminance and sync information.

(Audio may or may not be included per the relevant broadcast standard.) Most often, a "composite video" connection refers to a *baseband* signal, one which has not been placed on a higher-frequency carrier through modulation. However, many consumer products, especially television receivers, lack a separate input for such signals, and must accept all signals through the RF tuner/demodulator. Therefore, it is common for consumer-class video sources such as video-cassette recorders (VCRs) or DVD players to provide both a baseband composite video output, and the same signal modulated onto an RF carrier on a locally unused broadcast channel. In the US, VHF channels 3 and 4 (60–66 MHz and 66–72 MHz, respectively) are typically provided, and can be selected by the user.

Component video interfaces place the various components of the television signal on physically separate connections and cables. The primary advantage of this, at least in the consumer environment, is to ensure that these components do not interfere with one another. (This is, of course, of benefit only if these components have not previously been composited.) Also, since the component signals do not have to be carried within a limited bandwidth channel or comply with the other requirements of the "broadcast-style" composite signal, the bandwidth of these signals can be increased. The transmission channel therefore need not be the limiting factor in the quality of the displayed image.

One of the more common consumer video interfaces, provided by many different types of equipment, simply separates the luminance (Y, with syncs) and the combined chrominance or color-difference signals (C), placing them on physically separate channels. The chrominance signals are otherwise encoded and combined per the appropriate system specifications. This connection is generally referred to as a "Y/C" interface, although it is often mistakenly referred to in the generic sense as an "S-Video" connection. As will be discussed shortly, "S-Video" properly refers only to this form of interface using a specific physical connector. While not a purely composite interface, the Y/C form of connection generally is not referred to as "component" video either, as the chrominance signals are not separated into their most basic form.

8.11.2 The "RCA Phono" connector

A very common connector used for consumer-market baseband, RF, and component video connections is the "RCA" or "phono" connector, shown in Figure 8-11. This is a simple, inexpensive connector system which works reasonably well with small-diameter coaxial cabling. It is also, however, in common use in other consumer applications, especially for audio connections. Physically, the plug is characterized by a rounded-tip center pin, into which the center conductor of the cable may be inserted and soldered. This is surrounded by an

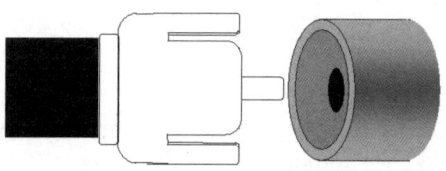

Figure 8-11 "RCA" or "phono" plug and jack.

insulating cylinder, over which the outer contact is provided in the form of a cylindrical shell with four lengthwise slots. The plug is held onto the jack solely by friction between this shell and the outer cylindrical surface of the jack, over which it fits. While this type provides an acceptable coaxial connection, it is not truly an impedance-matched connector system, and is therefore not the best choice for RF connections; it also can suffer from loosening of the physical connection with repeated insertions. Also, due to the other common uses of this type, especially as an audio connector, one must be careful not to use cable assemblies intended for less-critical application as video interconnects. (Audio cabling which uses this connector, for example, is almost certainly not constructed from coaxial cable at all, let alone being of the proper impedance.)

8.11.3 The "F" connector

A step up from the "RCA" type for high-frequency connections, if only due to being a better match to the RG-59 type coaxial cable common in consumer video interconnects, is the "F" connector type. This again provides a single "signal" contact, which in this case is generally the center conductor of the coaxial cable itself. The "shield" connection is provided via the outer cylindrical shell or barrel. The outer shell of the jack in this type is threaded, but the plug may provide either a friction-fit stationary outer shell or a threaded barrel. The connector is available in 75-Ω designs, and is typically a very inexpensive and simple connector to use; a crimp connection to the coaxial cable's shield is common. Due primarily to the use of the center conductor of the cable itself as the center pin, though, this type is not especially rugged and should be avoided in applications where frequent insertions and removals are expected. The "F" connector is quite common as an antenna input for television receivers, and as a composite video input or output connector for all consumer video equipment.

8.11.4 The BNC connector

The BNC ("Bayonet Neill–Concelmann", named for its designers) connector is a very rugged, compact, and relatively high-performance connector which provides a good match to coaxial cables up to the standard RG-59 size (special BNCs may also be found to accommodate larger cables). Shown in Figure 8-12, the plug provides a separate center pin, typically soldered or crimped to the coax center conductor, and an outer barrel for the "shield" contact. The rotating barrel provides a positive lock to the jack, by engaging two small pins protruding from the outer cylindrical portion of the jack (a "bayonet" connection, hence the name). While not as inexpensive as the "F" type, BNCs provide a much more positive connection, and are usable to very high frequencies. This type, due to its higher cost, is generally found only on high-end consumer equipment, but is common in professional video gear and has also been widely used in the computer industry. It is important to note, however, that the connector itself and standard cable assemblies using it are available in both 50-0hm and 75-ohm versions. Care must be taken to ensure that only 75-Ω types are used in standard video applications.

Figure 8-12 BNC connectors. (Photo courtesy of Don Chambers/Total Technologies, Inc. used by permission)

8.11.5 The N connector

Similar in appearance to the BNC, the somewhat-larger "N" connector differs in two significant points. First, the outer barrel of the plug, and the mating surface of the jack, employ a threaded connection rather than the bayonet type. The connector also is distinguished by a separate cylindrical contact between the center pin and the outer barrel, and so does not rely solely on the barrel to provide the "shield" or "return" connection. The "N" connector is a very high-performance, precision type, but relatively costly and more difficult to attach to the cable. Use of this type is restricted to professional-level studio and test equipment.

8.11.6 The SMA and SMC connector families

These connectors, generally found only in professional equipment and precision test and measurement gear, are precision coaxial connectors intended for use with miniature coax.

8.11.7 The "S-Video"/mini-DIN connector

A very popular connector in consumer video equipment, including television receivers, video-cassette recorders (VCRs), camcorders, etc., is the "S-Video" connector, based on the standard 4-pin miniature DIN connector (Figure 8-13). This provides separate luminance (Y) and chrominance (C) connections, and has become so associated with the Y/C interface that any such connection (even if physically separate connectors are used for the Y and C signals) is often referred to as "S-Video." The connector is relatively inexpensive and works reasonably well in this application, but does not provide a true coaxial, impedance-matched connection and so would not be suitable for similar use at higher frequencies. Alternate types, also confusingly referred to as "S-Video" connectors, use the 7-pin miniature DIN (the same size and overall shape as the 4-pin version), but carry either an I^2C interface (for control functions), or a separate composite video signal.

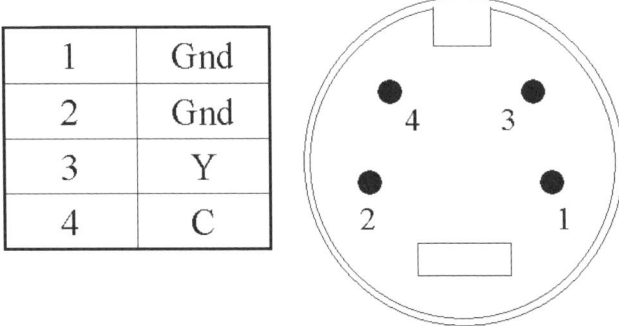

Figure 8-13 "S-Video" 4-pin mini-DIN connector (output jack) and pinout.

8.11.8 The SCART or "Peritel" connector

Possibly best viewed as the European counterpart to the S-Video connection, the SCART (Syndicat des Constructeurs d'Appareils Radiorécepteurs et Téléviseurs) connector (Figure 8-14), also known as the "Peritel" connector, also supports separate Y and C video connections. In addition, this connector provides stereo audio inputs and outputs and a separate composite video connection. An alternate pinout has also been defined for use in RGB systems. Status pins (both input and output) are used to define the active video signals at any given time, permitting devices to switch connections automatically.

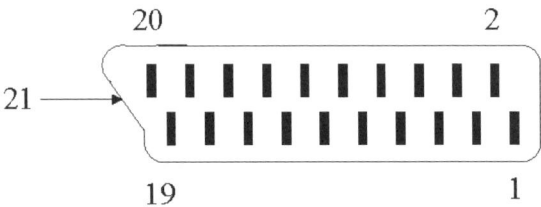

Figure 8-14 The SCART or "Peritel" connector. There are at least two pinouts in use for this connector, one intended for RGB video and the other intended for "S-Video" compatibility. These are as listed in Table 8-4.

Table 8-4 SCART/Peritel connector pinouts.

Pin no.	Type 1 assignment ("RGB")	Type 2 assignment ("S-Video")
1	Audio out B/right	Audio out B/right
2	Audio input B/right	Audio input B/right
3	Audio out A/left (or mono out)	Audio out A/left (or mono out)
4	Audio return	Audio return
5	Blue video return	Ground
6	Audio input A/left (or mono in)	Audio input A/left (or mono in)
7	Blue video input	–
8	Function select[a]	Function select[1]
9	Green video return	Ground
10	Data 2	Data 2
11	Green video input	–
12	Data 1	Data 1
13	Red video return	Chrominance return
14	Ground	Ground
15	Red video input	Chrominance (C) input
16	RGB/composite switching[b]	–
17	Ground	Luminance return
18	Ground	–
19	Video out (composite)	Video out (composite)
20	Video in (composite)	Luminance (Y) input
21(shell)	Common ground (shield)	Common ground (shield)

[a] The function select input switches the equipment between "standard TV" (0–2V), "widescreen" (5–8V), and "AV" (9.5–12V) modes. If the video source sets this pin to either of the latter two states, the SCART input is automatically used; the two differ only in that video processing to handle "anamorphic" 16:9 programming is enabled in the "widescreen" mode.

[b] The RGB/composite switch input determines the video inputs in use; when high (1–3V), the RGB video inputs are used, and when low (0–0.4V) the composite input is used.

9

Standards for Analog Video – Part II: The Personal Computer

9.1 Introduction

For over 30 years, broadcast television represented essentially the only electronic display system, and display interface, in truly widespread use. Various forms of electronic displays were in use in many applications, notably in test and measurement equipment and in early computing devices, but these were for the most part "embedded" display applications in which the display device was an integral part of the product. Such uses do not require much in the way of standardized external display interfaces. When remote location of the display itself was required, most often the connection was made using either an interface custom-designed for that product, or via television standards.

Through the 1960s and 1970s, the growing importance of electronic computing began to change this. Earlier analog computers had made use of existing "instrumentation-style" output devices – such as oscilloscopes and X–Y pen plotters – but the digital computer's ability to be programmed via something resembling "plain language" required a human interface more suited to text-based communications. At first, standard teletype units were adapted to this use. These were electromechanical devices that combined a keyboard and a printer with a simple electronic interface, capable of translating between these devices and a standard binary code for each character. (Direct descendants of this type of device are the modern character codes used to represent text information in digital form. Examples include the American Standard Code for Information Interchange (ASCII) and the various ISO character code definitions.

Teletype machines, while workable, had several significant disadvantages. Besides being large, slow, and noisy, they were greatly hampered by the need to print everything – both the text being entered by the operator, and any response from the computer – on paper. Editing text is particularly difficult with such a system, and any but the simplest interaction with the computer used a large amount of paper. The next logical step was to replace the printed page as the sole output device with an electronic display. As the CRT was the only practical option available, it became the heart of the first all-electronic computer terminals.

Still, these were little more than teletype machines with a "printer" that needed no paper. Input and output were still completely text-based, and the interface was essentially identical to that of the teletype – a digital connection, sometimes unique to the system in question, over which the operator and the computer communicated via character codes. (Computer printers, of course, continued to develop separately, but now as an output device used only when needed.)

9.2 Character-Generator Display Systems

The CRT terminal is of interest here, as it introduced the first use of character-generator text display. In this form of display, text is stored in the form of the ASCII or other code sets used by the terminal. The displayed image is produced by scanning through these codes in the order in which the text is to appear on the screen – usually in the form of lines running left to right and top to bottom – and the codes used as indices to a character read-only memory (ROM). This ROM stores the graphic representation of each character the terminal is capable of displaying. In this type of system (Figure 9-1), each character is generally produced within a fixed-size cell; the top line of each cell is read out, in order, for each character in a given line of text, and then the next line within each cell, and so forth. A line counter tracks the correct position vertically through the lines of text, as well as the appropriate line within each set of character cells. In this system, each visible line must be an integral number of character cell widths in length, and for simplicity, the blanking time and all subdivisions thereof are also counted in character widths. To this day, the use of the term "character" to denote the smallest increment of horizontal timing has been maintained.

The data from the character ROM is serialized, generally using a parallel-in, serial-out shift register, and could then be used (after amplification to the required level) to control the CRT electron beam. The same counters which track the line count within the displayed frame and the character count within each line are also used to produce the synchronization signals for the CRT, simply by starting and stopping these pulses at the appropriate count in each direction. Integrated, programmable control ICs to perform these functions were soon developed, requiring only the external ROM for character storage, a clock of the proper frequency to produce the desired timing, and a few other minor components.

Simple features such as underlining, inverse text display, and blinking were fairly simple to add to such a system. Blinking and/or inversion of the text (e.g., producing a black character within a white cell rather than vice-versa) are achieved simply by blocking or inverting the data from the character ROM, and gating this with a signal derived by dividing down the vertical sync signal to produce the desired blink rate. Underlining is easily achieved by forcing the video data line to the "on" state during the correct line of each character cell.

Simple color can also be achieved in such a system, by storing additional information with each character. For example, the character code itself might be stored as eight bits, plus three added bits, one for each of the primary colors. This permits each character to be as-

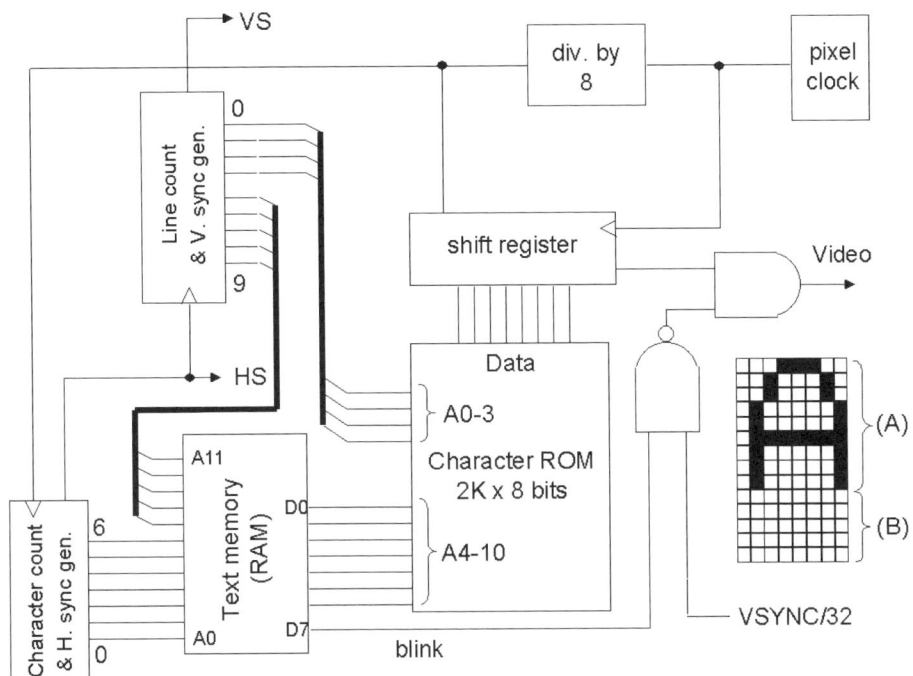

Figure 9-1 Character-generator video system. The image of each possible character is stored in a read-only memory (ROM), and read out under the control of a line counter and the output of text memory (RAM), which contains the codes corresponding to the text to be displayed on the screen. In this example, each character occupies an 8 pixel by 14 line cell; each cell may be viewed as having space for the main body of each character (A), plus additional lines for descenders, line-to-line spacing, etc. (B). The characters are selected in the proper sequence by addressing the text memory via character and line counters. Not shown is the data path from the CPU to the text memory.

signed any of eight colors, and the color information is simply gated by the character ROM data line to control the three beams of a color CRT.

9.3 Graphics

While the character-generator CRT terminal was a major improvement over the teletype machine, there was still no means for producing graphical output other than electromechanical systems such as pen plotters. As noted in Chapter 1, the earliest graphics displays were vector-scan devices; CRT-based displays in which images are literally "drawn" on the screen through direct control of the beam deflection. However, these were very limited in their capability, and never achieved a level of success which required the development of standard interfaces.

The logical development from the character-generator system, in terms of increasing the graphics capability, was to permit the images to be drawn in memory rather than using the permanent storage of a ROM. This brings us to the concept of a frame buffer, as was introduced in Chapter 1, in which images may be stored or synthesized. With the computer now

166 STANDARDS FOR ANALOG VIDEO – PART II: THE PERSONAL COMPUTER

able to create images in this manner, the concept of a pixel broadened from its meaning of simply a point sample of an existing image. Computer synthesis of imagery led to the "pixel" being thought of as simply the building block out of which images could be made. The concept of the pixel as literally a "little square of color," while technically incorrect, has become very deep-seated in the minds of many working in the computer graphics field. This unfortunately leads to some misconceptions, and conflicts with the proper point-sample model. Despite this conceptual problem, this usage of "pixel" has become very pervasive, and is practically impossible to avoid in discussions of computer graphics and display-system issues.

Regardless of what we define as a "pixel," however, the contents of the frame buffer must still be transferred in some way to the display device, and so we now move to consideration of the development of the display interface itself, as used in the computer industry.

9.4 Early Personal Computer Displays

The first personal computers had to rely either on existing terminal-type displays, communicating via standard serial or parallel digital interfaces, or on the one standard display device available – the television. With the PC market in its infancy, no standard computer monitor products yet existed. The original Apple II, Atari, and Commodore VIC PCs are all examples of early computers designed to use standard television receivers or video monitors as their primary display. A very few of the early PCs, notably the Commodore PET (1978), provided an integrated CRT display, although these were still generally based on existing television components.

But the limitations of such displays quickly became apparent. Character-based terminals were incapable of providing the graphics that PC users increasingly demanded, and more capable terminals were not an economical alternative for the home user. Televisions could not provide the resolution required for any but the simplest PC applications, especially when viewed from typical "desktop display" distances. Using television receivers as computer displays also required the use of TV-style color encoding, and the resulting loss of color resolution and overall quality. Display products specifically intended for computer use, and with them standardized monitor connections, were needed. These came with the introduction of the original IBM Personal Computer in 1981.

It should be noted at this point that many higher-end computers, such as those intended specifically for the scientific and engineering markets, retained the integrated-display model for some time. These more expensive products could afford displays specifically designed as high-resolution (for the time), high-quality imaging systems. Note, however, that no real "display interface" standardization had occurred in such systems. The display was directly connected to, and in such cases commonly in the same physical package as, the hardware that produced the images in the first place. When connections to external display devices were provided, they were either special, proprietary designs, or used existing standards from the television industry. The reliance on standards originally developed for television use, as in the case of the "lower-end" personal computer market, again was a major factor in shaping the signal standards and practices used in these systems. Such "scientific" desktop computer systems later developed into the "engineering workstation" market, which progressed separately and along a somewhat different path than the more common "personal computer" (or "PC").

9.5 The IBM PC

The IBM PC (and the "clones" of this system which quickly followed) was among the first, and certainly was the most successful, personal computer system to use the "two-box" (separate monitor and CPU/system box) model with a display specifically designed for the computer. Rather than using a standard "television" video output, and a display which was essentially a repackaged TV set (or even a standard portable television), IBM offered a series of display products as part of the complete PC system. To drive these, several varieties of video cards, or "graphics adapters" in the terminology introduced with this system, were provided. This model quickly became the standard of the industry, and at least the connector introduced with one of the later systems remains the de facto standard analog video output for PCs to this day.

These original IBM designs were commonly referred to by a three- or four-letter abbreviation which always included "GA", for "graphics adapter" or later "graphics array." The first generation included the Monochrome Display Adapter ("MDA", the one example which did not include "graphics" as it had nothing in the way of graphics capabilities), the Color Graphics Adapter (CGA), and the Enhanced Graphics Adapter (EGA). Later additions to the "GA" family included the Video Graphics Array, the Professional Graphics Adapter, the Extended Graphics Array (VGA, PGA, and XGA, respectively) and so forth. Today, only the "VGA" name continues in widespread use, at least in reference to a standard connector, although some of the others (notably VGA, SVGA, XGA, and SXGA) continue to be used in reference to display formats originally introduced with that hardware. (For example, "XGA" today almost always refers to the 1024 × 768 format, not to the original XGA hardware.)

9.6 MDA/Hercules

The original MDA adapter was a simple monochrome-only card, intended for use with a fixed-frequency display and providing what was effectively a 720 × 350 image format, although it was capable only of producing a text display using the ROM-based character generator technique described above. This system used a fixed character "cell" of 9 × 14 pixels, and so the image format produced can more properly be described as 25 lines of 80 characters each. (This "80 × 25" text format is a de facto standard for such displays.) The connection to the display was via a 9-pin D-subminiature connector, whose pinout is shown in Figure 9-2. Note that this might be considered a "digital" output, although if so it is of the very simplest variety. The primary video signal provided was a single TTL-level output, which simply switches the CRT's beam on and off to create the characters (although a separate "intensity" output was also provided, which could be used to change the brightness on a character-by-character basis). The MDA output also provided separate TTL-level horizontal and vertical synchronization ("sync") signals, a system which has been retained in PC standards to this day.

A similar video card of the same vintage was the "Hercules" graphics adapter, a name which is still heard in discussions of this early hardware. The Hercules card (named for the company which produced it) used essentially the same output as the IBM MDA, but a slightly different image format.

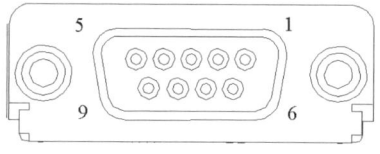

Pin	Signal	Pin	Signal
1	Ground	6	Intensity
2	n/c	7	Video (TTL)
3	n/c	8	H sync
4	n/c	9	V sync
5	n/c		

Figure 9-2 The MDA video output connector and pinout. This started the use of the 9-pin D-subminiature connector as a video output, which continued with the CGA and EGA designs.

9.7 CGA and EGA

A step up from the MDA was the Color Graphics Adapter, or "CGA" card. This permitted the PC user to add a color display to the system, albeit one that could provide only four different colors simultaneously, and that only at a relatively low-quality display format of 320 × 200 pixels. CGA also provided the option of monochrome operation at 640 × 200, mimicking the MDA format but with a slightly smaller character cell. Again, the 9-pin D-subminiature connector type was used (Figure 9-3), with previously unused pins now providing the additional outputs required for color operation.

A further increase in capabilities could be had by upgrading to the Enhanced Graphics Adapter, or EGA. This supported 16 different colors simultaneously, with a format of 640 × 350 pixels, in both graphics and text modes. The EGA retained the 9-pin connector of the CGA and MDA types, but with a slightly different pinout (also listed in Figure 9-3) as needed to support the increased color capabilities.

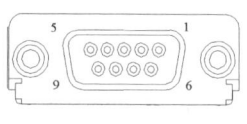

Pin	CGA	EGA	Pin	CGA	EGA
1	Ground	Ground	6	Intensity	Int/Grn 0
2	n/c	Red 0	7	n/c	Blue 0
3	Red	Red 1	8	H. sync	H. sync
4	Green	Green 1	9	V. sync	V. sync
5	Blue	Blue 1			

Figure 9-3 The revised pinout of the 9-pin connector for the Color Graphics Adapter (CGA) and Enhanced Graphics Adapter (EGA) products.

9.8 VGA – The Video Graphics Array

With the introduction of the VGA hardware and software definitions by IBM in 1987, the stage was set for PC video and graphics systems to come into their own as useful tools for both the home and professional user. Later products would build on VGA by increasing the pixel counts supported, adding new features, etc., but the basic VGA interface standards remain to this day. (A separate, lower-capability system introduced at the same time as VGA –

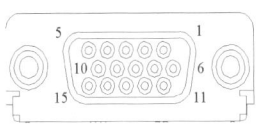

Pin	Original	DDC	Pin	Original	DDC
1	Red video		9	NC (key)	+5 VDC
2	Green video		10	Sync. return	
3	Blue video		11	ID 0	Unused
4	ID 2	Unused	12	ID 1	Data
5	Test	Return	13	Horizontal sync	
6	Red return		14	Vertical sync	
7	Green return		15	ID 3	Data clk.
8	Blue return				

Figure 9-4 The VGA video connector. Both the original pin assignments and those defined by the VESA Display Data Channel (DDC; see Chapter 11) are shown. Note that the original pinout is now obsolete and almost never found in current use. Used by permission of VESA.

the "MultiColor Graphics Array" or "MCGA" – never achieved the widespread acceptance of VGA and was soon abandoned.)

Among the most significant contributions of the VGA definition were a new output connector standard and a new display format, both of which are still referred to as "VGA."

The new connector kept the same physical dimensions as the 9-pin D-subminiature of the earlier designs, but placed 15 pins (in 3 rows of 5 pins each) within the connector shell (Figure 9-4). This is referred to as a "high-density" D-subminiature connector, and common names used for this design (in addition to simply "VGA") include "15-HD" or simply "15-pin D-sub." The new connector also supported, for the first time, "full analog" video, using a signal definition based loosely on the RS-343 television standard (roughly 0.7 V p-p video with a 75-Ω system impedance). However, separate TTL sync signals, as used in the earlier MDA/CGA/EGA systems, were retained instead of switching to the composite sync-on-video scheme common in television practice.

The new "VGA" timing, at 640 × 480 pixels and 60 Hz refresh, was also a tie to the television world, being in both format and timing essentially a non-interlaced version of the US TV standard (see the VGA timing details in chapter 7, and contrast these with the television standards discussed in Chapter 8). While this level of compatibility with television would very soon be abandoned by the PC industry in the move to ever-increasing pixel formats and refresh rates, this idea would later be revisited as the television and computer display markets converge. The VGA hardware still supported the earlier 720 × 350 and 640 × 200 formats (and in fact these remain in use as "boot" mode formats in modern PCs), but the 640 × 480 mode was intended to be the one used more often in normal operation, and provides a "square-pixel" format with an aspect ratio matching that of standard CRTs (4:3).

The VGA system also introduced, for the first time, a simple system for identifying the display in use. By the time of VGA's introduction, it was clear that the PC could be connected to any of a number of possible monitors, of varying capabilities. In order to permit the system to determine which of these were in use, and thereby configure itself properly, four pins of the connector were dedicated as "ID bits". The monitor, or at least its video cable and connector, could ground or leave floating various combinations of these and thereby identify itself as any of 16 possible types. This limitation to a relatively few predefined displays would soon prove to be unacceptable, however, and would be replaced by more sophisticated display identification systems (see Chapter 11).

The basic VGA connector would be retained by later, more capable "graphics adapter" hardware, such as the "Super VGA" (SVGA) and "extended VGA" (XGA) designs, and these add not only new features to the system but also support for higher and higher pixel counts, or "higher resolutions," to use the common PC terminology. SVGA introduced 800 × 600 pixels to the standard set, again a 4:3 square-pixel format, and was followed by the XGA 1024 × 768 format. As noted earlier, both of these names are now use to refer to the formats themselves almost exclusively, and the set has grown to include "Super XGA" (SXGA), at 1280 × 1024 pixels (the one 5:4 format in common use), and "Ultra XGA" (UXGA), or 1600 × 1200 pixels. With each increase in pixel count, however, the hardware retained support for the original "VGA" formats, as this was required in order to provide a common "boot-up" environment, and so *multifrequency* monitors became the norm in PC displays. This term refers to those displays which automatically adapt to any of a wide range of possible input timings; in the PC market, such will always at least support down to the 31.5 kHz horizontal rate required for the standard VGA modes. The development of such capability in the display was one of the primary factors driving the need for better display ID capability, such that the system could determine the capabilities of the display in use at any given time.

9.9 Signal Standards for PC Video

As noted above, the analog signal definitions used by the VGA system were loosely based on the RS-343 amplitude standards, or more correctly the European standards which had be developed from the earlier American practices. The reader may recall from the previous chapter that one significant difference between the American and European television standards was the absence of "setup", an amplitude distinction between the "blank" and "black" states, in the latter case. The original IBM VGA hardware provided the same 0.7 V p-p signal, without setup, as was common in European television, but again was distinguished from the TV standards in that the PC relied on the simpler, separate TTL-level sync signals. Both industries, at least, kept 75 Ω as the standard for the video interconnect system's characteristic impedance.

It is important to note, however, that the VGA specifications were never truly an industry standard in the sense of being formally reviewed and adopted by any standards organization or consortium. Manufacturers wishing to produce "VGA-compatible" hardware did so essentially by "doing what IBM had done," based on the products already brought to market, along with whatever guidance was to be found in the specifications themselves. Significantly, there was never a formal definition released for the video signal requirements under the VGA "standard," and this did lead to some confusion and compatibility problems, especially in recent years as video frequencies have increased and users have become more demanding in their image quality expectations. Only recently has a formal set of signal specifications been released, by the Video Electronics Standards Association (VESA).

The lack of formal standards and the slight difference between the various existing video standards made for some confusion in establishing the "correct" signal amplitudes in many systems. Video output circuits, typically either a separate "RAMDAC" IC (a device including both color look-up tables, in random-access memory or RAM, plus a digital-to-analog converter, as shown in Figure 9-5), or as part of an integrated graphics control IC, most often can be set to any desired signal amplitude (within limits), through the selection or adjustment

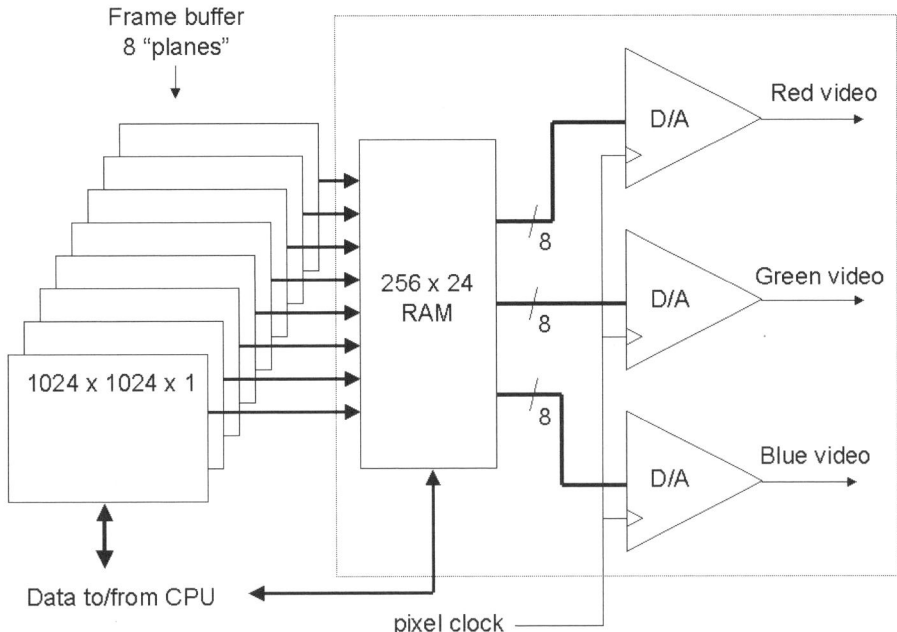

Figure 9-5 "RAMDAC" PC graphics output. In order to provide maximum flexibility within limited memory space, PC graphics systems began employing "color map" memory stages, coupled to digital-to-analog converters (DACs) to produce the video output. In this example, 1 Mbyte of frame buffer storage, organized as 1k × 1k pixels, each 8 bits "deep," feeds a 256 location by 24 bit RAM. This memory, whose contents are also written by the host CPU, maps the 8-bit values for each pixel to any of 2^{24}, or approximately 16.7 million, possible output values, or 8 bits for each primary color. The color-map memory and output DACs are often integrated into a single component, referred to as a RAMDAC. This technique remains in common use today, even with frame buffer systems providing far more than 24 bits per pixel, as it simplifies the implementation of numerous features.

of external components. In addition, such devices could be obtained in versions which did or did not include "setup", or offered a programmable choice here. The nominal VGA video signal level was, as mentioned above, 0.7 V p-p, approximately the same as both the European television video standard as well as the RS-343 definition of 0.714 V p-p. However, setting an output up to deliver the specified RS-343 signal exactly and then turning off "setup" would often simply drop the peak level by the setup or "pedestal" amplitude (0.054 V), resulting in a signal which peaked at only 0.660 V above the blank level. Thus, the nominal level of PC video signals could vary between 0.660, 0.700, and 0.714 V p-p, and for most PC graphics cards there is a considerable tolerance (±10% is typical). While such a wide range of possible amplitudes may not affect the basic operation of a CRT display (at least not in any way readily noticeable to the casual user), it does cause problems in critical imaging applications, and especially in those display types which require analog-to-digital conversion of such signals (as in common in many LCD and other non-CRT-based monitors).

A separate but equally serious problem for the non-CRT types results from the fact that there was no specification for the stability or skew of the sync signals, especially the horizontal sync, with respect to the video. Unlike systems that provide synchronization information as part of the video signal (as in the standard television definitions), the separate TTL syncs common in PC practice are not generally well controlled in this regard. While both the generation of the video signal itself and the sync signals are typically controlled by the same clock signal, there is most often no provision for further controlling the position of the sync edges with respect to the video signal. Skew and jitter of these signals, particularly with lower-quality video cabling (in which the sync signals are often carried on simple twisted pairs vs. the coaxial cable used for the video) can become a significant problem. Again, this is almost never a concern with CRT displays, owing to the stability of the phase-locked loop circuits used to "lock" the horizontal and vertical deflection timing. But displays which rely on "digitizing" analog video signals (again, analog-input LCD monitors are the most common example) must derive their sampling clocks from the sync signals, usually the horizontal sync. Skew and jitter in these signals make for considerable difficulty in maintaining a stable image in such cases.

As noted above, these concerns have recently begun to be addressed through the development of true industry standards for PC video signals. The VESA Video Signal Standard, released in 2000, represents the best attempt to date to provide a tighter specification of such signals; the basic requirements of this standard are outlined in Table 9-1. VESA had earlier produced the only true industry standard which documented the pinout of the VGA connector, as part of the Display Data Channel (DDC) standard released in 1994. Originally, this standard was intended as a replacement for the original four-bit VGA display ID scheme, and reassigned several of the ID pins. This standard is covered in detail in Chapter 11, but the final VESA-standard pinout for the "VGA" 15-pin connector is also given here in Figure 9-4. (VESA's "Plug & Display" standard, discussed later in this chapter, also attempted to establish specifications for the analog video signals carried by that connector, but these were not widely accepted. These represent an interesting footnote in PC video standards development, though, as the P&D specified a 0.700 Vp-p, zero-setup, and *DC-referenced* video signal. The blank level was set to zero volts, as reference to the analog signal ground.)

Table 9-1 A summary of the VESA Video Signal Standard requirements. Used by permission of VESA.

Parameter	Specification/comments
Max. luminance voltage	0.700 VDC, + 0.07V/–0.035 V; DC with respect to Return
Video rise/fall time	Max.: 50% of min. pixel clock period; min. 10% of min. pixel clock period; measured at 10-90% points
Video settling time	30% of min. pixel clock period to 5% final full-scale value
Video amplitude mismatch	Max. 6% channel-to-channel over full voltage range.
Video noise injection ratio	+/- 2.5% of maximum luminance voltage
Video channel/channel skew	Maximum of 25% of minimum pixel clock period
Video overshoot/undershoot	Max. of +/- 12% of step voltage, over full voltage range
Sync signal rise/fall time	Max. of 80% of minimum pixel clock period
Sync signal over/undershoot	Max. of 30% of high; no ringing into 0.5-2.4V range
Jitter (between H sync pulses)	15% of pk-pk or 6 sigma min. pixel clock period, 0 Hz to max. horizontal rate, over a minimum of 100K samples.

9.10 Workstation Display Standards

The engineering workstation industry, as previously noted, developed its own de facto standards separately from those of the PC market. As was the case with the PC, there were still no true industry standards, although there was some degree of commonality in the design choices made by the manufacturers of these systems. "Workstation" computers differ from those traditionally considered "PCs" in several ways impacting the display and the display interface options. First, since these are generally more expensive products, they are not under quite as much pressure to minimize costs; more expensive, higher-performance connectors, cabling, etc., are viable options. In addition, the workstation market has traditionally been dominated by the "bundled system" model, in which all parts of the system – the CPU, monitor, keyboard, and all peripherals – are supplied by the same manufacturer, often under a single product number. This makes for a very high "connect rate" (the percentage of systems in which the peripherals used are those made by the same manufacturer as the computer itself) for the display; unique and even proprietary interface designs are not a significant disadvantage in such a market. In fact, it has been very uncommon in the workstation market for two manufacturers to use identical display interfaces. The last difference from the PC industry comes from the operating system choice. Rather than a single, dominant OS used across the industry – as in the case with the various forms of Microsoft's "Windows" system in the PC market – workstation products have generally used proprietary operating systems, often some version of the Unix OS. Under this model, there is little need for industry-standard display formats or timings, nor are displays or graphics systems required to support the common VGA "boot" modes described earlier. This made for the workstation industry generally using fixed-timing designs, including single-frequency high-resolution displays. This has only recently changed, as a convergence of the typical workstation display requirements with the high end of the PC market has made it more economical for workstations to use the same displays as the PC. Even when using what are essentially multifrequency "PC" monitors, however, workstation systems commonly remain set to provide a single format and timing throughout normal operation.

In the early-to-mid-1980s, as the workstation industry began to develop physical system configurations similar to those of the PC (a processor or system box separate from the display), the most common video output connectors were BNCs, as described in the previous chapter (see Chapter 8, Figure 10), carrying separate RGB analog video signals. As with the PC, signal amplitude standards were approximately those of RS-343 (0.714 V p-p), again with very little standardization between the various possibilities (0.660, 0.700, and 0.714 V p-p) which might be produced. In a major departure from the typical PC practice, however, many workstation manufacturers chose to use composite sync-on-video, generally supplied on the green signal only (and thus making only this a nominal 1.000 V p-p, including the sync pulses).

Here again, the computer industry chose to do things somewhat differently from the standard practices established for television. In the television standards, the sync signals are always composited such that horizontal sync pulses are provided during the vertical sync pulse, although inverted from their normal sense (Figure 9-6); such pulses are referred to as *serration* of the vertical sync pulse. Further, television standards require that the reference edge of the horizontal sync pulse (generally, the trailing edge of this pulse is defined as the point from which all other timing within the line is defined) be maintained at a fixed position in time, regardless of the sense of the pulse. (The effect of this is that the sync pulse itself shifts by its own width between the "normal" version, and the "inverted" pulses occurring during the vertical sync interval.) In those computer display systems that used composite

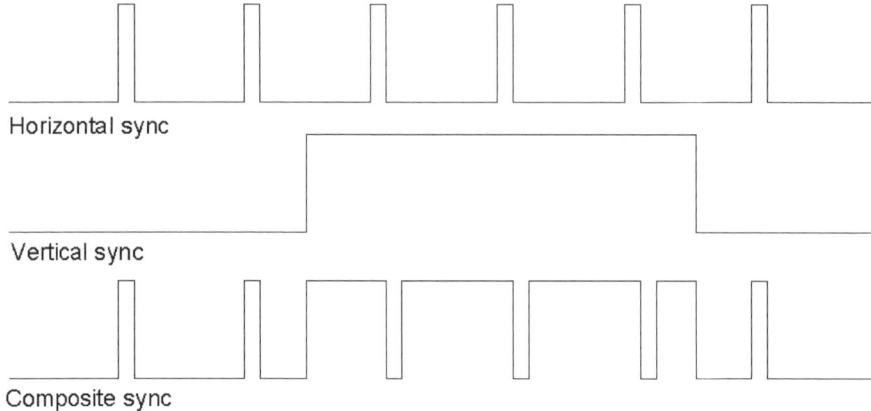

Figure 9-6 Sync compositing. If the horizontal and vertical sync pulses are combined as in standard television practice, the position of the falling edge of horizontal sync is preserved in the composite result, as shown.

sync, either as a separate signal or provided on the green video, this requirement has only rarely been met. Most often, computer display systems produce composite sync either by simply performing a logical OR of the separate horizontal and vertical sync signals, or (in a slightly more sophisticated system) performing an exclusive-OR of these. The former produces a composite sync signal in which the vertical sync pulse is *not* serrated (Figure 9-7a); this is sometimes referred to as a "block sync" system. If the syncs are exclusive-ORed, a serrated V. sync pulse results in the composited signal, but with a potential problem. Depending on the alignment of the horizontal sync pulse with the leading and trailing edges of the vertical sync, a spurious transition may be generated at either or both of these in the composited signal (Figure 9-7b).

The purpose of serration of the vertical sync pulse in a composite-sync system is to ensure that the display continues to receive horizontal timing information during the vertical sync (since the vertical sync pulse is commonly several line-times in duration). This prevents the phase-locked loop of the typical CRT monitor's horizontal deflection circuit from drifting off frequency. Should this PLL drift during the vertical sync, and not recover sufficiently during the vertical "back porch" time (that portion of vertical blanking during which horizontal sync pulses are again supplied normally), the result can be a visible distortion of the lines at the top of the displayed image (as in Figure 9-8). This is commonly known as "flagging," due to the appearance of the sides of the image near the top of the display. Serration of the vertical sync can eliminate this problem, but if done via the simple exclusive-OR method described above, the spurious transitions on either edge of the V. sync pulse can again cause stability problems in the display's horizontal timing.

As noted, the earliest workstation systems typically used BNC connectors to supply an RGB analog video output; various systems used three, four, or five BNCs, supporting either sync-on-green, a separate (typically TTL-level) composite sync output, or separate horizontal and vertical syncs (again commonly TTL). Again reflecting the relatively lower cost pressures in workstation systems vs. the PC market, the video cables for such outputs were commonly fairly high-quality coaxial cabling for both the video and sync signals.

WORKSTATION DISPLAY STANDARDS   175

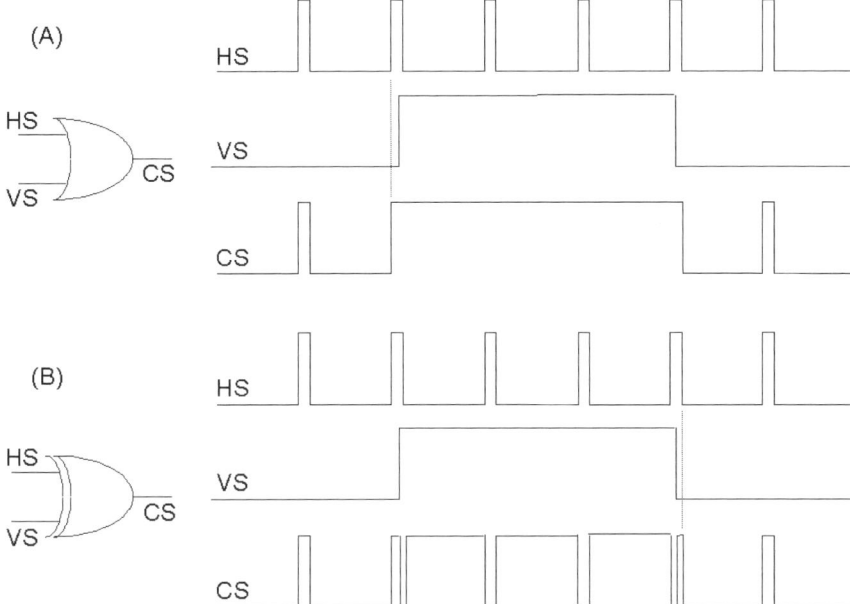

Figure 9-7 Sync compositing in the computer industry. As these signals are normally generated by digital logic circuits in computer systems, the composite sync signal has also commonly been produced simply by OR-ing (a) or excluding-OR-ing (b) the horizontal and vertical sync signals. The former practice is commonly known as "block sync," and no horizontal pulses occur during the composited vertical sync pulse. The XOR compositing produces serrations similar to those produced in television practice, but without preserving the falling edge position. Note also that, since the vertical sync in typically produced by counting horizontal sync pulses, the V. sync leading and trailing edges usually occur slightly after the H. sync leading edge; this can result in the spurious transitions shown in the composite sync output if the two are simply XORed, as in (b).

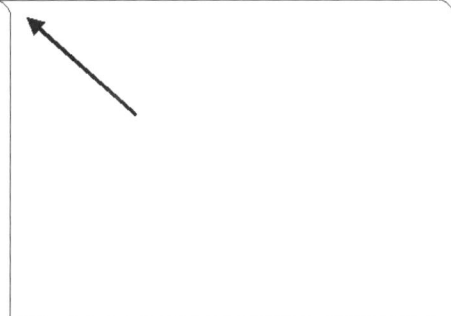

Figure 9-8 "Flagging" distortion in a CRT monitor. The top of the displayed image or raster shows instability and distortion as shown, due to the horizontal deflection circuits losing synchronization during the vertical sync period (when the horizontal sync pulses are missing or incorrect). This problem is aggravated by a too-short vertical "back porch" period, a horizontal phase-locked loop which takes too long to "lock" to the correct horizontal sync once it is acquired, or both.

9.11 The "13W3" Connector

While the BNC connector provides very good video signal performance, its size and cost make it less than ideal for most volume applications. As workstation designs faced increasing cost and, more significantly, panel-space constraints, the need arose for a smaller, lower-cost interface system. Some manufacturers, notably Hewlett-Packard, opted to use the same 15-pin high-density D-subminiature connector as was becoming common in the PC industry (although with a slightly modified pinout; for example, many of HP's designs using this connector provided a "stereo sync" signal in addition to the standard H and V syncs, and were capable of providing either separate TTL syncs or sync-on-green video). Most, however, chose to use a connector which became an "almost" standard in the workstation industry: the "13W3", which combined pins similar to those used in the 15HD with three miniature coaxial connections, again in a "D"-type shell (Figure 9-9).

While many workstation manufacturers – including Sun Microsystems, Silicon Graphics (later SGI), IBM, and Intergraph – used the same physical connector, pinouts for the 13W3 varied considerably between them (and hence the "almost" qualifier in the above). In some cases (notably IBM's use of this connector), there were different pinouts used even within a single manufacturer's product line. Many of these are described in Figure 9-10 (although this is not guaranteed to be a comprehensive list of all 13W3 pinouts ever used!).

While the closed, bundled nature of workstation systems permitted such variations without major problems for most customers, it has made for some complication in the market for cables and other accessories provided by third-party manufacturers.

Figure 9-9 The 13W3 connector. This type, common in the workstation market, also uses a "D"-shaped shell, but has three true coaxial connections for the RGB video signals, in addition to 10 general-purpose pins. Unfortunately, the pin assignments for this connector were never formally standardized. Used by permission of Don Chambers/Total Technologies.

EVC – THE VESA ENHANCED VIDEO CONNECTOR

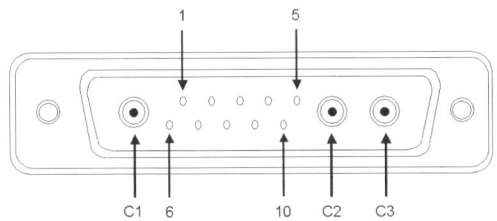

Pin	Sun	SGI	IBM
C1	Red video/Red return		
C2	Green video/Green return		
C3	Blue video/Blue return		
1	n.c./DDC SCL	MT 3/ SCL	ID 2
2	n.c./DDC +5V	MT 0/SDA	ID 3
3	Sense 2/n.c.	Comp. sync	n.c.
4	Sense/DDC rtn.	Horiz. sync	dig. ground
5	Comp. sync/HS	Vert. sync	H. sync
6	n.c./DDC SDA	MT 1/DDC +5	ID 0
7	n.c./V. sync	MT 2/DDC gnd	ID 1
8	Sense 1/n.c.	ground	n.c.
9	Sense 0/n.c.	ground	V. sync
10	gnd/sync return	sync 2/gnd	dig. ground

Figure 9-10 13W3 pinouts. This table is by no means comprehensive, but should give some idea of the various pinouts used with this connector, which has been popular in the workstation market. Note that alternate pinouts are shown for the Sun and Silicon Graphics (SGI) implementations; these are the original (or at least the most popular pre-DDC pinout) and the connector as it is used by that company with the VESA DDC standard (see Chapter 11). ("MT", "Monitor Type"; both these and the "Sense" pins in the Sun definition serve the same basic function as monitor ID pins in the IBM pinout here, and in the original IBM definition of the "VGA" connector.)

9.12 EVC – The VESA Enhanced Video Connector

By the early-to mid-1990s, the video performance requirements of the PC market were clearly running into limitations imposed by the VGA connector and typical cabling. In addition, there was an expectation that the display would become the logical point in the computer system at which to locate many other human-interface functions – such as audio inputs and outputs, keyboard connections, etc. These concerns led to the start of an effort, again within the Video Electronics Standards Association (VESA), to develop a new display interface standard intended to replace the ubiquitous 15-pin "VGA." A work group was formed in 1994 to develop an Enhanced Video Connector, and the VESA EVC standard was released in late 1995.

The EVC was based on a new connector design created by Molex, Inc. This featured a set of "pseudo-coaxial" contacts, basically four pins arranged in a square and separated by a crossed-ground-plane structure (in the mated connector pair) for impedance control and crosstalk reduction. The performance of this structure, dubbed a "Microcross™" by Molex, approximates that of the miniature coaxial connections of the 13W3 connector, while permitting much simpler termination of the coaxial lines and a smaller, lower-cost solution. In addition to the four "Microcross" contacts, which were defined as carrying the red, green, and blue video signals and an optional pixel clock, the EVC provided 30 general-purpose connections in three rows of ten pins each. These supported the usual sync and display ID functions, but also added audio input and output capability, pins for an "S-Video" like Y/C video input, and two general-purpose digital interfaces: the Universal Serial Bus (USB) and the IEEE-1394 high-speed serial channel, also known as "Firewire™" (Details on both of these interfaces are given in Chapter 11.) The EVC pinout is shown in Figure 9-11.

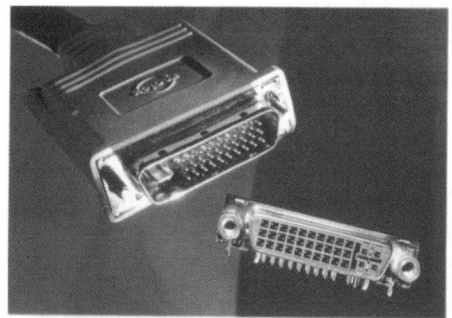
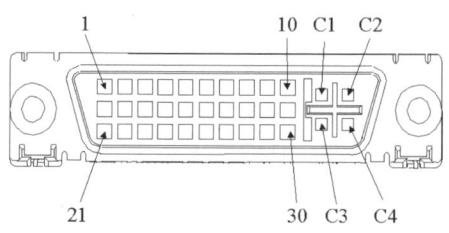

Pin	Signal	Pin	Signal
C1	Red video	C3	Pixel clock (optional)
C2	Green video	C4	Blue video
		C5*	Common video return
1	Audio output, right	16	USB data +
2	Audio output, left	17	USB data -
3	Audio output return	18	1394 shield/chg. pwr. return
4	Sync return	19	1394 Vg
5	Horiz./comp. sync (TTL)	20	1394 Vp
6	Vertical sync (TTL)	21	Audio input, left
7	Unused	22	Audio input, right
8	Charge power	23	Audio input return
9	1394 TPA -	24	Stereo sync (TTL)
10	1394 TPA +	25	DDC return/stereo return
11	Reserved	26	DDC data (SDA)
12	Reserved	27	DDC clock (SCL)
13	Video input, Y or composite	28	+5 VDC (DDC/USB)
14	Video input return	29	1394 TPB +
15	Video input, C	30	1394 TPB -

Figure 9-11 The VESA Enhanced Video Connector and pinout. Note: "C5" is the crossed ground plane connection in the C1–C4 "MicroCross™" area. *(Photograph courtesy of Molex Corp.; used by permission. "MicroCross™" is a trademark of Molex Corporation.)*

While the EVC design did provide significantly greater signal performance than the VGA connector, the somewhat higher cost of this connector, and especially the difficulty of making the transition away from the de facto standard VGA (which by this time boasted an enormous installed base), resulted in this new standard being largely ignored by the industry. Despite numerous favorable reviews, EVC was used in very few production designs, notably late-1990s workstation products from Hewlett-Packard. But while the EVC standard may from one perspective be considered a commercial failure, it did set the stage for two later standards, including one which now appears poised to finally displace the aging VGA.

9.13 The Transition to Digital Interfaces

Soon after the release of the EVC standard, several companies began to discuss requirements for yet another display interface, to address what was seen as the growing need to support displays via a digital connection. During the mid- to late-1990s, several non-CRT-based monitors, and especially LCD-based desktop displays, began to be commercially viable, and many suppliers were looking forward to the time where such products would represent a sizable fraction of the overall PC monitor market. While such displays do not necessarily require a digital interface (and in fact, as of this writing most of the products of this type currently on the market provide a standard "VGA" analog interface), there are certain advantages to the digital connection that are expected to ultimately lead to an all-digital standard.

The result of these discussions was a new standards effort by VESA, and ultimately the release of a new standard in 1997. This was named the "Plug & Display" interface (a play on the "plug & play" catchphrase of the PC industry), and was not so much a replacement for the EVC as an extension of it. In fact, EVC was soon renamed to become an "official" part of the Plug & Display (or "P&D") standard, as the "P&D-A" (for "analog only") connector. The P&D specification actually defined three semi-compatible connectors; including the EVC or "P&D-A", this system provided the option of an analog-only output, a combined analog/digital connection ("P&D-A/D"), or a digital-only output ("P&D-D").

The new "P&D" connector shape itself is shown in Chapter 10. This was intentionally made slightly different from the original EVC design; this, plus the absence of the "Microcross™" section in the digital-only version, ensures that analog-input and digital-input monitors (distinguished by the plug used) would only connect to host outputs that would support that type.

The basis of the change from EVC to P&D was the idea that future digital systems would not require the dedicated analog connections for audio I/O and video input of the EVC.

Support for these functions would be expected to be via the general-purpose digital channels (USB and/or IEEE-1394) already present on the connector. This freed a number of pins in the 30-pin field of that connector for use in supporting a dedicated digital display interface, which permitting the analog video outputs of EVC to be retained. The digital interface chosen for this was the "PanelLink™" high-speed serial channel, originally developed by Silicon Image, Inc. as an LCD panel interface, and which was seen to be the only design providing both the necessary capacity and the characteristics required for an extended-length desktop display connection. This was renamed the "TMDS™" interface, for "Transition Minimized Differential Signalling," in order to distinguish its generic use in the VESA stan-

180 STANDARDS FOR ANALOG VIDEO – PART II: THE PERSONAL COMPUTER

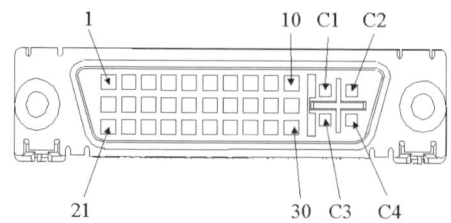

Pin	P&D-A (EVC)	P&D-A/D	P&D-D (digital only)
C1	Red or composite video		The P&D-D connector has nothing in this area; this prevents a P&D-D receptacle from accepting a P&D-A plug.
C2	Green or luminance (Y) video		
C3	Pixel clock (optional)		
C4	Blue or chrominance (C) video		
C5	Common video return		
1	Audio output, right	TMDS Data 2 +	TMDS Data 2 +
2	Audio output, left	TMDS Data 2 -	TMDS Data 2 -
3	Audio output, return	TMDS Data 2 return	TMDS Data 2 return
4	Sync return	Sync return	Unused
5	Horiz./comp. sync (TTL)	Horiz./comp. sync (TTL)	Unused
6	Vertical sync (TTL)	Vertical sync (TTL)	Unused
7	Unused	TMDS clock return	TMDS clock return
8	Charge power	Charge power	Charge power
9	1394 TPA -	1394 TPA -	1394 TPA -
10	1394 TPA +	1394 TPA +	1394 TPA +
11	Reserved	TMDS Data 1 +	TMDS Data 1 +
12	Reserved	TMDS Data 1 -	TMDS Data 1 -
13	Video input, Y or comp.	TMDS Data 1 return	TMDS Data 1 return
14	Video input return	TMDS Clock +	TMDS Clock +
15	Video input, C	TMDS Clock -	TMDS Clock -
16	USB data +	USB data +	USB data +
17	USB data -	USB data -	USB data -
18	1394 shield/chg. pwr. rtn.	1394 shield/chg. pwr. rtn.	1394 shield/chg. pwr. rtn.
19	1394 Vg	1394 Vg	1394 Vg
20	1394 Vp	1394 Vp	1394 Vp
21	Audio input, left	TMDS Data 0 +	TMDS Data 0 +
22	Audio input, right	TMDS Data 0 -	TMDS Data 0 -
23	Audio input, return	TMDS Data 0 return	TMDS Data 0 return
24	Stereo sync (TTL)	Stereo sync (TTL)	Unused
25	DDC/stereo return	DDC/stereo return	DDC return
26	DDC data (SDA)	DDC data (SDA)	DDC data (SDA)
27	DDC clock (SCL)	DDC clock (SCL)	DDC clock (SCL)
28	+5 VDC (USB/DDC)	+5 VDC (USB/DDC)	+5 VDC (USB/DDC)
29	1394 TPB +	1394 TPB +	1394 TPB +
30	1394 TPB -	1394 TPB -	1394 TPB -

Figure 9-12 A comparison of pinouts across the entire P&D family, including the P&D-A (formerly EVC). Used by permission of VESA.

dard from the Silicon Image implementations. The details of TMDS and other similar interface types are covered in the following chapter.

The various pinouts of the full P&D system, including the EVC for comparison, are shown in Figure 9-12. Note that the analog-only and analog/digital versions remain compatible, at least to the degree that the display itself would operate normally when connected to either. (Again, a digital-input display would not connect to the analog-only EVC.) All P&D-compatible displays and host systems were required to implement the VESA EDID/DDC display identification system (described in detail in Chapter 11), such that the system could determine the interface to be used and the capabilities of the display and configure itself appropriately. The P&D standard also introduced the concept of "hot plug detection," via a dedicated pin. This permits the host system to determine when and if a display has been connected or disconnected after system boot-up, such that the display identification information can be re-read and the system reconfigured as needed to support a new display. (Previously, display IDs were read only at system power-up, and the system would then assume that the same display remained in use until the next reboot.)

While the P&D standard greatly extended the capabilities of the original EVC definition, it saw only slightly greater acceptance in the market. Several products, notably PCs and displays from IBM, were introduced using the P&D connector system, but again the need for these new features was not yet sufficient to outweigh the increased cost over the VGA connector, and again the inertia represented by the huge installed base of that standard. There was also a concern raised at this time regarding the various optional interfaces supported by the system. While the P&D definition provided for use of the Universal Serial Bus and IEEE-1394 interfaces, these were not *required* to be used by any P&D-compatible host, and several manufacturers expressed the concern that this would lead to compatibility issues between different supplier's products. (This concern was a major driving force behind a later digital-only interface, the Digital Flat Panel or "DFP" connector, and ultimately the latest combined analog/digital definition, the Digital Visual Interface or "DVI". Both of these are covered in the following chapter.)

9.14 The Future of Analog Display Interfaces

At present, the end may seem to be in sight for all forms of analog display interfaces. Both television and the computer industry have developed all-digital systems which initially gained some acceptance in their respective markets, and which do have some significant advantages over the analog standards which they will, admittedly, ultimately replace. However, this replacement may not happen as rapidly as some have predicted. For now, the CRT remains the dominant display technology in both markets, and digital interfaces really provide no significant advantage for this type if all they do is to duplicate the functioning of the previous analog standards. And, as has been the case with many of the interface standards discussed in this chapter, the fact that analog video in general represents a huge installed product base makes a transition away from such systems difficult. A change to digital standards for all forms of display interfaces is not likely to be achieved until such systems provide a clear and significant advantage, in cost, performance, or supported features, over the analog connections. Digital interface standards have yet to realize this necessary level of distinction over their analog predecessors, but as will be seen in the following chapter, they are rapidly developing in this direction.

10

Digital Display Interface Standards

10.1 Introduction

While the history of analog display interfaces can be viewed as starting with television and then moving into the computer field, digital display interfaces took the opposite course. This is not surprising; while computing systems may have first been developed in the analog domain, the digital computer very quickly came to be the only serious contender. The age of digital television, however, did not truly begin until technologies and hardware developed for computing began to be adapted for TV use.

Digital interfaces might seem to be a natural for the computer industry, and in fact (as was shown in the previous chapter) the first display interfaces were of a very simple "digital" type. But the CRT display is not especially well-suited to a digital interfaces, and realizes few if any benefits from such (see Chapter 7). Discounting the general-purpose interfaces used with the early terminal-based systems, and the first crude CRT connections, the first widely successful digital display interfaces were those used with non-CRT types, in applications not readily supported by the CRT. The most obvious example of these are the displays used in calculators, "notebook" or "laptop" computers, and other such portable devices. These were for the most part, limited to embedded displays – those applications in which the display device itself is an integral component of the product, rather than being a physical separate peripheral. So the review of digital interface standards for displays begins with component-level interfaces. Later, however, as these alternative display technologies began to challenge the CRT on its traditional turf – desktop displays, such as computer monitors, and even larger non-portable devices such as television receivers – even the mainstream display interfaces have been forced to "go digital".

10.2 Panel Interface Standards

Early interfaces to LCD panels, as well of those for other technologies, were by no means standardized. To a large extent, this is true even today, although several industry standards have been developed, along with some de facto standardization around a particularly popular design or product line. Such interfaces were simply designed along with the panel and its integral driver ICs, to support the particular needs or capabilities of that design. There are, however, some features that these basic digital interfaces have in common. Almost always, the video data is organized by color, with only as many bits of input provided as are directly supported by the panel. In addition, the interface will typically provide a clock input, and the equivalent of the "horizontal and vertical sync" signals of the CRT interface – signals which, under whatever name, signify the start of a new line or new frame of data. The equivalent of a "blanking" signal, usually labelled "data valid", "display enable", or similar, is also provided, so that the loading of data into the panel can be inhibited without shutting down the input clock. It is also important to note that such interfaces do not always load the data into the panel in what might be the expected manner – one pixel per clock, scanning through the image in the conventional directions. Owing to the requirements of the various panel technologies, and to the desire for lower input data rates, simple digital panel interfaces often accept more than one pixel's worth of data per clock (requiring more data connections). It is also common, particularly in "passive-matrix" LCD panels (see Chapter 4), for the panel to be divided into two or more areas which will be loaded simultaneously (i.e., pixels will be loaded alternately into each area, scanning through these in the normal fashion, rather than proceeding strictly from left to right and top to bottom through the entire images. One of the earliest attempts to bring some standardization to this part of the industry – the first Flat Panel Display Interface standards (FPDI-I), introduced by the Video Electronics Standards Association (VESA) in 1992 – is shown in Figure 10-1. Note that in addition to the basic data and control interface, the physical connectors also typically provide lines intended for connection to external controls, such as "contrast" or "brightness" adjustments using off-

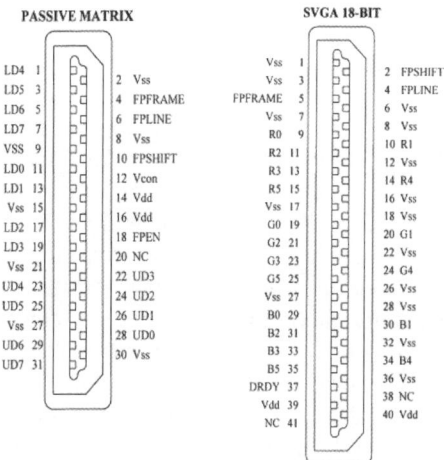

Figure 10-1 The VESA Flat Panel Display Interface (FPDI) standard connectors. This represents one of the earliest attempts to standardize the digital interface to LCD panels. Used by permission of VESA.

panel potentiometers. Panel power is also typically provided via the same connector, although backlight power (generally requiring a much higher voltage, provided via a separate inverter) is often assigned a separate connection.

10.3 LVDS/EIA-644

One of the most successful attempts to bring order at the panel interface level began with the introduction of the Low Voltage Differential Signalling technology by National Semiconductor Corp. in the early 1990s. Commonly referred to as simply "LVDS," it was also adopted by other manufacturers (notably Texas Instruments, which was the first second source of the product line), and was adopted as a standard by the Electronic Industries Association (later the Electronic Industries Alliance) as EIA-644.

While viewed by many as only a display interface, LVDS is actually best viewed as a general-purpose, unidirectional digital data connection. In its most popular form for panel use, an LVDS transmitter IC is used to encode up to 24 bits of data per input clock onto four differential serial pairs. A slightly different version, commonly referred to as "OpenLDI" (for "LVDS Display Interface), was later introduced for monitor use, and basically just adds four additional data pairs for increased capacity (Figure 10-2). In either case, LVDS involves serialization of the input data, distributing it among the four (or eight) serial pairs, and transmitting it at a clock rate seven times the original. The pixel clock is also transmitted via a separate differential pair. All pairs, both data and clock, operate in a true voltage-differential mode, with a swing of 355 mV on each line; the system impedance is a nominal 100 Ω. The additional four bits per clock (4 data pairs times 7 bits/clock provides 28 bits on each clock pulse) are used to add four general-purpose control bits to the data transmission;

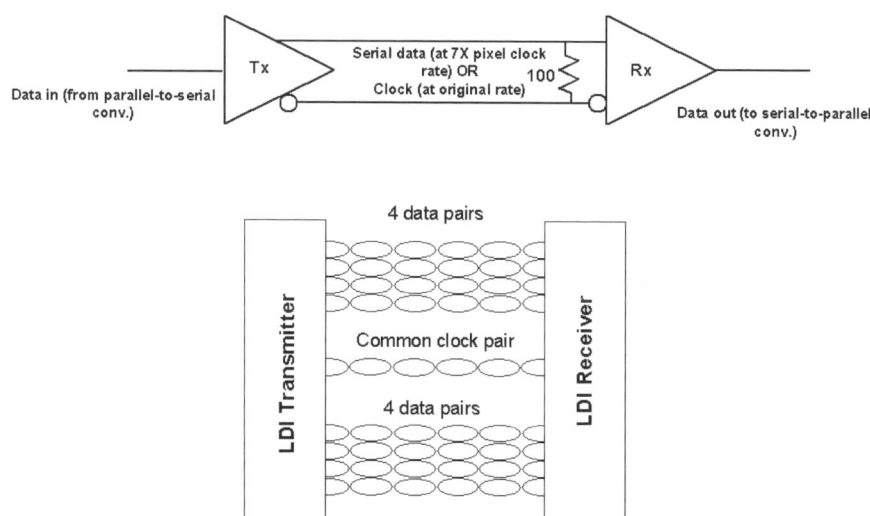

Figure 10-2 The Low-Voltage Differential Signalling (LVDS) system. LVDS in its basic form essentially serializes and distributes incoming data among four differential pairs, along with a clock signal transmitted on a similar pair. The "OpenLDI" version of this interface (bottom) adds four additional data pairs (which share the original clock pair) for additional data capacity.

in display applications, these are typically used to convey the line and frame sync signals, the display enable, plus one "custom" control signal to be used as required for a given application. An LVDS receiver accepts the data and clock pairs, uses the clock to both deserialize the data and to regenerate the original-rate pixel clock, and provides the video data, control signals, and clock as separated outputs.

When first introduced, the LVDS system (also referred to as "Flat-Link™" by National Semiconductor, and "FPD-Link™" by Texas Instruments) was capable of operating with pixel clocks up to 40 MHz. This was soon extended to 65 MHz, then 85 MHz, and ultimately to 112 MHz, which is the current upper limit of any LVDS devices. In the basic 24-bit version, a 65 MHz connection will support a 1024×768 panel at a 60 Hz refresh rate (assuming a "CRT-like" timing, with lengthy blanking periods; if blanking can be reduced, up to an 80 Hz rate could be supported at this format). It is also, of course, possible to employ different data mappings with this link (such as 12 bits per pixel, 2 pixels per clock) which can potentially increase the supported frame rate. However, a more popular means of adding capacity to the LVDS system is simply to add additional data pairs. So-called "dual channel" LVDS inputs have become popular for large-format LCD panels, in which eight data pairs (normally used to convey two 24-bit pixels) are provided using a single, common clock pair.

LVDS provides a relatively simple, efficient, and easy-to-use electrical interface, which has become extremely popular for flat-panel displays intended for embedded or integrated applications such as notebook computers. As such, it was also used as the basis for what has become the most popular industry standard for that market, the specifications published by the Standard Panels Working Group (SPWG). Formed in 1999 by seven notebook computer and display manufacturers (Compaq Computer Corp., Fujitsu, Hewlett-Packard Co., Hitachi, IBM Corp., NEC, and Toshiba), the SPWG's intent was to standardize not only the electrical interface and physical connector, but also the panel dimensions and mounting hardware. Conformance to the SPWG specifications permit notebook computer manufacturers to use multiple sources for a given display. (However, the SPWG does not set standards for display performance, colorimetry, etc., so some care must still be exercised to ensure that displays from different sources are truly interchangeable.) The first SPWG specification set standards for medium-sized, medium-format notebook displays: 10.4 inch, 12.1 inch, and 13.3 inch diagonal panels, of the "SVGA" and "XGA" (800×600 and 1024×768, respectively) formats, using a single (4 data pairs) LVDS channel. The SPWG 2.0 specification, released in 2000, provides similar standards for larger panels (up to 15.0 inch diagonal, and up to the "UXGA", or 1600×1200, format) using a dual-channel (8 data pairs) interface. A summary of the SPWG specifications is given in Figures 10-3 and 10-4; the complete specifications are available directly from the Standard Panels Working Group (www.displaysearch.com/SWPG).

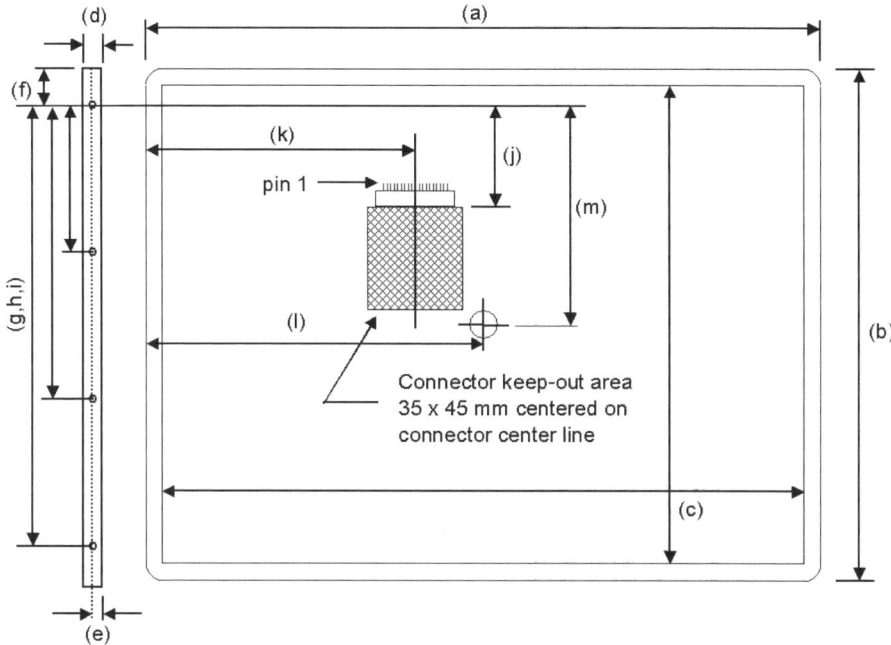

Ref.	Item	13.3" panel	14.1" panel	15.0" panel
a	Overall width	284.0 ± 0.5	299.0 ± 0.5	317.3 ± 0.5
b	Overall height	216.5 typ. 217.1 max.	228.0 typ. 228.6 max.	242.0 typ. 242.6 max.
c	Active area	270.3 × 202.8 ± 0.1	285.7 × 214.3 ± 0.1	304.1 × 228.1 ± 0.1
d	Maximum thickness	7.0	7.0	7.5
e	Mtg. hole C/L to surface	A: 3.7 ± 0.3 B: 2.8 ± 0.3	A: 3.7 ± 0.3 B: 2.8 ± 0.3	A: 4.1 ± 0.3 B: 3.1 ± 0.3
f	1st mtg. hole offset	13.6 +0.0/-2.5	15.3 +0.0/-2.5	12.8 +0.0/-2.5
g	2nd mtg. hole from (f)	68.8	54.0	56.9
h	3rd mtg. hole from (f)	121.4	144.3	160.4
I	4th mtg. hole from (f)	190.3	198.0	217.2
j	Connector mating surface from (f)	A: 26.4 ± 0.5 B: 24.0 ± 0.5	A: 29.7 ± 0.5 B: 28.0 ± 0.5	A: 47.2 ± 0.5 B: 30.5 ± 0.5
k	Connector C/L from edge	117.5 ± 0.5	125.1 ± 0.5	134.4 ± 0.5
l	Panel C/L to edge	142.5	150.15	159.35
m	Panel C/L to 1st mtg. ref.	95.1	99.15	108.65

Figure 10-3 Summary of SPWG 2.0 mechanical specifications. All dimensions in mm. Tolerances ± 0.3 mm unless otherwise indicated. Note: The SPWG 2.0 specification established two different panel types, referred to as "A" and "B", with differences as noted above. The "B" style, which is proposed for all designs from 2003 on, is intended to encourage a move toward thinner panels.

188 DIGITAL DISPLAY INTERFACE STANDARDS

Pin	Signal, 20-pin connector	Signal, 30-pin connector
1	Power supply, 3.3V (typ.)	Vss (ground)
2	Power supply, 3.3V (typ.)	Power supply, 3.3V (typ.)
3	Ground	Power supply, 3.3V (typ.)
4	Ground	DDC power (+3.3V)
5	Rin0 − (LVDS in, R0-R5, G0)	NC (reserved for supplier test)
6	Rin0 +	DDC clock (SCL)
7	Vss (ground)	DDC data (SDA)
8	Rin1− (LVDS in, G1-G5, B0-1)	Odd_Rin0 −
9	Rin1+	Odd_Rin0 +
10	Vss (ground)	Vss (ground)
11	Rin2− (LVDS in, B2-5, HS, VS, DE)	Odd_Rin1 −
12	Rin2+	Odd_Rin1 +
13	Ground	Vss (ground)
14	LVDS Clock −	Odd_Rin2 −
15	LVDS Clock +	Odd_Rin2 +
16	Vss (ground)	Vss (ground)
17	DDC power (+3.3V)	Odd_Clock −
18	NC (reserved for supplier test)	Odd_Clock +
19	DDC clock (SCL)	Vss (ground)
20	DDC data (SDA)	Even_Rin0 −
21		Even_Rin0 +
22		Vss (ground)
23		Even_Rin1 −
24		Even_Rin1 +
25		Vss (ground)
26		Even_Rin2 −
27		Even_Rin2 +
28		Vss (ground)
29		Even_Clock −
30		Even_Clock +

Figure 10.4 SPWG pinouts (per SPWG 2.0). The 2.0 version of the SPWG specification defines two connectors, as shown here. The 20-pin connector is for Style A XGA panels only; the 30-pin connector is used for SXGA+ (and above) panels of either style, and all Style B panels.

10.4 PanelLink™ and TMDS™

In the mid-1990s, a small Silicon Valley company, Silicon Image, Inc., introduced its "PanelLink™" interface system, intended as an alternative to the LVDS interface of National Semiconductor and Texas Instruments. Conceptually, PanelLink was similar to LVDS – a flat-panel interface system which serialized the data to be transmitted onto several differential data pairs, and sent this data along with a separate clock (on its own differential pair) to the receiver. However, PanelLink differed from LVDS in several significant aspects.

The most obvious change is a reduction in the number of data pairs; the basic PanelLink interface uses three data pairs rather than four, while still retaining the capacity for carrying 24 bits of video data per clock. Unlike the LVDS system, then, PanelLink isolates the three primary color channels, assigning one to each data pair. Next, the data is encoded prior to serialization, using a Silicon Image proprietary technique. This is an 8–10-bit encoding which is designed to both minimize the number of transitions on the serial lines, while also DC-balancing these lines. This encoding results in the data pair bit rate being 10× the original pixel clock rate, as opposed to the 7× rate of the original LVDS design. At the receiver, another difference between the two systems is found. In the PanelLink™ receiver, the clock is used to generate the 10× clock needed to recover the data, but the 10× clock is produced in several versions with different phase relationships to the original. These are used by the data receivers to independently recover and deserialize the data, permitting each data pair to in effect be independently resynchronized to the clock. This gives the PanelLink system considerable tolerance to data-to-clock or data-to-data (between data pairs) skew, typically up to ±½ the original pixel clock period. Finally, the PanelLink interface is not truly a voltage-differential system. As shown in Figure 10-5, this interface actually operates by "steering" a fixed current between the two lines of the pair. Using a fixed current rather than a fixed voltage provides some obvious advantages for a long-distance interconnect, but results in another subtle distinction. The PanelLink system requires an additional physical connection for each pair, for the return current path. (In practice, many standards using this system now define shared return paths for multiple data lines.) As originally commercialized by Silicon Image, the exact current level – and therefore the voltage swing across the standard 100-Ω terminating impedance – was not fixed, but rather would be set in each application through external components. Later, however, the demands of monitor interface standards lead to the current being set at 12 mA (the maximum allowed in the original transmitters) in most specifications. This results in a nominal 500 mV swing on the signal lines (1.0V differential).

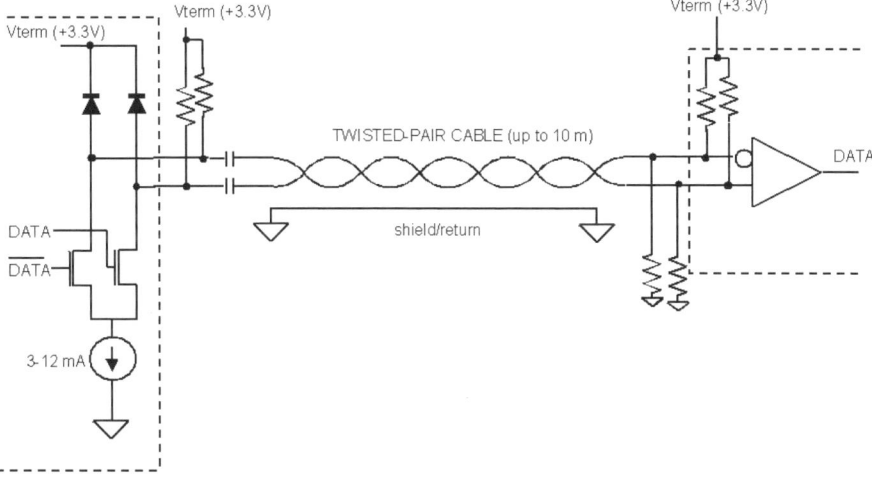

Figure 10-5 The Transition-Minimized Differential Signalling (TMDS) interface. This simplified view of the driver and receiver circuits, shown as used in a monitor interface application, illustrates how the electrical connection operates by steering current between the two conductors of the pair.

190 DIGITAL DISPLAY INTERFACE STANDARDS

In 1995, the PanelLink interface came to the attention of a VESA committee working on the problem of standardizing a digital interface for monitors. Due to its advantages for longer interconnects, as needed for monitor applications, the system was chosen as the basis for the new monitor standard, and the basics of the PanelLink system included as a part of the VESA specification. To distinguish the standard version of the interface from the products offered by Silicon Image, the name "Transition Minimized Differential Signalling," or TMDS, was adopted. VESA soon published two standards based on TMDS: the "Plug & Display," or "P&D," monitor interface (see Chapter 9), and a new panel-level interface standard, FPDI-II. The FPDI-II standard was never widely adopted, and standardization of panel-level interfaces had to wait for the later SPWG specifications (above). But PanelLink/TMDS' use in the P&D standard set the stage for further development of digital monitor standards based on this system.

Use of TMDS as a monitor connection raised an additional concern. In the case of an interface between physically separate products, as with a monitor and its host system, there is no guarantee that the reference or "ground" potential will be the same, and therefore no guarantee that supply voltages of even the same nominal level will actually be compatible. This results in the possibility of problems for interfaces using DC connections; in the case of TMDS, there is a possible problem if the transmitter and receiver supply voltages differ by more than two diode drops in many designs. This would be a relatively rare occurrence in most cases, and systems have been successfully built using a DC-connected TMDS interface (the difference in transmitter and receiver supply voltages is generally limited by the specifications applicable to such systems, typically to not more than 0.6 V). However, the VESA P&D and later monitor-oriented specifications suggested the use of either capacitive or inductive AC coupling at the receiver to avoid any possible problems. In notebook applications, where a common power supply could be assumed for both the receiver and transmitter, there is no potential for this problem, and so direct coupling is the norm (Figure 10-6).

In the first products provided by Silicon Image, the PanelLink interface supported pixel clocks up to 65 MHz, competing with the LVDS components available at the time. Later

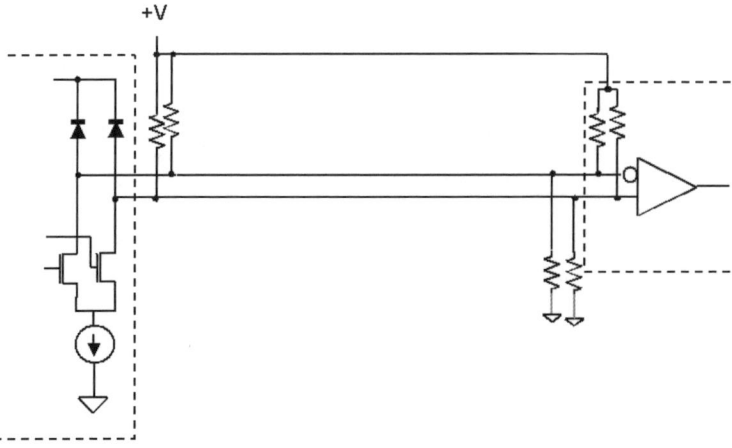

Figure 10-6 When used in a notebook application, or a similar situation in which the receiver and transmitter share the same power supply, the TMDS interface has no problem with potential voltage mismatches between the two, and direct coupling of the data and clock pairs may safely be used.

speed increases raised the upper limit to 85 MHz, then 112 MHz, and finally 165 MHz, in the fastest TMDS products currently available. There has been some indication that further increases in speed may be possible, to rates above 200 MHz, but for now interface standards using TMDS also allow for additional data pairs, as was done with LVDS, when increased data capacity is required. Silicon Image, Inc., has also licensed the basic intellectual property needed for the implementation of the interface to numerous other companies, so that compatible transmitters and receivers are now available from multiple sources.

10.5 GVIF™

A relative newcomer to the market, Sony's Gigabit Video Interface (GVIF) product line is in its present form aimed at the lower end of the computer market and at digital television. GVIF is also a differential, serial digital interface, but one which uses only one pair of physical conductors. In this system, 24-bit (per pixel) data is encoded (using a proprietary algorithm) and serialized by the transmitter. The data encoding is such that the transmitted data stream is "self-clocking", i.e., the pixel clock can be derived from the serial stream at the receiver, and used to deserialize and decode the data, and present it at the receiver outputs. The maximum 65 MHz pixel clock rate limits the current system not higher than approximately the 1024×768 format at CRT-like timings, although there is certainly sufficient capacity for consumer digital television or digital HDTV, at least in compressed form. To date, no industry standards have been written around the GVIF system, although there has been some interest in basing a standard for "head-mounted" or "eyeglass" displays on it.

10.6 Digital Monitor Interface Standards

Through the 1990s, the growing interest in non-CRT displays as desktop PC monitors – and particularly the increasing importance of the LCD monitor – led to several attempts at digital interface standards for this market. While only one of these is currently seeing any significant degree of success, a look at the history of digital monitor interface development through the past decade is very useful in order to understand how this latest standard was shaped.

10.7 The VESA Plug & Display™ Standard

Soon after the introduction of the Enhanced Video Connector (EVC) by VESA in 1995, several of the companies which had been involved in the development of that standard began to discuss the possibility of extending the EVC concept to support a digital display interface. What resulted was the VESA "Plug & Display" standard (the name being a play on the "Plug & Play" concept being promoted at the time), which was first released in 1997. Plug & Display, or "P&D" as it was more commonly called, retained a significant degree of compatibility with the original EVC design, and in fact the original EVC was later incorporated into the Plug & Display standard as "P&D-A" (for analog-only).

The P&D connector (Figure 10-7; pinout shown in Figure 9-13) retained the same size and basic shape of EVC, and in one version (P&D-A/D) retained the "Microcross™" analog video section. The 3-row, 10-column field of pins was also retained, but with the analog audio I/O and video input pins now redefined for the support of a single channel (3 data pairs

192 DIGITAL DISPLAY INTERFACE STANDARDS

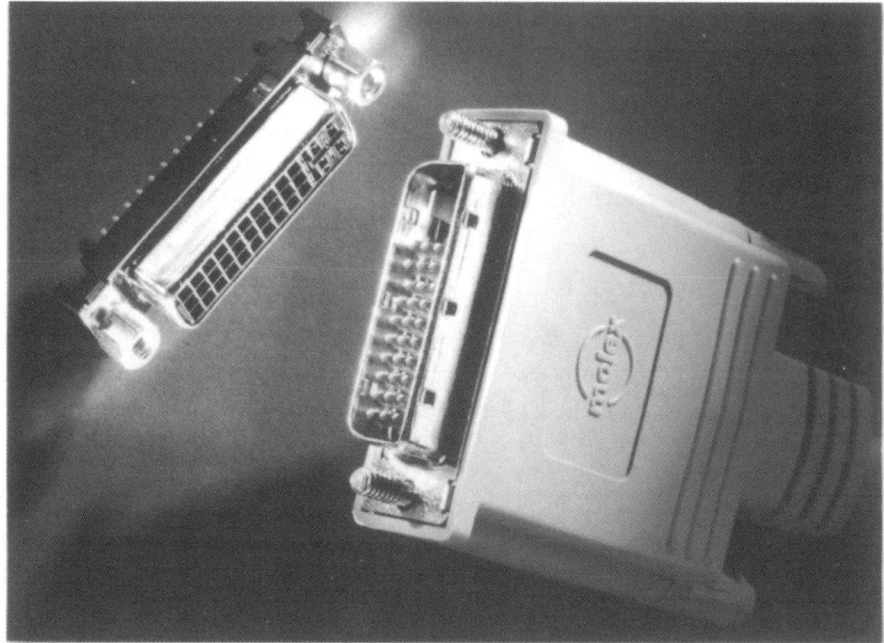

Figure 10-7 The VESA Plug & Display Connector, or "P&D," was a modification of the original VESA Enhanced Video Connector, with the intention of supporting both analog and digital display interfaces on a single physical connector. *(Photograph courtesy of Molex Corp., used by permission.)*

plus one clock pair) of the TMDS digital interface. (The PanelLink interface from Silicon Image, Inc., was standardized and renamed "TMDS" by VESA as part of the P&D effort.) The P&D standard also introduced the concept of "hot plug detection," whereby the host system could detect the disconnection and reconnection of a display at this connector. This was required by the dual video interfaces supported; without such a scheme, the host would have no way of knowing, for example, that an analog-input display had been disconnected and replaced by one using the digital interface. Detecting such an event permits the host to ID the display upon connection, rather than simply at system power-up, and re-set the graphics system outputs to drive the new display.

The Plug & Display standard also made a slight change to the connector shell design from the original EVC. This simple change made P&D a full connector system, permitting hosts to readily support either the digital interface, analog, or both, simply by using the proper connector. Analog-input displays, using the original EVC plug, would connect only to host providing the EVC (now "P&D-A") or the combined-output "P&D-A/D" receptacles. Similarly, digital-input displays, using the new shape plug, would connect to either a P&D-A/D or the new digital-only ("P&D-D") receptacles. This ensured that a display could only physically connect to a host capable of supporting it. (A proposed extended P&D, which would add a separate section to the connector for extending the TMDS support to two channels, was never developed.)

Despite these new features, the P&D system saw only limited acceptance. The biggest concern expressed by most potential users was the optional support for the IEEE-1394 and USB interfaces. Display manufacturers could not be sure that either would be supported on any given host system, and similarly system manufacturers were not willing to design support for either in without suitable displays being available. While the industry struggled to resolve this, two new options were developed – and interest in the P&D system declined.

10.8 The Compaq/VESA Digital Flat Panel Connector – DFP

While discussion continued regarding the use of P&D's "optional" interfaces, there was still a need for a simple digital interface that could be easily implemented to support non-CRT displays. Compaq Computer introduced what it called the "DFP" connector – for "Digital Flat Panel" – on several PC products in the summer of 1997. DFP was intended as a bare-minimum implementation of a digital display interface, one which could easily be used in addition to the existing analog interface provided by the "VGA" connector.

Using a 20-pin "micro delta ribbon," or "MDR" connector (from a family developed by 3M), the DFP specification supported a single channel of the TMDS interface, the "hot plug" system of P&D, and the basic VESA Display Data Channel (DDC) connection for display ID. DFP saw some success as a standard connection for LCD monitors, but was seen by many as only a short-term solution to be used alongside the VGA connector – until both were replaced by P&D, or by whatever the industry finally determined would be the long-term solution. DFP was later adopted as a VESA standard, in essentially the same form as originally introduced (Figure 10-8).

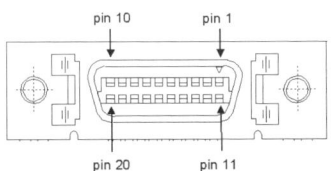

Pin	Signal	Pin	Signal
1	TX1 + (TMDS pair 1, +)	11	TX2 +
2	TX1 -	12	TX2 -
3	TMDS 1 shield/return	13	TMDS 2 shield/return
4	TMDS clock shield/return	14	TMDS 0 shield/return
5	TMDS Clock +	15	TX0 +
6	TMDS Clock -	16	TX0 -
7	Logic ground	17	No connect
8	+5 VDC (from host)	18	Hot plug detection
9	No connect	19	DDC data (SDA)
10	No connect	20	DDC clock (SCL)

Figure 10-8 The Compaq (later VESA) Digital Flat Panel (DFP) connector. Used by permission of VESA.

10.9 The Digital Visual Interface™

In order to finally establish a new video interface standard which would be acceptable to both the major systems manufacturers and display makers, the Digital Display Working Group (DDWG) was formed in 1999. The core members, known as the DDWG Promoters' Group, was made up of seven of these companies: Compaq, Fujitsu, Hewlett-Packard, IBM, Intel, NEC, and Silicon Image. The new standard, called the Digital Visual Interface (DVI™ 1.0) was also to be based on Silicon Image's PanelLink or TMDS technology, and the connector chosen was very similar to the VESA P&D. The major changes from P&D to DVI included the deletion of the optional IEEE-1394 and USB interfaces, and the additional of a second TMDS data channel (three data pairs), sharing the same clock pair as the basic channel. (The additional data pairs in some cases also share the ground/return connections of the original set.) DVI also raised the question of digital content protection for the first time; while not required under the 1.0 specification, the use of an Intel-proprietary encryption system (High-Definition Content Protection, or HDCP) is officially recognized under DVI.

Physically, the DVI connectors resemble the VESA P&D, although with two fewer columns of pins and a slightly different shell design. This prevents direct physical compatibility between the two, although it is possible to connect P&D and single-channel or analog DVI with the appropriate adapters. Like P&D, DVI defined both a digital-only version (DVI-D), and one which supports both analog and digital interfaces (DVI-I), again via the Microcross™ pseudo-coaxial connector design originated by Molex. Pinouts for both DVI versions are shown in Figure 10-9.

As of this writing, DVI has begun to see fairly widespread adoption as an LCD monitor connection, although it has yet to significantly displace the VGA connector or other options for CRT monitors, either in analog or digital form. There is also considerable interest in the standard as a possible solution for consumer television applications, for example as an interconnect between digital HDTV decoders ("set-top boxes") and digital-input receivers. The support for digital content protection provided by DVI (the HDCP encryption system) is of particular interest in such applications. However, there remain some open issues within the DVI specification, which are being addressed by a joint effort between the DDWG and several consumer electronics manufacturers and their industry association, the CEA. Among these are the need for audio support and the possibility of alternate color encoding methods, such as a "YUV" or similar encoding rather than the DVI-standard RGB. The DDWG is also working on some implementation concerns which have been raised by the computer industry, such as the use of DVI as a display input connector and the standard means for transitioning between single- and dual-channel TMDS support. The former is a concern due to the possibility of different display capabilities being available depending on whether the analog or digital interface is in use – yet the display can only provide a single set of ID information at a time. The use of the second TMDS channel is also problematic under the current specification. While the DVI 1.0 standard set 165 MHz as the limit for the basic single-TMDS version, it did not set explicit guidelines for managing both single- and dual-channel operation within a given system. The second channel could potentially be used for supporting larger display/image formats, increased "color depth" or both – but how this is to be negotiated between the display and host has not yet been well defined.

Still, even with these minor concerns, DVI at this point represents the interface most likely to win widespread adoption within not only the computer display industry, but beyond it to consumer applications as well. Development of the standard will no doubt continue for

THE DIGITAL VISUAL INTERFACE™   195

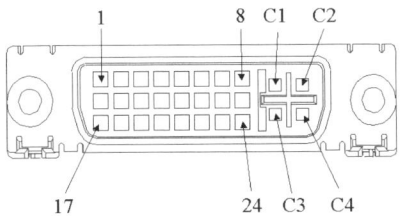

Pin	DVI-I	DVI-D
C1	Red video (analog)	Like P&D-D, the DVI-D connector has nothing in this area.
C2	Green video (analog)	
C3	Blue video (analog)	
C4	Horizontal sync (TTL)	
C5	Common return	
1	TMDS Data 2 −	TMDS Data 2 −
2	TMDS Data 2 +	TMDS Data 2 +
3	TMDS Data 2/4 shield	TMDS Data 2/4 shield
4	TMDS Data 4 −	TMDS Data 4 −
5	TMDS Data 4 +	TMDS Data 4 +
6	DDC clock (SCL)	DDC clock (SCL)
7	DDC data (SDA)	DDC data (SDA)
8	Vertical sync (TTL)[1]	Unused
9	TMDS Data 1 −	TMDS Data 1 −
10	TMDS Data 1 +	TMDS Data 1 +
11	TMDS Data 1/3 shield	TMDS Data 1/3 shield
12	TMDS Data 3 −	TMDS Data 3 −
13	TMDS Data 3 +	TMDS Data 3 +
14	+5 VDC	+5 VDC
15	Ground/Sync return	Ground
16	Hot plug detect	Hot plug detect
17	TMDS Data 0 −	TMDS Data 0 −
18	TMDS Data 0 +	TMDS Data 0 +
19	TMDS Data 0/5 shield	TMDS Data 0/5 shield
20	TMDS Data 5 −	TMDS Data 5 −
21	TMDS Data 5 +	TMDS Data 5 +
22	TMDS Clock shield	TMDS Clock shield
23	TMDS Clock +	TMDS Clock +
24	TMDS Clock −	TMDS Clock −

Figure 10-9 The Digital Visual Interface. Both DVI-I and DVI-D pinouts are shown; the connectors are identical, except that the DVI-D is blank in the area of the "MicroCross™" analog connections. Note: the "vertical sync" (pin 8) and "horizontal sync" (pin C4) signals of the DVI-I are for use only by displays using the analog connection; they cannot be used with the digital interface. TMDS pairs 3-5 comprise a second data channel, for added capacity; its use is optional.

196 DIGITAL DISPLAY INTERFACE STANDARDS

some time, but there seems to be enough momentum building behind this latest standard to ensure at least success beyond that achieved by the earlier attempts.

10.10 The Apple Display Connector

Some mention should also be made of the Apple Display Connector (ADC), although this is a proprietary design used (to date) only in Apple Computer Corp. systems. In many ways, the ADC resembles both the VESA Plug & Display connector (in the "P&D-A/D" form) and the Digital Visual Interface standard. Like them, it is also based on the Molex "Microcross" connector family, and physically resembles a P&D-A/D connector with a slightly modified shell shape. Also like the P&D and DVI standards, ADC supports both analog and digital outputs in a single physical connector, and again uses the "TMDS" electrical interface standard. As in the DVI connector, up to two TMDS data channels (comprising three data pairs

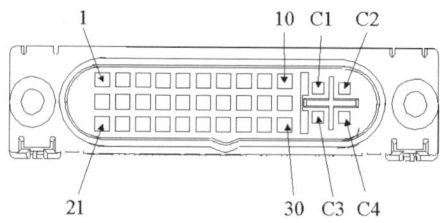

Pin	Signal	Pin	Signal
C1	Blue video (analog)	C3	Horizontal sync (TTL)
C2	Green video (analog)	C4	Red video (analog)
		C5	Analog video/DDC return
1	+28 VDC	16	TMDS Data 1/3 shield
2	+28 VDC	17	TMDS Data 3 -
3	LED	18	TMDS Data 3 +
4	TMDS Data 0 -	19	DDC clock (SCL)
5	TMDS Data 0 +	20	TMDS Clock shield
6	TMDS Data 0/5 shield	21	USB data +
7	TMDS Data 5 -	22	USB data -
8	TMDS Data 5 +	23	USB return
9	DDC data (SDA)	24	TMDS Data 2 -
10	Vertical sync	25	TMDS Data 2 +
11	28V return	26	TMDS Data 2/4 shield
12	28 V return	27	TMDS Data 4 -
13	Soft Power	28	TMDS Data 4 +
14	TMDS Data 1 -	29	TMDS Clock +
15	TMDS Data 1 +	30	TMDS Clock -

Figure 10-10 The Apple Display Connector and its pinout.

each) are supported, and the ADC again relies on the VESA DDC and EDID standards for display identification and control.

In addition to the analog video, TMDS, and DDC interfaces supported by DVI, the ADC connector adds a power supply (two pins carrying +28 VDC, along with two dedicated return pins) and the USB interface. There is also a "soft power" signal (pin 13), which can be used to place the monitor into a low-power mode (and thereby providing Apple monitors with a power-management system that is independent of the PC-standard VESA DPMS). The pinout for the ADC is shown in Figure 10-10.

10.11 Digital Television

As mentioned at the beginning of the chapter, the development of "digital" television was to a great extent driven by the development of the computer industry, the opposite of the course of analog video interfaces. In fact, to this point there is still not a widespread, consumer-level digital interface standard for television; the first such may come through the consumer industry's adoption of DVI, as mentioned above. Digital television began first as a production or broadcast studio technology, permitting a wider range of storage, editing, and effects options than had been available with the earlier analog-only systems. As these applications did not involve the development of significant new interface standards specifically oriented toward displays, they are beyond the scope of this work and will not be examined in detail here. However, certain developments within the "digital" realm are discussed in Chapter 12, as part of our discussion of high-definition television standards and their impact on the display interface.

From the standpoint of the requirements on the video interface itself, television in either analog or digital form generally represents a less-demanding application than does computer video, solely due to the much lower data rates required. However, getting "TV" and "computer" signals to co-exist in a single system can be a challenging problem, due to a number of factors. For one thing, the data rates required for digital television can actually be *below* the limits of many computer-oriented interfaces. As an example, the pixel clocks normally used for standard-definition television (usually represented using either 720×480 or 720×576 image formats, or similar) fall under the typical lower limit for the digital interfaces discussed above, if transmitted in their usual interlaced form. (TMDS, for instance, typically has a lower pixel clock limit of 25 MHz.) In addition, the different color encoding methods used for television, along with the need to carry synchronized supplemental data such as audio, further complicates the compatibility issue. Finally, while most computer graphics systems are designed around the assumption of "square" pixels (equal numbers of sample per unit distance in both horizontal and vertical directions), this is not the case in most digital television standards.

10.12 General-Purpose Digital Interfaces and Video

While not in general used as display interfaces per se, two popular digital interface standards were designed with the transmission of digital video in mind, and deserve some mention here. They have not to date seen widespread use as display connections, but at least have some potential here, especially in consumer-entertainment applications.

The Universal Serial Bus, or USB, was first introduced by a consortium of seven companies (Intel, IBM, NEC, Compaq, Digital Equipment Corp., Microsoft, and Northern Telecom) in 1995. It was intended as a general-purpose, low-to-medium speed, low-cost connection for desktop PC devices and other applications requiring only a short-distance interconnect. USB 1.0 defined a very flexible, "self-configuring" desktop connection system, with two levels of performance: a 1.5 Mbps link for keyboards and other low-speed devices, and a higher-performance 12 Mbps link with the potential for supporting high-quality digital audio and even some basic digital video devices. The USB connector/cable system uses just four wires: a +5 V power connection and its association ground, plus a bidirectional differential pair for the data signals. The data transmission format is specified such that the serial data stream is "self-clocking", i.e., the timing information required to properly recover the data may be derived from the data stream itself.

A single USB host controller can support up to 127 peripheral devices simultaneously, although it must allocate the available channel capacity among these. Typically, practical USB installations will have perhaps a half-dozen devices on the interface at once, especially if any of these have relatively high data-rate requirements, such as digital audio units or a video camera.

The USB 1.0 specification has so far seen most of its acceptance in the expected markets; human-input devices for PCs, such as keyboards, mice, trackballs, etc. – and significantly for low-to-medium resolution video devices such as simple cameras. However, the recent development of a much more powerful version of the standard may increase its acceptance in the video/display areas. USB 2.0, in its initial release, defined performance levels up to 480 Mbps, more than sufficient for the support of standard compressed HDTV data and even the transmission of uncompressed video at "standard definition" levels. USB was defined from its inception to permit the transmission of "isochronous" data, meaning types such as audio or video in which the timing of each data packet within the overall stream must be maintained for proper recovery at the receiver.

At this level of performance, USB 2.0 may be a serious competitor for the other widely used general-purpose digital interface, the IEEE-1394 standard. The "1394" system is also often referred to as "FireWire™," although properly speaking that name should be used only for the implementations of the system by Apple Computer, which originated the technology in 1986. IEEE-1394 is conceptually similar to the USB system – a point-to-point general purpose interconnect using a small, simple connector – but supported much higher data rates at its first introduction. Like USB, the 1394 interface supports isochronous data transmission, and so has been widely accepted in digital audio and video applications. The standard "1394" connector has six contacts – one pair for power and ground, as in USB (although typically using higher voltages) – and two pairs which make up the data channel.

The IEEE-1394/"FireWire" interface has been introduced on some products as a consumer-video connection, although not as a display interface per se. In its original form, this standard defined several levels of performance, up to a maximum of 400 Mbits/s. This (per the capacity requirement analyses of Chapter 5) is more than sufficient for standard television-quality video (roughly 640 × 480 to 720 × 576 pixels at 60 fields/s, 2:1 interlaced), but not for typical computer-display video formats and timings. The "1394" interface is very likely to see increases in capacity, to as much as 3.2 Gbits/s, but even this is still somewhat low for use with today's "high-resolution" displays and timings. Both IEEE-1394 and USB 2.0, therefore, may become widely used for consumer and even professional video-editing and similar applications, but neither is likely to see any serious use as a high-resolution display interface.

10.13 Future Directions for Digital Display Interfaces

To date, digital interfaces as used for display devices themselves (as opposed to the more general usage of digital techniques for video, as in the television industry) have more or less simply duplicated the functions of the earlier analog systems. In the computer industry, for example, analog RGB video connections have begun to give way to digital, but these new interfaces still provide video information in parallel RGB form, with regular refreshes of the entire display. The encoding of the information has changed form, but nothing else in the operation of the display system.

In future standards, including several that are currently under development, this situation is expected to change. With the image information remaining in digital form from generation through to the input of the display, new models of operation become possible. Most of these rely on the assumption that the display itself will, in an "all-digital" world, contain a frame buffer, a means of storing at least one full frame of video information. This is actually quite common today even in many analog-input, non-CRT displays, as it is required for frame-rate conversion.

Frame storage within the display enables much more efficient usage of the available interface capacity. First, there is no longer any need for the image data to be updated at the refresh rate required for the display technology being used. CRT displays, for example, will generally require a refresh rate of 75–85 Hz to appear "flicker-free" to most users, but providing this frame rate over a product-level (i.e., between physically separate products, as opposed to the interface which might exist within a monitor to the display device itself) can be extremely challenging, especially over long distances. If a frame buffer is placed into the monitor, between the external interface and the display device itself, *frame-rate conversion* may be performed within the monitor, permitting the display to be refreshed at a much higher rate than is now required on the interface. Further, since the display timing is now decoupled to a large degree from the timing of the incoming display data, the external interface need not lose capacity to "overhead" losses such as the blanking intervals (which represent idle time for the interface, using the traditional model). (Frame-rate conversion is already used in many non-CRT displays, such as LCD-based monitors, but in the other direction – to convert the wide range of refresh rates used in "CRT" video to the relatively narrow range usable by most LCD panels.) The data rate required on the interface is now determined solely by the pixel format and the frame rate needed for acceptable motion rendition.

Further improvements in the efficiency of the interface may be obtained by realizing that the typical video transmission contains a huge amount of redundant information. If the image being displayed, for example, is text being typed on a plain white background – very common in computer applications – repeatedly sending the information representing any part of the image but the *new* text in each frame is a waste of capacity. Again assuming that the display contains sufficient frame storage and "intelligence," it will be far more efficient to permit *conditional update* of the display. Using this method, only those portions of the image which have changed from frame to frame need to be transmitted, greatly reducing the data rates required of the interface. (Note, however, that if full-screen, full-motion video is possible, the peak capacity required to handle this must still be available – although only at the rate required for convincing rendition of the motion, as noted above.) As with any reduction in information redundancy – as was previously discussed relative to data compression techniques – the possibility of errors in the transmitted data becoming a problem for the user is increased. However, this can be addressed by including some form of error detection and/or

200 DIGITAL DISPLAY INTERFACE STANDARDS

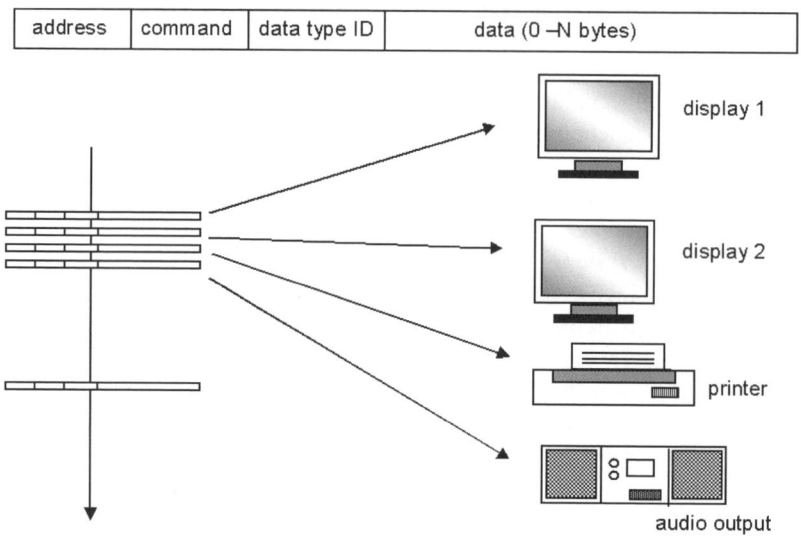

Figure 10-11 "Packet" video. In a digital transmission system, video data may be "packetized" into blocks of a predefined format. In this hypothetical example, data packets have been defined which include the address of the intended recipient device, commands for that device, identification of the type of data (if any) the packet carries, and the data itself (which may be of variable length). Such a system would permit the addressing of multiple display devices over a single physical connection, and even the transmission of data types besides video – such as text or digital audio – with each type being properly routed only to devices capable of handling it.

correction into the system definition, without significantly affecting the improvement in efficiency.

The ability to make more efficient use of available interface capacity can be exploited in several ways. First, and probably most obviously, higher pixel counts – "higher resolution" formats – can be accommodated than would otherwise be the case. (Or, conversely, a given format can be supported by a lower-rate interface.) However, by adopting techniques from common digital-networking practice, new capabilities may be introduced that were never possible in previous display interfaces. By "packetizing" the data – sending bursts of information in a predefined format (Figure 10-11) – and by defining portions of these packets as containing address and other information besides the image data itself, several new modes of operation become possible. First, a single physical and electrical channel could be used to carry multiple types of information; besides the video data itself, audio and other supplemental data (such as teletext) can be carried simply by permitting each to be uniquely identified in the packet "header." With the packet also providing address information, multiple displays could be connected to a single host output. This would rely on sophisticated display identification and control systems, building on existing standards (such as the VESA DDC and EDID specifications, which is discussed in Chapter 11), to communicate the unique capabilities of each display in the system to the host. The ability to support multiple separate displays could also be extended to support arrays of physically separate display devices which are to be viewed as providing a single image – a "tiled" display system, as in Figure 10-12.

Such a "packet video" system is currently under development by the Video Electronics Standards Association (VESA), based on a system originally developed by Sharp Electron-

FUTURE DIRECTIONS FOR DIGITAL DISPLAY INTERFACES 201

Figure 10-12 A tiled display system. In this type of display system, multiple separate display devices are physically arranged so as to be viewed as a single image. Ideally, the borders between the individual screens are zero-width, or at least narrow enough so as to be invisible to the viewer. Such a system is difficult to manage with conventional interfaces and separate image sources, but becomes almost trivially simple when using a packetized data transmission system.

ics, Hitachi, Toshiba, and IBM Japan. The Digital Packet Video Link, or "DPVL," standard is expected to provide all of the functionality described above, and is currently planned to be released in two stages. The first, "DPVL-Light" may be published as a standard by late 2002 or early 2003, and will support some of the more basic functionality in a single-display version of the system. This first version will be capable of being added to existing systems with a minimum of hardware changes. Later, the full DPVL standard will enable full packet-video functionality in new designs.

DPVL and similar packet-video systems will not necessarily require new physical and electrical interface standards, at least initially. At first, they may be expected to use existing channels, such as the TMDS interface and the DVI physical connection, and could be viewed simply as a new command/data protocol layer. (Note that many of the electrical interfaces presented in this chapter, such as LVDS, TMDS, etc., can be used as general-purpose digital channels, despite being primary used in display applications at present.) Eventually, new physical and electrical standards may be required for a system which is truly optimized for packet-video transmission. At the very least, the display ID and control channel will likely need to be improved from current standards, to permit better support for multiple displays. A higher-speed "back channel," capable of carrying greater amounts of information from the display to the host (the current digital display interfaces are basically unidirectional) may also be required. Possible solutions for all of these currently exist, however, and so the development and acceptance of packet video as a standard display interface method will again be limited primarily by the difficulty and costs of making the transition from current standards.

11

Additional Interfaces to the Display

11.1 Introduction

While the subject of this book so far has primarily been the transmission of image information from any of a number of different sources to the display device, there is very often a need to convey additional types of data between the two. Examples of the functions enabled by additional or supplemental interfaces between the host system and the display device include identification of the display and its capabilities, control of the display by the host (or even vice versa, permitting user controls to be built into or attached to the display), and the transmission of other forms of information to and from the user (as in the case of audio input and output capabilities at the display).

Often, it would be possible to carry such information within the same channel as the image itself; in some situations, as in the case of broadcast television, this is mandatory, as only one channel is available. But in many applications, and especially in the case of point-to-point, wired interfaces, it is simpler and more efficient to provide a separate electrical connection within the same overall physical connector and cabling. This chapter discusses several of the more popular supplemental interfaces used in display systems.

11.2 Display Identification

In systems where the display choice is highly constrained by the nature of the system standards (or simply due to the dominance of the market by a single display type), there is little need for the image source to be concerned about the exact nature of the display in use; its characteristics are either mandated by the standards applicable to that situation. This is the situation which at least has been common in consumer television. The timing, format, and even to a great extent the characteristics of the display (assumed to be a CRT-based device of

204 ADDITIONAL INTERFACES TO THE DISPLAY

specified color and response characteristics) were fixed, and so no "custom" configuration of the source was required. (In a normal broadcast situation, custom configuration of the source is not even *possible*; the broadcaster must generate one signal for all receivers, regardless of their individual characteristics.)

As noted previously, early computer systems used fixed-frequency displays, which often were "bundled" as part of the complete product and so were under the control of the system manufacturer. With the introduction of the "VGA" and subsequent graphics systems in the PC market, it soon became possible for a mismatch to exist between the host graphics hardware and the display in use. Not all displays could support the fastest timings that were within the capabilities of the hardware. To ensure that the proper video mode was chosen so as to make best use the available display, the PC had to be able to identify the display used in that system.

Basic ID capability was first defined by IBM as part of the original VGA specification. In its original form, this functionality was provided simply by defining four pins on the "VGA" connector – the common 15-pin D-subminiature connector still found on practically every personal computer – as "ID pins" (Figure 11-1). They would be selectively grounded by the display, thus permitting up to 16 different displays (or more precisely, 15 different display products as well as the absence of any display) to be uniquely identified to the host system.

By the 1990s, it was apparent that this system was no longer useful, and in fact was not being used in the majority of situations due to its inherent limitations. Display capabilities had grown far beyond what could be identified in the simple four-bit system of the original VGA definition; new display technologies and more sophisticated software meant for a growing need for much more information to be provided for each display. But in the absence

Pin	Signal	Pin	Signal
1	Red video	9	N/C (mech. key)
2	Green video	10	Sync return
3	Blue video	11	Monitor ID Bit 0
4	Monitor ID Bit 2	12	Monitor ID Bit 1
5	Test (ground)	13	Horiz. Sync. (TTL)
6	Red return	14	Vert. Sync. (TTL)
7	Green return	15	Monitor ID Bit 3
8	Blue return		

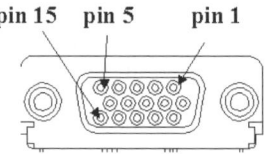

pin 15   pin 5   pin 1

0010 - Super VGA
0101 - Monochrome VGA
0110 - Color VGA
1111 - No monitor connected

Figure 11-1 Original VGA HD-15 connector pin assignment. This shows the "VGA" connector as originally defined, with its four ID pins. These could be selectively grounded (set to the "0" state) by the monitor (or its cable) to identify the type of monitor in use. Some of the possible monitor ID codes are listed. Note that this pinout should be considered obsolete today; for the current VGA pinout per industry standards, see Figure 9-7. As with all connector pinout diagrams in this book, the pin numbering shown is for the female connector, looking in from the "front" of the connector (as it would normally be viewed on the PC rear panel).

of a suitable channel for conveying such information via the standard display interface, the first means established for improved display ID had to use another route.

11.3 The VESA Display Information File (VDIF) Standard

With no standard hardware channel defined for the communication of large amounts of display ID information, the first standards to go beyond the VGA "ID pins" were simply definitions of standard file formats through which the display could be described. An example is the VESA Display Information File, or "VDIF" standard. This specification described an 8 kilobyte file that could provide information on the display type, supported timings, colorimetry, etc.. An overview of the contents of the VDIF file is given in Table 11-1. As there was no means for either storing such data in the display itself, or conveying the file to the host even if there were, it was expected that the display manufacturer would provide the VDIF on a physical storage medium such as a disc. The user would install the VDIF onto the PC from the disc, at the time the display was installed.

Such methods, of course, require that the file be provided to the computer system separately from the act of connecting the display itself, and the user must make sure that the proper file is installed for the display to be used. This still leaves open the possibility that the system will not have the proper information for the display in use at any given time, as it is certainly possible that a different display could be connected without at the same time updating the file in use. In any event, relying on user intervention is both cumbersome and unreliable. For this reason, VDIF and similar file-based display descriptors never achieved any significant acceptance in the industry. It was soon recognized that it would be far preferable to provide a standard, universal system through which the display itself could provide all the information required for the computer to be properly configured, simply by being connected to the computer. This was accomplished in the next generation of VESA display ID standards, with the definition of a completely new system which was rapidly adopted and is today used in practically all PC displays.

Table 11-1 Basic VDIF format.[a] Used by permission of VESA.

Section	Item	Comments
Version	Version	Major & minor VDIF version numbers
Monitor description	File date & revision	
	Manufacturer	
	Model number	
	Max. resolution	
	Version	Version number of this model
	Serial number	
	Date of manufacture	
	Monitor type	"monochrome" or "color"
	CRT size	Actual CRT size, in inches (diagonal)
	Phosphor decay	100–10% decay time in μs. Separate values for red, green, and blue if color.
	Border color	Recommended; separate 0–100 RGB values
	White point, luminance	CIE 1931 xy coordinates, followed by Y in cd/m^2

206 ADDITIONAL INTERFACES TO THE DISPLAY

Section	Item	Comments
	Chromaticity	CIE xy coord. for red, green, and blue, or a single pair for monochrome.
	Gamma	May provide separate values for red, green, and blue, or single overall.
	Video type	"Analog", "TTL", "ECL", "DECL", or "Other"
	Video term. resistance	Nominal termination of video inputs
	White level	These give levels, referenced to blank, for
	Black level	100% and 0% luminance; a non-zero value
	Blank level	for blank is used to define the DC offset in a DC-coupled system
	Sync level	Volts BELOW blanking level; zero if sync-on-video not used.
	Sync type	"Analog", "TTL", "ECL", "DECL", or "Other".
Operational limits	Sync configuration	Defines separate, composite, or sync-on-video (green) system.
	Min. horizontal frequency	In kHz
	Max. horizontal frequency	In kHz
	Min. vertical frequency	In Hz
	Max. vertical frequency	In Hz
	Max. pixel clock	In MHz
	Max. horizontal pixels	The maximum number of pixels which may be addressed horizontally
	Max. vertical pixels	As above, for the vertical axis
	Horizontal line dimension	Max. length of H. addressable line (mm)
	Vertical height dimension	Max. height of addressable image (mm)
	Minimum horiz. retrace	Min. time in μsec required for retrace
	Minimum vert. retrace	Min. time in msec required for retrace
	Pre-adjusted timing name	Mfg. assigned or standard timing name
	Horizontal pixels	Horizontal addressibility in pixels
	Vertical pixels	Vertical addressibility in lines
	Horizontal frequency	In kHz
	Vertical frequency	In Hz
	Pixel clock	In MHz
	Character width	In pixels
Pre-adjusted timings (multiples of these may be provided)	H. addressable line length	In mm
	V. addressable height	In mm
	Pixel aspect ratio	Given as two numbers, with pixel width first (e.g., "1,1" denotes a "square" pixel)
	Scan type	"Non-interlaced", "Interlaced", "Other"
	Sync. polarity	Separate values for H and V sync polarity
	Horizontal total time	Total H. period in microseconds.
	Horizontal addressable time	Time in microseconds during which active (addressable) video is displayed
	Horizontal blank start	The time between the beginning of the addressable line, and the beginning of the H. blanking time, in microseconds

Section	Item	Comments
	Horizontal blank time	Duration of H. blanking, in μs
	Horizontal sync start	The time between the beginning of the addressable line and the beginning of the H. sync pulse, in microseconds.
	Horizontal sync time	Duration of the H. sync pulse, in μs
	Vertical total time	Defined as for the corresponding horizontal values, but all vertical values are given in milliseconds.
	Vertical addressable time	
	Vertical blank start	
	Vertical blank time	
	Vertical sync start	
	Vertical sync time	
Gamma table (optional)	Number of entries	Either a single gamma table, or separate tables for each channel, may be provided.
	Table of relative luminance values	

[a] It should be noted that this format is essentially obsolete at this point, having been replaced by the much more widely used VESA Extended Display Identification Data standard ("EDID"), a 128-byte block (plus extensions) which is stored within the display itself and communicated to the host system via the VESA Display Data Channel (DDC) communications standard.

11.4 The VESA EDID and DDC Standards

First released in 1994, this new ID standard was actually defined in two separate documents: the VESA Display Data Channel (DDC) standard, and the accompanying Extended Display Identification Data (EDID) standard. The DDC specification established the standard electrical interface over which display identification information could be transmitted; EDID defined the format for that information, originally in the form of a single 128-byte file. This could be stored in a relatively inexpensive EEPROM (electrically erasable, programmable read-only memory) device, enabling the display itself to contain its own descriptor file.

The original Display Data Channel definition was based in part on the "I^2C™" (for "Inter-Integrated Circuit) interface, originally designed by Phillips Semiconductor as a channel for communications between different ICs within a single product. This is a two-wire connection, with one wire used as a bidirectional serial data line (SDA), and the other as a system clock (SCL). (This same hardware interface had previously been used as the basis for the ACCESS.bus standard, which was an early attempt at a "universal" connection for human-interface devices such as keyboards, etc..) However, as I^2C capability was not expected to quickly become standard in graphics hardware, the DDC standard also defined a simpler protocol that could be quickly implemented as a minor modification to existing designs. In this more basic mode of operation, called "DDC1," serial data was transmitted over the same physical line as in the I^2C mode, but was clocked by the vertical sync signal from the host system. (The standard permitted a temporary increase in the V. sync rate, with the video to the display blanked, for faster transmission of the data). Any modes using the actual I^2C interface were collectively referred to as "DDC2". In the original standard, these included "DDC2AB," in which the system was used as defined in the ACCESS.bus standard (anticipating the connection of keyboards and other input devices at the display), while the "DDC2B" mode was the simpler one-directional protocol in which the I^2C channel was used simply to read the ID data from the display (as an I^2C memory read from address A0h). The ACCESS.bus interface was soon abandoned by the industry (effectively replaced by the later Universal Serial Bus, or USB, standard), and the sync-driven DDC1 was not used beyond some very early implementation of the system. Today, both the DDC1 and DDC2AB modes have been dropped from the standard, and virtually all computer systems and displays obtain display ID information via the DDC2B mode.

Use of the I^2C electrical interface limits the maximum transmission rate under DDC to approx. 100 kbits/s, and the maximum distance between the display and the host system to 10 m (at least without intermediate "repeater" devices in the channel), but neither of these places an unreasonable limitation on the use of this system. To place the DDC interface on the existing de facto standard "VGA" connector (the 15-pin high-density D-subminiature discussed in Chapter 9), VESA redefined what had previously been the "ID" pins of this connector (see Figure 9-7). Pins 12 and 15 were assigned to the SCL and SDA lines of the interface, respectively. Two additional pins (5 and 9) were assigned as a dedicated return for the interface and a +5 VDC supply, such that the DDC hardware in the display could be supplied with power and read by the host even if the display was in a reduced-power state or off.

Regardless of the DDC mode used, the information transmitted from the display to the host system is the 128-byte EDID file. This provides data on the timings supported by the display, its basic characteristics (the default white point, chromaticity of the three primary colors, and the basic response or "gamma" number), the display model and serial number, and similar information. An overview of the EDID structure, per its current definition, is given in Table 11-2. Note the 8-byte header at the start of the table; this is a holdover from the DDC1 mode, in which the data was transmitted continuously over the serial data line and some method was required for recognition of the start of the file.

The 128-byte size of the basic EDID structure was originally chosen to match that of readily available and inexpensive I^2C-compatible EEPROMs. However, the standard recognized that additional information might be required in future applications of the system. For this reason, a means of adding additional 128-byte "extension blocks" was also defined under the standard. To date, several extensions have either been defined as additions to the EDID standard or are going through the standardization process, including additional timing blocks and information needed for digital-input displays and microdisplay-based products. In additional, the 72-byte "detailed timings" section of the original EDID was later redefined to permit the last three of its 18-byte blocks to be optionally used as "monitor descriptors". Several of these descriptors and extensions are summarized in Table 11-3. It should also be noted that at one point, a "second-generation" EDID definition was released – the 256-byte "EDID 2.0" – which was intended to cover both analog- and digital-input displays in a single basic format. This was developed primarily for use with the VESA "Plug & Display" interface standard (see Chapter 10). As the P&D interface has today largely been abandoned in favor of the later Digital Visual Interface (DVI) standard, and the 128-byte EDID format was already being provided in a large number of different display models, the EDID 2.0 definition has basically been discarded.

Table 11-2 Basic EDID file format (base 128-byte EDID)[a] Used by permission of VESA.

Item	No. of bytes	Description
Header	8	Hex. values: 00, FF, FF, FF, FF, FF, FF, 00
Product ID	2	Manufacturer name (3-character EISA ID code)
	2	Product code (manufacturer's assigned model number)
	4	32-bit serial number (manufacturer defined)
	1	Week of manufacture (week number)
	1	Year of manufacture (Year – 1990; e.g., 1998 = "8")
EDID vers./revision	2	One byte each for VESA version and revision number
Basic parameters	1	Video input definition byte (one-bit flags)
	1	Maximum horizontal image size, in centimeters
	1	Maximum vertical image size, in centimeters
	1	Gamma, stored as ($\gamma \times 100$) – 100; e.g., $\gamma = 2.5$ is "150d"
	1	Feature support (one-bit flags)
Color characteristics (1931 CIE xy coordinates)	2	Lower 2 bits of xy values for red, green, blue, and white
	2	Upper 8 bits of red x, then upper 8 bits of red y
	2	Upper 8 bits of green x, then upper 8 bits of green y
	2	Upper 8 bits of blue x, then upper 8 bits of blue y
	2	Upper 8 bits of white x, then upper 8 bits of white y
Established timings supported	3	24 single-bit flags indicating support for basic VESA timings, including VGA/SVGA/XGA/SXGA
Standard timing codes	16	Indicates support for up to 8 timings, identified by a standard 2-byte name encoded as follows: First byte: (horizontal active pixels/8) – 31 Second byte: first 2 bits give aspect ratio (00=16:10; 01= 4:3; 10=:4; 11=16:9); last 6 bits = refresh rate – 60 (Hz)
Detailed timings (Note: As of the EDID 1.1 release, the last three blocks in this section may also be used as monitor descriptors; see the current EDID standard for details.)	72	Provides detailed information on up to 4 timings, stored as 18-byte blocks; all information stored 1 byte each unless otherwise noted: pixel clock (MHz; 2 bytes, LSB first); H. active (lower 8 bits); H. blanking (lower 8); H active/H blanking (upper 4 for each, in order); 3 bytes for V. active/blanking, same coding; H. sync offset/pulse width (one byte each, lower 8 bits); V. sync offset/pulse width (lower 4 bits of each); 1 bytes with upper 2 bits of H. sync offset/width, V. sync offset/width; 2 bytes for H and V image size in mm, lower 8 bits each; one byte with upper 4 bits for both H and V image size; one byte each for H and V border width in pixels; 1 byte flags (interlace, stereo, etc.). The first 18-byte block must be used as a detailed timing, assumed to be the display's preferred timing.
Ext. flag/checksum	2	Number of ext. blocks to follow, plus one byte checksum (such that the sum over the entire 128 bytes is 00h)

[a] See current VESA "Extended Display Identification Data" standard for details.

Table 11-3 ID codes for monitor descriptor and extension blocks under the VESA EDID standard. Used by permission of VESA.

ID code (hex)	Name	Description
Monitor Descriptors (may be in last 3 18-byte blocks of base EDID)		
00-0F		Manufacturer-defined descriptor
10	Dummy	Used to identify an unused block
11-F9	(undefined)	Unassigned at present
FA	Standard timings	Stores additional 2-byte codes identifying supported timings, using same coding system as base EDID (see Table 11-2)
FB	Additional color data	Gives chromaticity and gamma data on additional white points, if needed
FC	Monitor name	ASCII code, up to 13 bytes
FD	Monitor range limits	Defined by VESA EDID standard
FE	ASCII string	Any desired ASCII information, up to 13 bytes
FF	Monitor serial number	Stores ASCII code, up to 13 bytes
EDID Extension Blocks (additional 128-byte blocks which may follow the base EDID)		
30	"Consumer"	Reserved code for proposed "consumer electronics" (television, etc.) extension; under development
40	DI-EXT	VESA Display Information Extension block. Released in 2001, this extension provides 128 bytes of additional data, including added color and gamma data and information specifically for digital/dual-input and non-CRT display types
60	HMD-EXT	VESA Head-Mounted Display Extension block (expected to be released in 2002)

The DDC and EDID standards are among the most successful in the personal-computer display industry, and today are found in practically ever PC and many higher-end systems. They have enabled "plug and play" use of various display types and technologies in this market, and have become increasingly important with the introduction of mixed analog/digital interface standards (as described in Chapters 9 and 10) and new technologies which do not resemble the CRT in their behavior.

11.5 ICC Profiles and the sRGB Standard

Before moving on to other physical and electrical interfaces often used with displays, we should mention two additional standards that relate to the problem of display identification. Prior to the introduction of the DDC and EDID standards, as noted above, the host system had no practical means of determining the capabilities and characteristics of the display in use. This was not only a problem from the standpoint of choosing the correct timing and image format for optimum use of the system, but also in terms of ensuring reasonable accurate color in displayed images. The original broadcast television standards (and to some degree, the newer HDTV standards) avoid color-matching problems to a large degree by specifying the relevant performance parameters of the television receiver. The colors of the phosphors, the white point, and the response or "gamma curve" of the displays to be used within each television system are controlled by the standards defining that system. Broadcasters can transmit programming under the assumption that the display will perform per these requirements, and that colors will therefore be properly rendered.

This has never been the case in the computer industry. While many of the display designs used were originally derived from common television practices, there are truly no comparable standards in this industry. Even in a given product, there are often user adjustments for white point and other factors which can affect the displayed color. This situation requires that the host system also have information on the color aspects of the display, at least the generic capabilities for the model in question and preferably measured data from the specific unit in use.

In the absence of standard channels over which such information could be communicated, the industry employed two approaches. First, a standardized file format was developed, specifically for the purpose of conveying color-related information. Defined by the International Color Consortium, "ICC profiles" may be produced not only for electronic displays, but also printers, cameras, and other image input and output devices. Profile information may be attached to image files when a given input device creates them, and this, along with the corresponding output-device profile, may be used by the system to correct or compensate the video data for proper color rendering. Such profiles are, like the VDIF description mentioned above, generally assumed to be supplied by the device manufacturer separate from the device itself. Unlike the DDC/EDID system, there is generally no standard system for obtaining ICC profile information directly from the peripheral itself, and so this method suffers from the same problems as the VDIF. In the case of displays, however, the current definitions of the EDID file and its extensions provide sufficient information so that a reasonably complete ICC profile could be derived from this data. To date, such a translation has not commonly been implemented in either standard operating systems or applications programs.

The ICC profile specification is a fairly complex and flexible standard, and attempting to summarize a typical display profile here would be both difficult and likely misleading. For the purposes of this discussion, it is suffice to note that a complete profile provides the same sort of information as the formats detailed above, such as the color space used by the device, its chromaticity characteristics, response curve(s), etc., but in much more detailed and precise form than could be conveyed in a space-constrained representation such as the EDID file. The full ICC profile specification may be downloaded from the ICC web site (www.color.org), and is recommended if more detailed information is desired.

An alternative, which may be used in the absence of device-specific color information such as an ICC profile, is provided by the "sRGB" standard. This is a specification developed by researchers at Hewlett-Packard Co. and Microsoft (and now standardized as IEC 61966, Part 2.1), and which defines a standard model for a color-output device. The sRGB model establishes norms for the primary colors (based on typical CRT phosphor values), the "white point" (6500K, specifically the CIE D65 illuminant), the response curve or "gamma," and several other key parameters. Rather than providing for a means of conveying device-specific information, the sRGB system assumes that standard devices will be built to conform to (or at least can be set up to) these specifications. The operating system or application program therefore can perform color transformations, etc., assuming this standardized output device. Table 11-4 summarizes the key features of the sRGB specification.

The sRGB model can provide significantly improved color performance over the common PC environment in which no information at all about the output device's characteristics is available or used in the generation of images. It also has the advantage of simplicity; given a single standard definition, color information can be properly generated without the need for

212 ADDITIONAL INTERFACES TO THE DISPLAY

Table 11-4 Basic sRGB definitions.[a]

Parameter	Definition
Display luminance	80 cd/m^2 (white; 100% drive on R, G, and B inputs)
Display white point	D65; $x = 0.3127$, $y = 0.3291$
Primary chromaticities	ITU-R BT 709-2 primaries:
	Red: $x = 0.6400$, $y = 0.3300$
	Green: $x = 0.3000$, $y = 0.6000$
	Blue: $x = 0.1500$, $y = 0.0600$
Gamma	2.4
Offset (R, G, and B)	0.055
Background	20% of white luminance (for background areas on screen)
Surround	Reflectance of 20% of reference ambient illuminance
Ref. amb. illuminance	64 lx
Ambient white point	D50; $x = 0.3457$, $y = 0.3585$
Veiling glare	1.0%

[a] All chromaticity coordinates provided are per the 1931 CIE xy system. The "gamma" and "offset" values define the display response curve as: $V = [(V' + 0.055)/1.055]^{2.4}$ for all channels.

complex, device-specific calculation. However, being a generic color system – the definition of a standard "color space" – sRGB lacks flexibility and cannot provide quite the optimum performance for all devices. It is a very CRT-centric definition; the primary colors, response curve, and other factors in effect describe a typical color CRT display, and may not be a good match to the actual characteristics of other types. For critical applications, it is still better to use product-specific information such as an ICC profile or the color information from EDID, or better yet to have such information generated for the particular unit in question.

11.6 Display Control

Beyond merely identifying and describing the display to the host system, there has also been the desire to permit the adjustments and controls commonly provided to the user to be accessed by the host. Such capability cannot only allow the development of "soft" control panels (those which duplicate the front-panel controls of the monitor through software), but also can provide for fully automated adjustment of the display by the video source (as might be the case in automated color correction and management). Such functionality could be implemented via any of a number of general-purpose interfaces already included in many display connector standards, but in several cases new systems intended specifically for the control of various display functions have been defined. Much of this activity has again been performed under the auspices of the Video Electronics Standards Association (VESA), and driven by the needs of the personal computer market. However, many of these standards have attracted the attention of other industries, such as the consumer-television, and are likely to be adopted (and possibly adapted) for their use.

Initially, display control functionality was added to the various interface standards in the form of definitions intended to control specific functions only, such as control of the power-management features. Later, however, as more capable channels have been developed and supported within the interface standards, display control has become more generalized. Using a general-purpose digital interface, commands may be defined for the control of a wide range of disparate display functions. As this trend continues, display control will likely become just another function of a general-purpose, system-wide control and data interface (such as USB), and eventually may come to be transmitted on the same physical and electrical interface as the video data itself, along with other types of supplemental data and commands.

11.7 Power Management

As CRT displays represent a significant portion of the electrical power required in most personal computer systems, the development of PC power management techniques in the late 1980s and early 1990s could hardly succeed without attention to these products. Several products and systems were introduced which included the capability for reducing the display power or shutting the display off completely during idle periods, including some which relied on the host simply blanking the video signal for a specified period of time to trigger display shutdown. However, in 1993, VESA released the Display Power Management Signalling (DPMS) standard, and this method was quickly adopted by the industry.

Basically, DPMS allows the host to place the display in any of four defined power states by enabling or disabling the synchronization signals. If either or both sync signals are determined to be inactive by the display, it leaves the normal, "active" mode of operation and enters one of the reduced-power states as shown in Table 11-5. Note that either sync may be considered "inactive" if its rate falls below a defined frequency (40 Hz for the vertical signal, and 10 kHz for the horizontal). This was done to permit host systems that did not have the ability to completely shut off the sync signals to still implement DPMS control, simply by reprogramming the video timing. The four power states were defined to correspond to the "Advanced Power Management" definitions, originally created by Intel and Microsoft for PC systems. DPMS does not specify the actual power levels or recovery times for any of the power states, nor the specific means of implementing the reduced-power states; only the signalling used to control them. However, minimum levels of power-reduction have been specified by various regulatory bodies, such as the Environmental Protection Agency in the US (through their "Energy Star" program). It is assumed, that in general the "deeper" reduced power states will require longer recovery times. In a CRT display, for example, the lowest level of power reduction (the "standby" state) might be implemented by shutting down the deflection circuits, but leaving the CRT filament powered up for a relatively rapid recovery. The deeper "suspend" state might turn the filament off or at least operate it at a greatly reduced power level.

Table 11-5 VESA DPMS power management states. Used by permission of VESA.

DPMS state	Horizontal sync	Vertical sync
On (normal operation)	Active	Active
Standby	Inactive	Active
Suspend	Active	Inactive
Off	Inactive	Inactive

[a] Note that all video signals must be blanked prior to entering any state below "On", and should not be restored until after the return to the "On" state.

214 ADDITIONAL INTERFACES TO THE DISPLAY

It is important to note that while the DPMS system permits the display to be put into an "off" state by the host, it is not required that the display be capable of recovering from that state without user intervention. Therefore, it is permissible for the display to be turned off under DPMS control and truly be disconnected from the AC line completely, requiring the user to press the power button to re-enable the display. Most manufacturers have chosen to implement an "active off" state as the lowest-power DPMS level. In such a design, the display continues to consume a very small amount of power even when "off," but only enough to support continued monitoring of the sync signals and recovery from this state.

11.8 The VESA DDC-CI and MCCS Standards

Due to the success of the I^2C-based Display Data Channel standard, it was logical to extend the use of this from simply a display-ID channel to a fully bidirectional command and control interface. This was achieved with the release of the VESA "DDC-CI" (Display Data Channel – Command Interface) standard in 1998. The nomenclature of the overall DDC/CI system is somewhat confusing. This standard replaced and expanded upon the earlier "DDC2B+" definition, which was part of the original DDC standard and which permitted the I2C channel to be used in a bidirectional master/slave mode, using an ACCESS.bus host driver. (The difference between this and the full "DDC2AB" mode was that DDC2B+ was a single-device implementation, while "DDC2AB" indicated support for the full ACCESS.bus specification.) The DDC/CI standard redefined this as "DDC2Bi," and abandoned the ACCESS.bus driver model for a new DDC-specific driver (DDC.DLL at the host PC), although much of the ACCESS.bus protocol definition is retained.

DDC-CI exploits the existing bidirectional capabilities of the I^2C interface, adding a new command protocol to the original ID-oriented DDC system. The display device itself operates as an I2C slave device, at address 6E/6F (the EDID information remains accessible as an I2C memory device at A0/A1). The specification also permits a limited number of additional functions or devices to be connected via the DDC2Bi channel, such as pointers (touchscreen, trackball, etc.), audio devices, and so forth. Each is assigned a predefined I^2C address by the DDC/CI standard, all in the F0-FF address space. (Other functions have been defined and assigned standard addresses by Philips or other companies/organizations using the I^2C system, in other address ranges.)

A separate standard, the VESA Monitor Control Command Set (MCCS), was released along with DDC-CI, and provided a fairly complete set of standard command functions that could be implemented in a CI-capable display. An overview of these is given in Table 11-6. Note that these standards also permit manufacturer-specific "custom" commands to be defined, permitting extension of the basic set as required for a given display product.

Table 11-6 VESA Monitor Control Command Set (MCCS) summary.

Code (hex)	Description
01	Degauss; causes display to execute a degaussing operation. No additional value
10	Brightness; increased values increase the black level luminance (cutoff drops relative to the video signal)
12	Contrast; increased values increase the ratio between max. and min. displayed luminance (i.e., video signal gain)
16	Red video gain
18	Green video gain
1A	Blue video gain
6C	Red video black level
6E	Green video black level
70	Blue video black level
1C	Focus; varies apparent spot size
20	Horizontal position. Increasing values shift the image to the right
22	Horizontal size
24	Horizontal pincushion. Increasing causes L and R sides to become more convex
26	Horizontal pincushion balance
28	Horizontal convergence. Increasing moves red right and blue left (w.r.t. green)
2A	Horizontal linearity
2C	Horizontal linearity balance
30	Vertical position. Increasing values move the image up
32	Vertical size
34	Vertical pincushion. Increasing causes top/bottom to become more convex
36	Vertical pincushion balance
38	Vertical convergence. Increasing moves red up and blue down (w.r.t. green)
3A	Vertical linearity
3C	Vertical linearity balance
40	Parallelogram (Key balance) Increasing shift top right with respect to bottom
42	Trapezoid (Key) Increasing lengthens the top edge relative to the bottom
44	Tilt (Rotation) Increasing rotates the image clockwise about its center
46	Top corner distortion
48	Top corner distortion balance
4A	Bottom corner distortion
4C	Bottom corner distortion balance
56	Horizontal moiré cancellation
58	Vertical moiré cancellation
5E	Input level select; selects from among predefined video signal amplitude standards; see MCCS standard for details
60	Video source select; selects from among predefined input sources/connectors; see MCCS standard for details
7A	Adjust focal plane. Adjusts focus of optics (projection displays)
7C	Adjust zoom. Adjusts optical zoom (projection displays)
7E	Trapezoid. Adjusts trapezoid optically (projection displays)
80	Keystone. Adjusts keystone optically (projection displays)
8A	TV saturation. Increasing values increase the saturation of colors (TV inputs)
8C	TV sharpness. Increases amplitude of high-frequency video (TV inputs only)
8E	TV contrast. Affects "TV" video inputs only
90	TV hue (tint). Increasing shifts tint or hue toward red (TV inputs only)
A2	Auto size/center. 0 – none selected; 1 – disabled; 2 – enabled

216 ADDITIONAL INTERFACES TO THE DISPLAY

AC	Read back horizontal frequency, 3 bytes; returns FFFFFF if incapable
AE	Read back vertical frequency; returns FFFF if incapable
B0	Value =1: Store current settings; 2 = restore factory default settings
CE	Secondary input source select; see 60 above
D0	Output source select 1. See MCCS standard for details
D2	Output source select 2. See MCCS standard for details
D4	Stereo mode select. See MCCS standard for details
D6	DPMS mode select; 0 – none; 1 – on; 2 – standby; 3 – suspend; 4 – off
D8	Color temp. preset select. 0 – none; all others use index from EDID file
DA	Scan format. 0 – none selected; 1 – underscan; 2 – overscan; 3 – 16:9 letterbox
E0-FF	Manufacturer-defined

[a] Unless otherwise noted, all commands are followed by a two-byte value. Codes in the A0-AF range are "read-only"; when given, the display returns a 2-byte value unless otherwise noted. Note: This is an overview, and does not list all codes defined by the MCCS standard. See that standard for details.

11.9 Supplemental General-Purpose Interfaces

As noted previously, the recent trend has been for more general-purpose interfaces to be included in the display interface definition, along with the primary video channel. This permits a wide variety of additional data and control functions to be communicated to the display. In addition to providing identification and control of the display itself, as covered in the previous sections, the addition of such channels to the display interface permits the display to host numerous other devices and peripherals. In many applications, for instance, the display is a logical point at which to centralize those functions through which the user interacts with the system. It is the one major system component which can be counted on to be located in the immediate vicinity of the user, and also can be used to provide power to additional devices at only a slight incremental cost in the display power supply. Therefore, many display designs also incorporate input devices (such as touchscreens, trackballs, keyboards, etc.), removable-media mass storage, audio input and/or output, or at least the connectors necessary for the attachment of these.

Until recently, the primary distinguishing features of such additional interfaces, as compared with the video channel itself, were that they were relatively low-capacity, bidirectional, and digital connections. As the display interface proper changes from the analog standards of the past to the new digital types, these distinctions become less important and the need for such physically separate general-purpose interfaces less clear. The sole additional requirement for supporting these functions which has not yet generally been provided by the display interface is a "back channel," some means for communicating data from the display to the host system. This is also likely to change, given the needs of more "network-like" systems such as the Digital Packet Video Link discussed in the previous chapter. As display systems move to this model, it is likely that all communications between the host and display – including all command, ID, and supplemental functions, as well as the video data itself – will be carried over a single common channel, supported by a somewhat lower-capacity back channel.

Until such a system is widely adopted, several general-purpose digital interfaces will see continued support on standard display connector definitions. The I^2C interface, in the form of the VESA Display Data Channel standard, has already been discussed, as has the now-obsolete ACCESS.bus specification which was based upon it. At this point, there are two such systems in widespread use: the Universal Serial Bus, or USB, and the IEEE-1394 standard, also known by the trade names "Firewire™" and "i-Link™", which refer specifically to the implementations of this standard by Apple Computer and Sony Corp., respectively. No other challengers have appeared, and it seems likely that these will continue to be the dominant standards for such applications. While these have been mentioned earlier in relation to the recent display interface standards that support them, they will now be covered in greater detail here.

11.10 The Universal Serial Bus

The Universal Serial Bus, or "USB" as it is commonly known, was originally defined as a low-to-medium speed interface intended primarily for the connection of human input devices (such as keyboards, mice, etc.) and digital audio peripherals. Using a simple, four-wire physical connection, comprising a single bidirectional data pair, a +5 VDC power line, and a ground, the USB 1.0 specification permitted two modes of operation to meet the needs of these peripherals. In the "low-speed" mode, a peak raw data rate of 1.5 Mbits/s is supported; in the faster, standard mode, USB provides a peak rate of 12 MBits/s. (In realistic situations, accounting for the overhead of the system and management of multiple devices on a given physical interface, more typical observed data rates are in the range of 700–800 kbytes/s.) This is sufficient to support simultaneous operation of multiple low-speed devices such as keyboards, etc., along with perhaps one or two peripherals requiring higher data rates, such as digital audio (speakers) or a low-to-medium resolution video camera.

Despite its name, USB is not truly a "bus" system. Rather, it employs a "tiered-star" topology, in which a single controller manages all peripheral devices as the "root" of a tree-like arrangement (Figure 11-2). From the root controller, devices are connected via "hubs" which provide for branching at each level. The specification permits tiering up to five levels

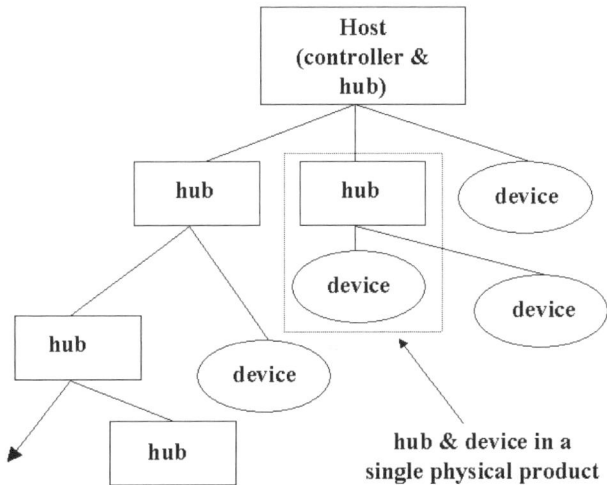

Figure 11-2 USB "tiered-star" topology. A single controller may support up to 127 devices, but the system is limited to five levels of tiering. This is only the logical topology; physical products may combine hubs and logical USB devices as desired.

deep, although there is no set limit on the number of separate "branches" ("downstream" ports) which may be supported by each hub. Hubs may be either "powered" or "unpowered"; as the name would imply, a powered hub has its own power supply, and does not rely on the +5 VDC line provided by its "upstream" connection. In either case, all USB "downstream" ports (such as those provided by the host or a hub for the connection of peripherals) must provide up to 500 mW of power (100 mA) on the +5 VDC line when a peripheral is connected and initialized. Under host control, a given peripheral may then be provided with up to 2.5 W (500 mA), if this remains within the limitations of the host or hub.

USB was designed to permit "hot plugging" of all peripheral devices, meaning that any device may be connected to or disconnected while the rest of the system remains powered up. Upon connection of each device (or at initialization of a previously connected set of devices at system power-up), the device is identified by the host controller and assigned an address within the current structure. The speed at which the device operates is also identified, and the port to which it is connected is then configured to provide the desired level of service (either the low-speed 1.5 Mbit/s mode or the full 12 Mbit/s). As each device is identified, the proper driver for that device or device class may be loaded by the host system. The USB specifications provide for sufficient standardization of device operation that a "generic" USB driver, at least for a given device class (such as human-input devices), may typically be used with all devices of that class connected to the system.

USB provides for a mix of isochronous and non-isochronous (or "asynchronous") data streams to be supported simultaneously; as noted in the previous chapter, an "isochronous" data transmission is one in which the receipt of the data by the receiving device is time-critical. An example of this is the transmission of digital audio, in which the data representing each sample must be received within the defined sample period. To support such transmissions, the USB controller allocates the available capacity on the interface as best it can, giving priority to devices requiring such isochronous flow. Thus, it is possible for one or more isochronous devices to "hog the bus" at the expense of others.

For any transmission type or rate, the USB interface transmits data serial on a single differential pair, in a "half-duplex" mode. All operations must be initiated by the host controller; no direct communications (i.e., "peer-to-peer" transmissions) are permitted between peripherals. Owing to the transmission protocol and format, the maximum length of any USB cable is strictly limited to a maximum of 5 m (slightly under 16.5 feet); to extend any branch beyond this limit requires at least one intermediate hub. A given physical product may contain both device and hub functions, although these will still be treated as logically separate blocks by the controller. As a device is a separate function from a hub, such configurations can appear to provide "daisy-chain" connections, although the limitation to five levels in the tiered-star topology represents a hard limit on the length of such a "chain." For maximum flexibility, then, most hubs will provide multiple downstream ports.

The specification defines two physical connectors, shown in Figure 11-3. The "type A" connector, the more rectangular, "flatter" of the two, is used as the "downstream" connection from hosts and hubs. The "type B" connector, which is more nearly square in cross-section, is used on USB peripherals.

In April, 2000, a new USB consortium led by Compaq, Hewlett-Packard, Lucent, NEC, Intel, and Microsoft released the USB 2.0 specifications, which greatly increased the data rate supported by the system. USB 2.0 remains completely compatible with the earlier specification (which in its latest revision was "USB 1.1"), but permits data rates up to 480 Mbits/s. USB 1.1 peripherals can be used in a USB 2.0 system, although the port to which

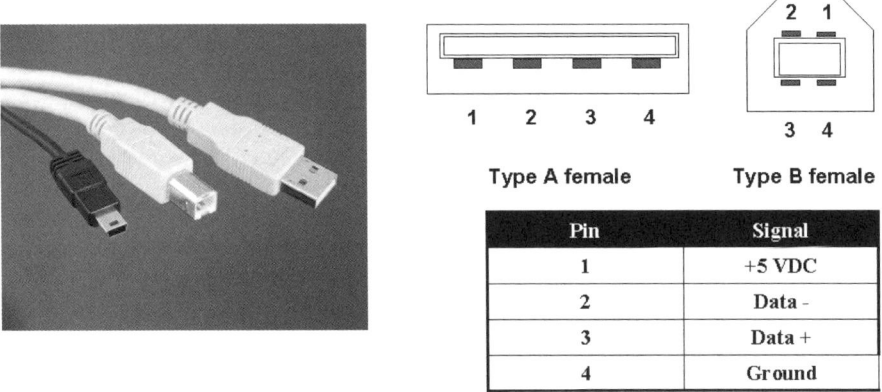

Figure 11-3 USB connectors and pinouts. As shown in the photograph, USB connectors come in several styles and sizes; this picture shows the "A" and "B" type standard-sized connectors (in white), plus an example of a mini-USB connector. The "A" connector, the flatter of the two, is typically the "upstream" connection – the output of a hub or controller, for example. The more squared-off "B" type connector is the "downstream" connection, used at the input to USB devices. (*Photograph courtesy of Don Chambers/Total Technologies, Inc.; used by permission.*)

they are connected will then be configured by the host controller to either the 12 Mbit/s or 1.5 Mbit/s modes, prohibiting higher-speed devices to be connected downstream of that port. USB 2.0 and USB 1.1 hubs may be mixed in a given system, but the user must be aware of which is which such that the higher-speed USB 2.0 devices are not connected to a hub which cannot support them. Existing USB cabling, if compliant with the original standard, is completely capable of supporting the faster rate.

11.11 IEEE-1394/"FireWire™"

In 1986, Apple Computer developed a new high-speed serial interface which was introduced under the name "FireWire™." The specifications for this interface were later standardized by the Institute of Electrical and Electronic Engineers as IEEE-1394, released in its original form in 1995. Currently, this interface is commonly known under both the "1394" and "FireWire™" names, although properly the latter refers only to Apple's implementation (and is a trademark of that company). Other proprietary names for this interface have also been used by various companies, such as "i.Link™" for Sony's implementation. We refer to it here simply as "1394." This standard has become the digital interface of choice in several markets, especially in both consumer and professional digital audio/video (A/V) systems. IEEE-1394 has also been selected as the underlying physical/electrical interface standard for the VESA/CEA Home Network standard.

In many respects, 1394 resembles USB; both are serial interfaces intended for the easy connection of various types of devices in a "networked" manner. Both permit "hot-plugging" (connection and disconnection of devices at any time), and both provide for power to be carried by the physical interface along with data. And, with the recent introduction of the USB

2.0 specification, the data rates supported are at least comparable. The original IEEE-1394-1995 standard defined operation at 100, 200, and 400 Mbits/s.

However, there are significant differences between the two that continue to distinguish them in the market. The most obvious is in the topology of the system. Unlike USB, which relies on a single controller as the master of the entire "tree" of hubs and devices, 1394 is based on peer-to-peer communications. Any device connected to the interface may request control of the bus, under defined protocols and capacity-allocation limits; there is no single master controller. 1394 has a cable length limit similar to USB's – in this case, 4.5 m under the original specification – but permits the use of "repeaters" to extend the connection up to 16 times between devices. Note that longer cable lengths may be possible at lower data rates, or using expected future specifications for improved cabling. 1394 also lends itself to optical connections, which may permit repeaterless connections of 50 to 100 m or more. Each 1394 bus can support up to 63 separate devices, but there is also a provision for "bridges" between buses. A maximum of 1,023 buses may be interconnected via 1394 bridges, for an ultimate limit of 64,449 interconnected devices.

Physically, the standard 1394 connector provides six contacts in a rectangular shell which has a slight resemblance to the USB Type A connector. This same connector type, however, is used for all 1394 ports. The connector and its pinout are shown in Figure 11-4. A 1394 cable comprises two twisted pairs, individually shielded, for the data channels; these are crossed between the data contacts at each end, such that each device sees a separate "transmit" and "receive" pair. The remaining two contacts are for power and ground, in this case a DC supply of up to 1.5 A at 8 to 40 VDC. An overall shield is also specified, covering all six conductors. A smaller 4-pin connector has also been used in some products, which deletes the power connection and its return.

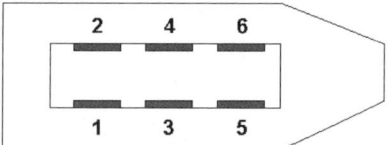

Pin	Signal
1	Power (8–40 VDC, 1.5 A)
2	Ground
3	Twisted pair B -
4	Twisted pair B +
5	Twisted pair A -
6	Twisted pair A +

Figure 11-4 IEEE-1394/"FireWire™" connector and pinout. Unlike USB, the IEEE-1394 standard uses the same connector at both ends of the cable. The interface is based on two twisted-pair connections, both carrying data and data strobe signals, but these are switched from one end of the cable assembly to the other such that each device sees a "transmit" and "receive" pair. The pinout shown here is for the male connector. (*Photograph courtesy of Don Chambers/Total Technologies, Inc.; used by permission. "FireWire" is a trademark of Apple Computer Corp.*)

Like USB, the 1394 standard defines support for both isochronous and asynchronous communications, and so also lends itself well to supporting timing-critical, streaming-data applications such as digital video or audio transmission. Any single device may request up to 65% of the available capacity of the bus, and a maximum of 85% of the total capacity may be allocated to such requests across all devices on the bus. This ensures that some capacity will always be available for asynchronous communications, preventing such from being completely shut down through an "overload" of isochronous allocations.

Also like USB, the 1394 standard continues to be developed. A recently completed revision to the original specifications will extend the capacity of this system beyond the original 100/200/400 Mbit/s levels, adding 800, 1600, and 3200 Mbit/s as supported rates. With this increase in capacity, 1394 can continue to support the high-speed connections required by mass storage devices, high-definition television, and multiple digital audio streams.

The slightly more complex cabling design, and especially the more elaborate protocol and the requirement that all devices be peers (and so effectively have "controller" capability) result in the 1394 bus being somewhat more costly to implement, overall, than the USB system. However, this increased cost does buy increased capabilities, as described above. Both systems are likely to continue to co-exist, with USB being the more attractive choice as a "desktop" PC peripheral interconnect, and 1394 remaining the standard for digital A/V system connections and "home network" and similar applications.

12

The Impact of Digital Television and HDTV

12.1 Introduction

As was noted in Chapter 8, broadcast television represents a particularly interesting case of a "display interface," as it requires the transmission of a fairly high-resolution, full-color, moving picture, and a significant amount of supplemental information (audio, for instance) over a very limited channel. This is especially true when considering the case of *high-definition television,* or "HDTV," as it has developed over the last several decades.

HDTV began simply as an effort to bring significantly higher image quality to the television consumer, through an increase in the "resolution" (line count and effective video bandwidth, at least) of the transmitted imagery. However, as should be apparent from the earlier discussion of the original broadcast TV standards, making a significant increase in the information content of the television signal over these original standards (now referred to as "standard definition television", or "SDTV") is not an easy task. Television channels, as used by any of the standard systems, do not have much in the way of readily apparent additional capacity available. Further, constraints imposed by the desired broadcast characteristics (range, power required for the transmission, etc.) and the desired cost of consumer receivers limits the range of practical choices for the broadcast television spectrum. And, of course, the existing channels were already allocated – and in some markets already filled to capacity. In the absence of any way to generate brand-new broadcast spectrum, the task of those designing the first HDTV systems seemed daunting indeed.

In a sense, new broadcast spectrum *was* being created, in the form of cable television systems and television broadcast via satellite. Initially, both of these new distribution models used essentially the same analog video standards as had been used by conventional over-the-air broadcasting. Satellite distribution, however, was not initially intended for direct reception by the home viewer; it was primarily for network feeds to local broadcast outlets, but individuals (and manufacturers of consumer equipment) soon discovered that these signals were relatively easy to receive. The equipment required was fairly expensive, and the receiving antenna (generally, a parabolic "dish" type) inconveniently large, but neither was outside the high end of the consumer television market. Moving satellite television in to the mainstream of this market, however, required the development of the "direct broadcast by satellite", or DBS, model. DBS, as exemplified in North America by such services as Dish Network or DirecTV, and in Europe by BSkyB, required more powerful, purpose-built satellites, so as to enable reception by small antennas and inexpensive receivers suited to the consumer market. But DBS also required a further development that would eventually have an enormous impact on all forms of television distribution – the use of digital encoding and processing techniques. More than any other single factor, the success of DBS in the consumer TV market changed "digital television" from being a means for professionals to achieve effects and store video in the studio, to displacing the earlier analog broadcast standards across the entire industry. And the introduction of digital technology also caused a rapid and significant change in the development of HDTV. No longer simply a higher-definition version of the existing systems, "HDTV" efforts transformed into the development of what is now more correctly referred to as "Digital Advanced Television" (DATV). It is in this form that a true revolution in television is now coming into being, one that will provide far more than just more pixels on the screen.

12.2 A Brief History of HDTV Development

Practical high-definition television was first introduced in Japan, through the efforts of the NHK (the state-owned Japan Broadcasting Corporation). Following a lengthy development begun in the early 1970s, NHK began its first satellite broadcasts using an analog HDTV system in 1989. Known to the Japanese public as "Hi-Vision", the NHK system is also commonly referred to within the TV industry as "MUSE," which properly refers to the encoding method (MUltiple Sub-Nyquist Encoding) employed. The Hi-Vision/MUSE system is based on a raster definition of 1125 total lines per frame (1035 of which are active), 2:1 interlaced with a 60.00 Hz field rate. If actually transmitted in its basic form, such a format would require a bandwidth well in excess of 20 MHz for the luminance components alone. However, the MUSE encoding used a fairly sophisticated combination of bandlimiting, subsampling of the analog signal, and "folding" of the signal spectrum (employing sampling/modulation to shift portions of the complete signal spectrum to other bands within the channel, similar to the methods used in NTSC and PAL to interleave the color and luminance information), resulting in a final transmitted bandwidth of slightly more than 8 MHz. The signal was transmitted from the broadcast satellite using conventional FM.

While technically an analog system, the MUSE broadcasting system also required the use of a significant amount of digital processing and some digital data transmission. Up to four channels of digital audio could be transmitted during the vertical blanking interval, and digitally calculated motion vector information was provided to the receiver. This permitted the receiver to compensate for blur introduced due to the fact that moving portions of the image are transmitted at a lower bandwidth than the stationary portions. (This technique was employed for additional bandwidth reduction, under the assumption that detail in a moving object is less visible, and therefore less important to the viewer, than detail that is stationary within the image.) An image format of 1440 × 1035 pixels was assumed for the MUSE HDTV signal.

While this did represent the first HDTV system actually deployed on a large scale, it also experienced many of the difficulties that hinder the acceptance of current HDTV standards. Both the studio equipment and home receivers were expensive – early Hi-Vision receivers in Japan cost in excess of ¥4.5 million (approx. US$35,000 at the time) – but really gave only one benefit: a higher-definition, wider-aspect-ratio image. Both consumers and broadcasters saw little reason to make the required investment until and unless more and clearer benefits to a new system became available. So while HDTV broadcasting did grow in Japan – under the auspices of the government-owned network – MUSE or Hi-Vision cannot be viewed as a true commercial success. The Japanese government eventually abandoned its plans for long-term support of this system, and instead announced in the 1990s that Japan would transition to the digital HDTV system being developed in the US.

In the meantime, beginning in the early 1980s, efforts were begun in both Europe and North America to develop successors to the existing television broadcast systems. This involved both HDTV and DBS development, although initially as quite separate things. Both markets had already seen the introduction of alternatives to conventional, over-the-air broadcasting, in the form of cable television systems and satellite receivers, as noted above. Additional "digital" services were also being introduced, such as the teletext systems that were became popular primarily in Europe. But all of these alternates or additions remained strongly tied to the analog transmission standards of conventional television. And, again due to the lack of perceived benefit vs. required investment for HDTV, there was initially very little enthusiasm from the broadcast community to support development of a new standard.

This began to change around the middle of the decade, as TV broadcasters perceived a growing threat from cable services and pre-recorded videotapes. As these were not constrained by the limitations of over-the-air broadcast, it was feared that they could offer the consumer significant improvements in picture quality and thereby divert additional market share from broadcasting. In the US, the Federal Communications Commission (FCC) was petitioned to initiate an effort to establish a broadcast HDTV standard, and in 1987 the FCC's Advisory Committee on Advanced Television Service (ACATS) was formed. In Europe, a similar effort was initiated in 1986 through the establishment of a program administered jointly by several governments and private companies (the "Eureka-95" project, which began with the stated goal of introducing HDTV to the European television consumer by 1992). Both received numerous proposals, which originally were either fully analog systems or analog/digital hybrids. Many of these involved some form of "augmentation" of the existing broadcast standards (the "NTSC" and "PAL" systems), or at least were compatible with them to a degree which would have permitted the continued support of these standards indefinitely.

However, in 1990 this situation changed dramatically. Digital video techniques, developed for computer (CD-ROMs, especially) and DBS use had advanced to the point where an all-digital HDTV system was clearly possible, and in June of that year just such a system was proposed to the FCC and ACATS. General Instrument Corp. presented the all-digital "Digi-Cipher" system, which (as initially proposed) was capable of sending a 1408 × 960 image, using a 2:1 interlaced scanning format with a 60 Hz field rate, over a standard 6 MHz television channel. The advantages of an all-digital approach, particularly in enabling the level of compression which would permit transmission in a standard channel, were readily apparent, and very quickly the remaining proponents of analog systems either changed their proposals

Table 12-1 The final four US digital HDTV proposals.[a]

Proponent	Name of proposal	Image/scan format	Additional comments
Zenith/AT&T	"Digital Spectrum Compatible"	1280 × 720 progressive scan	59.94 Hz frame rate 4-VSB modulation
MIT/General Instruments	ATVA	1280 × 720 progressive scan	59.94 Hz frame rate 16-QAM modulation
ATRC[b]	Advanced Digital Television	1440 × 960 2:1 interlaced	59.94 Hz field rate SS-QAM modulation
General Instruments	"DigiCipher"	1408 × 960 2:1 interlaced	59.94 Hz field rate 16-QAM modulation

[a] The proponents of these formed the "Grand Alliance" in 1993, which effectively merged these proposals and other input into the final US digital television proposal.
[b] "Advanced Television Research Consortium"; members included Thomson, Philips, NBC, and the David Sarnoff Research Center.

to an all-digital type or dropped out of consideration. By early 1992, the field of contenders in the US was down to just four proponent organizations (listed in Table 12-1), each with an all-digital proposal, and a version of the NHK MUSE system. One year later, the NHK system had also been eliminated from consideration, but a special subcommittee of the ACATS was unable to determine a clear winner from among the remaining four. Resubmission and retesting of the candidate systems was proposed, but the chairman of the ACATS also encouraged the four proponent organizations to attempt to come together to develop a single joint proposal. This was achieved in May, 1993, with the announcement of the formation of the so-called "Grand Alliance," made up of member companies from the original four proponent groups (AT&T, the David Sarnoff Research Center, General Instrument Corp., MIT, North American Philips, Thomson Consumer Electronics, and Zenith). This alliance finally produced what was to become the US HDTV standard (adopted in 1996), and which today is generally referred to as the "ATSC" (Advanced Television Systems Committee, the descendent of the original NTSC group) system.

Before examining the details of this system, we should review progress in Europe over this same period. The Eureka-95 project mentioned above can in many ways be seen as the European "Grand Alliance," as it represented a cooperative effort of many government and industry entities, but it was not as successful as its American counterpart. Europe had already developed an enhanced television system for direct satellite broadcast, in the form of a Multiplexed Analog Components (MAC) standard, and the Eureka program was determined to build on this with the development of a compatible, high-definition version (HD-MAC). A 1250-line, 2:1 interlaced, 50 Hz field rate production standard was developed for HD-MAC, with the resulting signal again fit into a standard DBS channel through techniques similar to those employed by MUSE. However, the basic MAC DBS system itself never became well established in Europe, as the vast majority of DBS systems instead used the conventional PAL system. The HD-MAC system was placed in further jeopardy by the introduction of an enhanced, widescreen augmentation of the existing PAL system. By early 1993, Thomson and Philips – two leading European manufacturers of both consumer television receivers and studio equipment – announced that they were discontinuing efforts in HD-MAC receivers, in favor of widescreen PAL. Both Eureka-95 and HD-MAC were dead.

As in the US, European efforts would now focus on the development of all-digital HDTV systems. In 1993, a new consortium, the Digital Video Broadcasting (DVB) Project, was formed to produce standards for both standard-definition and high-definition digital television transmission in Europe. The DVB effort has been very successful, and has generated numerous standards for broadcast, cable, and satellite transmission of digital television, supporting formats that span a range from somewhat below the resolution of the existing European analog systems, to high-definition video comparable to the US system. However, while there is significant similarity between the two, there are again sufficient differences so as to make the American and European standards incompatible at present. As discussed later, the most significant differences lie in the definitions of the broadcast encoding and modulation schemes used, and there remains some hope that these eventually will be harmonized. A worldwide digital television standard, while yet to be certain, remains the goal of many.

During the development of all of the digital standards, interest in these efforts from other industries – and especially the personal computer and film industries – was growing rapidly. Once it became clear that the future of broadcast television was digital, the computer industry saw a clear interest in this work. Computer graphics systems, already an important part of television and film production, would clearly continue to grow in these applications, and personal computers in general would also become new outlets for entertainment video. Many predicted the convergence of the television receiver and home computer markets, resulting in combination "digital appliances" which would handle the tasks of both. The film industry also had obvious interests and concerns relating to the development of HDTV. A large-screen, wide-screen, and high-definition system for the home viewer could potentially impact the market for traditional cinematic presentation. On the other hand, if its standards were sufficiently capable, a digital video production system could potentially augment or even replace traditional film production. Interested parties from both became vocal participants in the HDTV standards arena, although often pulling the effort in different directions.

12.3 HDTV Formats and Rates

One of the major areas of contention in the development of digital and high-definition television standards was the selection of the standard image formats and frame/field rates.

The only common goal of HDTV efforts in general was the delivery of a "higher resolution" image – the transmitted video would have to provide more scan lines, with a greater amount of detail in each, than was available with existing standards. In "digital" terms, more pixels were needed. But exactly how much of an increased was needed to make HDTV worthwhile, and how much could practically be achieved?

Using the existing broadcast standards – at roughly 500–600 scan lines per frame – as a starting point, the goal that most seemed to aim for was a doubling of this count. "HDTV" was assumed to refer to a system that would provide about 1000–1200 lines, and comparable resolution along the horizontal axis. However, many argued that this overlooked a simple fact of life for the existing standards: that they were not actually capable of delivering the assumed 500 or so scan lines worth of vertical resolution. As was covered in Chapter 8, several factors combine to reduce the delivered resolution of interlaced systems such as the traditional analog television standards. It was therefore argued that an "HD" system could provide the intended "doubling of resolution" by employing a progressive-scan format of between 700 and 800 scan lines per frame. These would also benefit from the other advantages of a progressive system over interlacing, such as the absence of line "twitter" and no need to provide for the proper interleaving of the fields. So two distinct classes of HDTV proposals became apparent – those using 2:1 interlaced formats of between roughly 1000 and 1200 active lines, and progressive-scan systems of about 750 active lines.

Further complicating the format definition problem were requirements, from various sources, for the line rate, frame or field rates, and/or pixel sampling rates to be compatible with existing standards. The computer industry also weighed in with the requirement that any digital HD formats selected as standards employ "square pixels" (see Chapter 1), so that they would be better suited to manipulation by computer-graphics systems. Besides the image format, the desired frame rate for the standard system was the other major point of concern. The computer industry had by this time already moved to display rates of 75 Hz and higher, to meet ergonomic requirements for "flicker-free" displays, and wanted an HDTV rate that would also meet these. The film industry had long before standardized on frame rates of 24 fps (in North America) and 25 fps (in Europe), and this also argued for a frame or field rate of 72 or 75 Hz for HDTV. Film producers were also concerned about the choice of image formats – HDTV was by this time assumed to be "widescreen," but using a 16:9 aspect ratio that did not match any widescreen film format. Alternatives up to 2:1 were proposed.

But the television industry also had legacies of its own: the 50 Hz field rate common to European standards, and the 60 (and then 59.94+) Hz rate of the North American system. The huge amount of existing source material using these standards, and the expectation that broadcasting under the existing systems would continue for some time following the introduction of HDTV, argued for HD's timing standards to be strongly tied to those of standard broadcast television. Discussions of these topics, throughout the late 1980s and early 1990s, often became quite heated arguments.

In the end, the approved standards in both the US and Europe represent a compromise between these many factors. All use – primarily – square-pixel formats. And, while addressing the desire for progressive-scan by many, the standards also permit the use of some interlaced formats as well, both as a matter of providing for compatibility with existing source material at the low end, and as a means of permitting greater pixel counts at the high. The recognized transmission formats and frame rates of the major HDTV standards in the world as of this writing are listed in Table 12-2. (It should be noted that, technically, there are no format and rate standards described in the official US HDTV broadcast specifications. The information shown in this table represents the last agreed-to set of format and rate standards proposed by the Grand Alliance and the ATSC. In a last-minute compromise, due to objections raised by certain computer-industry interests, this table was dropped from the proposed US rules. It is, however, still expected to be followed by US broadcasters for their HDTV systems.) Note that both systems include support for "standard definition" programming. In the DVB specifications, the 720 × 576 format is a common "square-pixel representation of 625/50 PAL/SECAM transmissions. The US standard supports two "SDTV" formats: 640 × 480, which is a standard "computer" format and one that represents a "square-pixel" version of 525/60 video, and 720 × 480. The latter format, while not using "square" pixels, is the standard for 525/60 DVD recordings, and may be used for either 4:3 (standard aspect ratio) or 16:9 "widescreen" material. It should also be noted that, while the formats shown in the table are those currently expected to be used under these systems, there is nothing fundamental to either system that would prevent the use of the other's formats and rates. Both systems employ a packetized data-transmission scheme and use very similar compression methods. (Again, the biggest technical difference is in the modulation system used for terrestrial

Table 12-2 Common "HDTV" broadcast standard formats.

Standard	Image format (H × V)	Rates/scan format	Comments
US "ATSC" HDTV broadcast standard	640 × 480	60/59.94 Hz; 2:1 interlaced	"Standard definition" TV. Displayed as 4:3 only
	720 × 480	60/59.94 Hz; 2:1 interlaced	SDTV; std. DVD format. Displayed as 4:3 or 16:9
	1280 × 720	24/30/60 Hz;[a] progressive	Square-pixel 16:9 format
	1920 × 1080	24/30/60 Hz;[a] progressive and 2:1 interlaced	Square-pixel 16:9 format. 2:1 interlaced at 59.94/60 Hz only
DVB (as used in existing 625/50 markets)	720 × 576	[b]	SDTV; std. DVD format for 625/50 systems
	1440 × 1152	[b]	2× SDTV format; non-square pixels
	1920 × 1152	[b]	1152-line version of common 1920 × 1080 format; non-square at 16:9
	2048 × 1152	[b]	Square-pixel 16:9 1152-line format
Japan/NHK "MUSE"	1440 × 1035 (effective)	59.94 Hz; 2:1 interlaced	Basically an analog system; will be made obsolete by adoption of an all-digital standard

[a] The ATSC proposal originally permitted transmission of these formats at 24.00, 30.00 and 60.00 frames or fields per second, as well as at the so-called "NTSC-compatible" (N/1.001) versions of these.

[b] In regions currently using analog systems based on the 625/50 format, DVB transmissions would likely use 25 or 50 frames or fields per second, either progressive-scan or 2:1 interlaced. However, the DVB standards are not strongly tied to a particular rate or scan format, and could, for example, readily be used at the "NTSC" 59.94+ Hz field rate.

broadcast.) There is still some hope for reconciliation of the two into a single worldwide HDTV broadcast standard.

12.4 Digital Video Sampling Standards

"Digital television" generally does not refer to a system that is completely digital, from image source to output at the display. Cameras remain, to a large degree, analog devices (even the CCD image sensors used in many video cameras are, despite being fixed-format, fundamentally analog in operation), and of course existing analog source material (such as video tape) is very often used as the input to "digital" television systems. To a large degree, then, digital television standards have been shaped by the specifications required for sampling analog video, the first step in the process of conversion to digital form.

Three parameters are commonly given to describe this aspect of digital video – the number and nature of the samples. First, the sampling clock selection is fundamental to any video digitization system. In digital video standards based on the sampling of analog signals, this clock generally must be related to the basic timing parameters of the original analog stan-

230 THE IMPACT OF DIGITAL TELEVISION AND HDTV

dard. This is needed both to keep the sampling grid synchronized with the analog video (such that the samples occur in repeatable locations within each frame), and so that the sampling clock itself may be easily generated from the analog timebase. With the clock selected, the resulting image format in pixels is set – the number of samples per line derives from the active line period divided by the sample period, and the number of active lines is presumably already fixed within the existing analog standard.

12.4.1 Sampling structure

Both of these will be familiar to those coming from a computer-graphics background, but the third parameter is often the source of some confusion. The *sampling structure* of a digital video system is generally given via a set of three numbers; these describe the relationship between the sampling clocks used for the various components of the video signal. As in analog television, many digital TV standards recognize that the information relating to color only does not need to be provided at the same bandwidth as the luminance channel, and therefore these "chroma" components will be *subsampled* relative to the luminance signal. An example of a sampling structure, stated in the conventional manner, is "YC_RC_B, 4:2:2", indicating that the color-difference signals "C_R" and "C_B" are each being sampled at one-half the rate of the luminance signal Y. (The reason for this being "4:2:2" and not "2:1:1" will become clear in a moment.) Note that "C_R" and "C_B" are commonly used to refer to the color difference signals in digital video practice, as in "YC_RC_B" rather than the "YUV" label common in analog video.

12.4.2 Selection of sampling rate

Per the Nyquist sampling theorem, any analog signal may be sampled and the original information fully recovered only if the sampling rate exceeds a lower limit of one-half the bandwidth of the original signal. (Note that the requirement *is* based on the bandwidth, and not the upper frequency limit of the original signal in the absolute sense.) If we were to sample, say, the luminance signal of a standard NTSC transmission (with a bandwidth restricted to 4.2 MHz), the minimum sampling rate would be 8.4 MHz. Or we might simply require that we properly sample the signal within the standard North American 6 MHz television channel, which then gives a lower limit on the sampling rate of 12 MHz. However, as noted above, it is also highly desirable that the selected sampling rate be related to the basic timing parameters of the analog system. This will result in a stable number and location of samples per line, etc.

A convenient reference frequency in the color television standards, and one that is already related to (and synchronous with) the line rate, is the color subcarrier frequency (commonly 3.579545+ MHz for "NTSC" system, or 4.433618+ MHz for PAL). These rates were commonly used as the basis for the sampling clock rate in the original digital television standards. To meet the requirements of the Nyquist theorem, a standard sampling rate of four times the color subcarrier was used, and so these are generally referred to as the "$4f_{sc}$" standards. The basic parameters for such systems for both "NTSC" (525 lines/frame at a 59.94 Hz field rate) and "PAL" (625/50) transmissions are given in Table 12-3. Note that these are intended to be used for sampling the composite analog video signal, rather than the separate components, and so no "sampling structure" information, per the above discussion, is given.

Table 12-3 $4f_{sc}$ sampling standard rates and the resulting digital image formats.

Parameter	525/60 "NTSC" video	625/50
Color subcarrier (typ.; MHz)	3.579545	4.433619
4 times color subcarrier (sample rate)	14.318182	17.734475
No. of samples per line (total)	910	1135
No. of samples per line (active)	768	948
No. of lines per frame (active)	485	575

12.4.3 The CCIR-601 standard

The $4f_{sc}$ standards suffer from being incompatible between NTSC and PAL versions. Efforts to develop a sampling specification that would be usable with both systems resulted in CCIR Recommendation 601, "Encoding Parameters for Digital Television for Studios." CCIR-601 is a component digital video standard which, among other things, establishes a single common sampling rate usable with all common analog television systems. It is based on the lowest common multiple of the line rate for both the 525/60 and 625/50 common timings. This is 2.25 MHz, which is 143 times the standard 525/60 line rate (15,734.26+ Hz), and 144 times the standard 625/50 rate (15,625 Hz). The minimum acceptable luminance sampling rate based on this least common multiple is 13.5 MHz (six times the 2.25 MHz LCM), and so this was selected as the common sampling rate under CCIR-601. (Note that as this is a component system, the minimum acceptable sampling rate is set by the luminance signal bandwidth, not the full channel width.) Acceptable sampling of the color-difference signals could be achieved at one-fourth this rate, or 3.375 MHz (1.5 times the LCM of the line rates). The 3.375 MHz sampling rate was therefore established as the actual reference frequency for CCIR-601 sampling, and the sampling structure descriptions are based on that rate. Common sampling structures used with this rate include:

- 4:1:1 sampling. The luminance signal is sampled at 13.5 MHz (four times the 3.375 MHz reference), while the color-difference signals are sampled at the reference rate.
- 4:2:2 sampling. The luminance signal is sampled at 13.5 MHz (four times the 3.375 MHz reference), but the color-difference signals are sampled at twice the reference rate (6.75 MHz). This is the most common structure for studio use, as it provides greater bandwidth for the color-difference signals. The use of 4:1:1 is generally limited to low-end applications for this reason.
- 4:4:4 sampling. All signals are sampled at the 13.5 MHz rate. This provides for equal bandwidth for all, and so may also be used for the base RGB signal set (which is assumed to require equal bandwidth for all three signals). It is also used in YC_RC_B applications for the highest possible quality.

The resulting parameters for CCIR-601 4:2:2 sampling for both the "NTSC" (525/60 scanning) and "PAL" (625/50 scanning) systems are given in Table 12-4.

232 THE IMPACT OF DIGITAL TELEVISION AND HDTV

Table 12-4 CCIR-601 sampling for the 525/60 and 625/50 systems and the resulting digital image formats.

Parameter	525/60 "NTSC" video	625/50
Sample rate (luminance channel)	13.5 MHz	13.5 MHz
No. of Y samples per line (total)	858	864
No. of Y samples per line (active)	720	720
No. of lines per frame (active)	480	576
Chrominance sampling rate (for 4:2:2)	6.75 MHz	6.75 MHz
Samples per line, C_R and C_B (total)	429	432
Samples per line, C_R and C_B (active)	360	360

12.4.4 4:2:0 Sampling

It should be noted that another sampling structure, referred to as "4:2:0," is also in common use, although it does not strictly fit into the nomenclature of the above. Many digital video standards, and especially those using the MPEG-2 compression technique (which will be covered in the following section) recognize that the same reasoning that applies to subsampling the color components in a given line – that the viewer will not see the results of limiting the chroma bandwidth – can also be applied in the vertical direction. "4:2:0" sampling refers to an encoding in which the color-difference signals are subsampled as in 4:2:2 (i.e., sampled at half the rate of the luminance signal), but then also averaged over multiple lines. The end result is to provide a single sample of each of the color signals for every four samples of luminance, but with those four samples representing a 2 × 2 array (Figure 12-1) rather

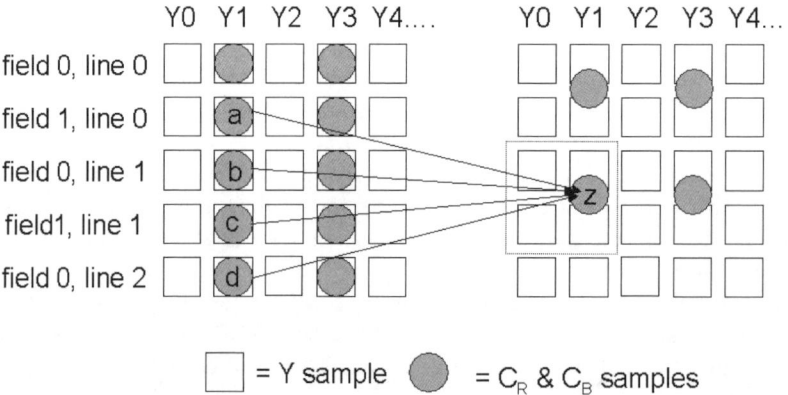

Figure 12-1 4:2:2 and 4:2:0 sampling structures. In 4:2:2 sampling, the color-difference signals C_R and C_B are sampled at half the rate of the luminance signal Y. So-called "4:2:0" sampling also reduces the effective bandwidth of the color-difference signals in the vertical direction, by creating a single set of C_R and C_B samples for each 2 × 2 block of Y samples. Each is derived by averaging data from four adjacent lines in the complete 4:2:2 structure (merging the two fields), as shown; the value of sample z in the 4:2:0 structure on the right is equal to $(a + 3b + 3c + d)/8$. The color-difference samples are normally considered as being located with the odd luminance sample points.

than four successive samples in the same line (as in 4:1:1 sampling). This is most often achieved by beginning with a 4:2:2 sampling structure, and averaging the color-difference samples over multiple lines as shown. (Note that in this example, the original video is assumed to be 2:1 interlace scanned – but the averaging is still performed over lines which are adjacent in the complete frame, meaning that the operation must span two successive fields.)

12.5 Video Compression Basics

As was noted in Chapter 6, one of the true advantages of a digital system over its analog counterpart is the ability to apply digital processing to the signal. In the case of broadcast television, digital processing was necessary in order to deliver the expected increase in resolution and overall quality, while still sending the transmission over a standard television channel. In short, it was the availability of practical and relatively low-cost digital compression and decompression hardware that made digital HDTV possible. This same capability can also be exploited in another way, and has in commercial systems – by reducing the channel capacity required for a given transmission, digital compression also permits multiple standard-definition transmissions, along with additional data, to be broadcast in a single standard channel.

Consider one frame in one standard "HD" format – a 1920 × 1080, 2:1 interlaced transmission. If we assume an RGB representation at 8 bits per color, each frame contains almost 50 million bits of data; transmitting this at a 60 Hz frame rate would require almost a 375 Mbytes/s sustained data rate. Clearly, the interlaced scanning format will help here, reducing the rate by a factor of two. We can also change to a more efficient representation; for example, a YC_RC_B signal set, with 8 bits/sample of each, and then apply the 4:2:2 subsampling described above. This would reduce the required rate by another third, to approximately 124 Mbytes/s, or just under 1 Gbit/s.

But by Shannon's theorem for the data capacity of a bandlimited, noisy channel, a 6 MHz TV channel is capable of carrying not more than about 20 Mbit/s. (This assumes a signal-to-noise ratio of 10 dB, not an unreasonable limit for television broadcast.) Transmitting the HDTV signal via this channel, as was the intention announced by the FCC in calling for an all-digital system, will require a further reduction or compression of the data by a factor of about 50:1! (Note that this can also be expressed as requiring that the transmitted data stream correspond to *less than one bit per pixel* of the original image.) Digital television, and especially digital HDTV, requires the use of sophisticated compression techniques.

There are many different techniques used to compress video transmissions. Entire books can, and have, been written to describe these in detail, and we will not be able to duplicate that depth of coverage here. However, a review of the basics of compression, and some specific information regarding how digital video data is compressed in the current DTV and HDTV standards, is needed here.

Recall (from Chapter 5) that compression techniques may be broadly divided into two categories: *lossless* and *lossy*, depending on whether or not the original data can be completely recovered from the compressed form of the information (assuming no losses due to noise in the transmission process itself). Lossless compression is possible only when redundancy exists in the original data. The redundant information may be removed without impacting the receiver's ability to recover the original, although any such process generally increases the sensitivity of the transmission to noise and distortion. (This must be true, since the redundant information is what gives the opportunity for "error correction" of any type.) Lossy compression methods are those which actually remove information from the transmission, relative to the original data, in addition to removing redundancy. In most cases, lossy compression methods are acceptable only when the original transmission can be analyzed, and classes of information within it distinguished according to how important they are to the receiver (or to the end user). For example, many digital audio systems employ a compression method in which information representing sounds at too low a level, relative to other tones that would be expected to "mask" them, is removed from the data stream.

It should be clear at this point that the practice of limiting the bandwidth of the color-difference signals (and subsampling them in the digital case) is an example of such a system. Information relating to color *is* being lost in order to "fit the signal into the channel". But this loss is accepted, as it is known that the eye places much more importance on the luminance information and will not miss the high-frequency portions of the color signals. Interlacing has also been cited as an example of a crude compression scheme in the analog domain, one which is lossless for static images but which gives up resolution along the vertical axis for moving objects (among other problems). Other techniques that may be seen as simple types of compression include:

- Full or partial suppression of sidebands (theoretically lossless; removes redundancy). This is used, as noted, in standard analog television broadcasting.
- Removal of portions of the transmission corresponding to "idle" periods of the original signal. For example, in a digital system, there is no need to transmit data corresponding to the blanking periods of the original signal, as the receiver can restore these as long as the individual lines/fields are still distinguishable.
- Run-length encoding of digital data. This is exactly what the name implies; in systems where significantly long runs of a steady value (either 1 or 0) may be expected to be produced, transmitting the length of these runs rather than the raw data will often save a significant amount of capacity. This is a lossless technique.
- Variable-length coding (VLC). Given that not all values are equally likely in a given system, a VLC scheme assigns short codes to the most likely of these values and longer codes to the less likely. An excellent example is "Morse" (actually "International") radio code, in which the patterns of dots and dashes were assigned roughly in accordance with the frequency of the letters in English. Thus, the letter "E" is represented by a single dot ("."), while a relatively uncommon letter such as "X" is represented by a longer pattern of dots and dashes ("–.–")
- Quantization. While the act of quantization is not necessarily lossy (if the quantization error is significantly smaller than the noise in the original signal), quantization can also be used as a lossy compression scheme. In this case, the loss occurs in the deletion of information that would otherwise correspond to lower-order bits. (The loss, or error, so introduced may be reduced through dithering the values of the remaining bits over multiple samples.)

In the compression method most commonly used in present-day digital video practice (the "MPEG-2" system; the abbreviation stands for the Motion Picture Experts Group), a combination of several of these, along with a transform intended to reduce the impact of the compression losses, are employed. The remainder of this section examines the details of this system as it is commonly implemented in current DTV/HDTV standards.

12.5.1 The discrete cosine transform (DCT)

The basis for the compression method used in current digital television standards, although not technically a compression method itself, is the Discrete Cosine Transform, or DCT. This might best be seen as a two-dimensional, discrete-sample analog to the familiar Fourier transform, which permits the conversion of time-domain signals to the frequency domain, and vice versa. The two-dimensional DCT, as used here, transforms spatial sample information from a fixed-size two-dimensional array into an identically sized array of coefficients. These coefficients indicate the relative content of the original sample set in terms of discrete spatial frequencies in X and Y. Mathematically, the discrete cosine transform of any $N \times N$ two-dimensional set of values $f(j,k)$ (where j and k may have values from 0 to $N-1$) is given by

$$F(u,v) = \frac{4C(u)4C(v)}{N^2} \sum_{j=C}^{N-1} \sum_{k=C}^{N-1} f(j,k) \cos\left(\frac{(2j+1)v\pi}{2N}\right) \cos\left(\frac{(2k+1)u\pi}{2N}\right)$$

where $C(x)$ is defined as $1/(\sqrt{2})$ for $x = 0$, and 1 for $x = 1,2,...,N-1$.

As commonly used in digital television standards, the DCT process operates by dividing the original set of samples – the pixels of the image – into blocks of 8 pixels by 8 lines each. (The 8×8 block size is a compromise between the desire for significant compression, the requirement to be able to perform this process at video-suitable rates, and the requirement to maintain an acceptable level of delivered image quality. It will be apparent that these same methods could be applied to differently sized blocks) These are transformed via the above to 8x8 blocks of DCT coefficients. Note that this is a fully reversible operation, assuming that arbitrary precision can be obtained throughout, and so the DCT itself does not involve a loss of information. (In practical terms, when this is implemented digitally, the calculations must be carried out using a greater number of bits than were present in the original samples, to avoid loss due to truncation or rounding. In the case of 8-bit input values, the coefficients must be allowed at least 12 bits (signed) each for the process to be reversible. If fewer than this number of bits is provided, the DCT becomes irreversible without loss.) Note also that the DCT can result in negative values for the coefficients (due to the use of the cosine function).

As the DCT operating on 8×8 blocks of pixels produces coefficients relating to 8 discrete spatial frequencies each in X and Y, the coefficients in each cell of the resulting 8×8 array are best viewed as giving the relative weight of each of 64 *basis functions*, as shown in Figure 12-2. These images show the appearance of the combination of the separate X and Y waveforms corresponding to each point in the 8x8 array. The original image therefore may be recovered by summing these basis functions in accordance with this relative weighting.

12.5.1.1 Weighting, Quantization, Normalization, and Thresholding
As noted, the DCT itself is not a compression technique, and (as long as sufficient accuracy is maintained in the coefficients) neither results in a loss of data or a reduction in the data rate. However, it places the image information into a form that is now easier to compress without significant impact in the final image quality. The array of coefficients so generated represents the spatial frequency content of the original block of pixels, from the DC component as the top-leftmost coefficient, to the highest frequency in both X and Y at the bottom

236 THE IMPACT OF DIGITAL TELEVISION AND HDTV

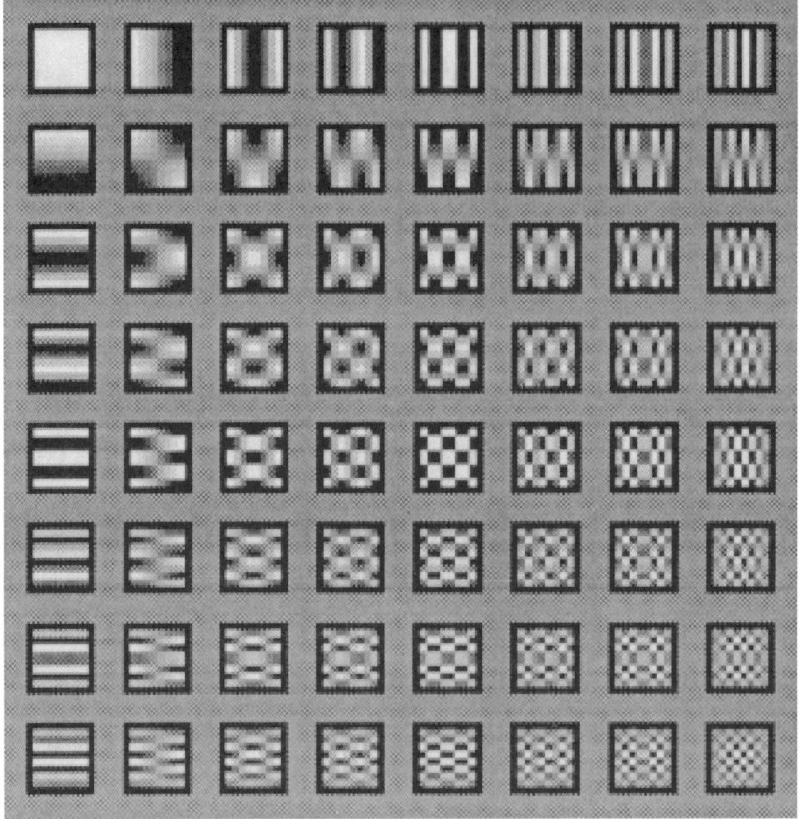

Figure 12-2 The 8 × 8 basis functions of the discrete cosine transform (DCT) compression method.

right. In terms of importance to the overall image, it can be assumed that the DC component is the most important value for this block, as the higher frequency coefficients represent finer detail. Thus, the first step following the transform in a DCT-based compression is to assign "weights" to the coefficients based on their relative importance. The weighting of each is generally specified in the particular compression standard in use. Following weighting of the coefficients, they will be quantized and truncated (converted to a specific bit length for each, which may not be constant across the full set), normalized per the requirements of the standard or system in use, and subjected to "thresholding." The thresholding step sets low-value coefficients – those below a specified threshold, which again may vary across the full set – to zero. These steps are illustrated in Figure 12-3.

12.5.1.2 Encoding

At this point, some compression of the original data may have been achieved, as information corresponding to the higher spatial frequencies may have been eliminated or reduced in importance (via the quantization and thresholding processes). Further compression may now be achieved by noting that the resulting coefficients for these upper frequencies are those most likely to have a zero value. Thus, the information in this array of coefficients may be most efficiently transmitted through a "zig-zag" ordering of the data (per Figure 12-3c), and apply-

COMPRESSION OF MOTION VIDEO 237

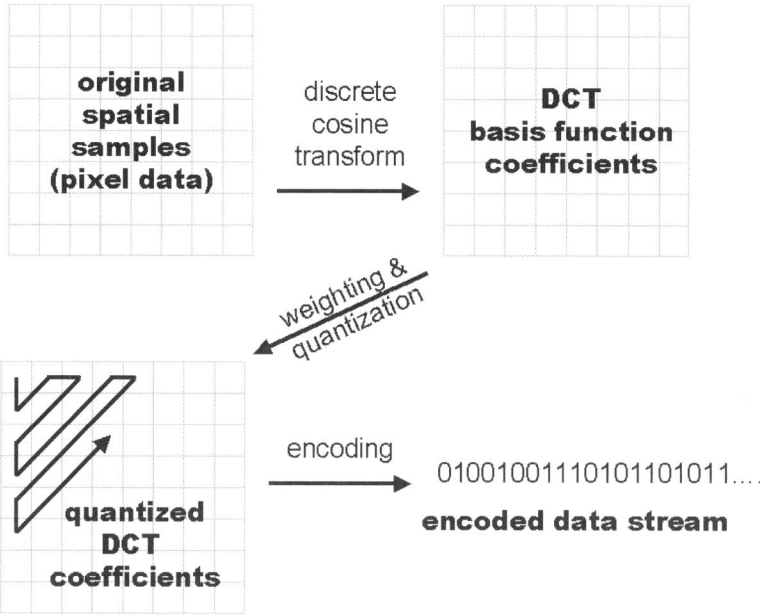

Figure 12-3 Outline of the complete DCT-based compression process. A block of pixels of a predefined size (in this case, 8 × 8) is transformed to an equivalent-sized array of DCT basis function coefficients. This step is lossless, assuming sufficient precision in the calculation of the coefficients. These are then weighted and quantized – a lossy step – and re-ordered in the "zig-zag" fashion shown. This maximizes the length of runs of zero values, as these are more likely in the higher-frequency coefficients. Finally, the sequence is run-length encoded for further compression.

ing a combination of run-length and variable-length coding techniques to the resulting series of values. The "zig-zag" ordering tends to maximize the length of zero runs, making for the most effective compression in this manner.

By itself, a DCT-based compression method as described here can result in a compression ratio of up to about 20:1. (It is important to note that the parameters of DCT compression, such as the weighting table, quantization, etc., may be adjusted to trade off image quality for compression, and vice versa. This can even be done "on the fly" to meet varying requirements of the transmission channel, source material, and so forth.) But so far, we have looked only at compressing the data in a given single image – which might be a still picture, at this point. Further compression is required for HDTV applications, and is achieved by noting the redundancies that exist in motion video.

12.6 Compression of Motion Video

The DCT-based compression methods described up to this point can achieve a significant savings for single, isolated images (and in fact are used in such applications, as in the JPEG – Joint Photographic Experts Group – format). However, television is a medium for transmitting *moving* images, as a series of successive stills, and we must also ask if further compression can be achieved by taking advantage of this. In the most extreme case, which is the television transmission of a still picture, the answer is clearly yes – conventional TV broadcasting practice would be sending the same picture over and over again, an obvious case of extreme redundancy. It would be far better to simply send the picture once, and not send anything further until a change in the image had occurred.

But even "moving" pictures exhibit this same sort of redundancy to a large degree. In the vast majority of cases, there is still considerable similarity between successive frames of a film or a video transmission. An object may move, but the appearance of the object itself has changed little if any from one frame to the next. In the simplest example of this, consider "motion" which occurs by panning a camera across a scene comprising only still objects. Some new information enters the frame at one side, while "old" information leaves at the other, but the majority of the image could easily be described as "same as before, but shifted by *this* amount in *that* direction." Sending such a description, plus the small amount of "new" data which has entered the image, is clearly more efficient that sending a complete new frame.

In practice, these concepts may be extended and generalized through the use of *motion estimation/prediction* techniques, applied to the same blocks of pixels as were used in the DCT transform above, and through the transmission of *difference* or *error* information to be used in correcting the resulting "predicted" frames. This technique is shown in the series of illustrations given in Figure 12-4.

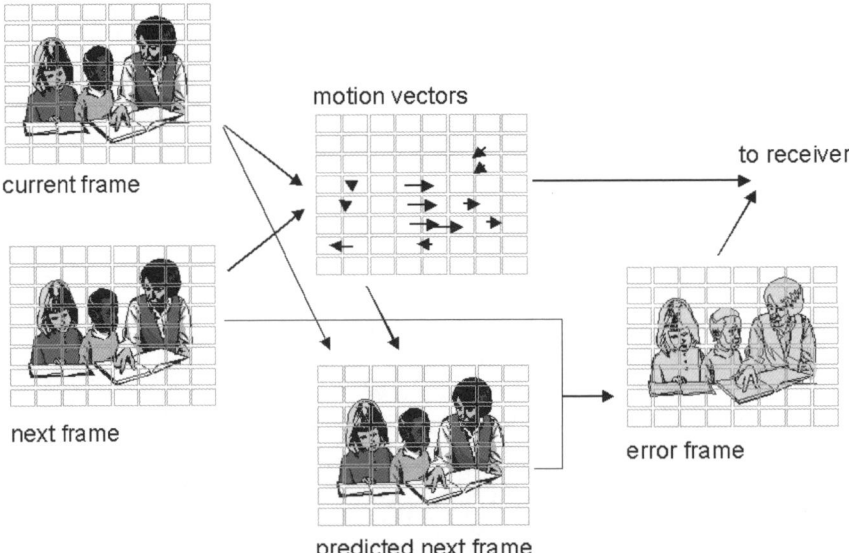

Figure 12-4 Motion prediction. Rather than compressing and transmitting each frame individually, digital television systems achieve further efficiency gains through the use of motion prediction. Here, *motion vectors* are calculated for each block of the original (as used in the DCT compression process described earlier) by comparing adjacent frames. These give the average motion for the pixels in that block. These vectors may be used to produce a predicted next frame, which is then compared to the actual next frame. The errors found in this comparison are themselves compressed and transmitted to the receiver along with the motion vectors, and together permit the receiver to generate an fairly accurate version of the next frame.

The process begins by assigning a *motion vector* to each of the previously determined blocks of pixels within the image. This is done by comparing successive frames or fields of the series being transmitted, and attempting to determine the new (or previous) location of each block. That determination is made by minimizing the error resulting from treating a given frame as only the translation of blocks in the preceding or following frame. Note the it is assumed that all pixels in a given block experience the same displacement, and that the motion of each block is assumed to consist only of translation in X or Y – the block size and shape does not change, nor are possible rotational motions considered. With a motion vector assigned to each block, it now becomes possible to create a *predicted frame* based solely on this information and the blocks from the previous frame. (Note: the MPEG compression system actually involves some *bidirectional* motion prediction, in which frames are predicted not only from the previous frame, but also from the next. Throughout this discussion, one should keep in mind that the techniques discussed are also often applied in the "reverse" direction as well as the forward.)

However, it is clear that such a predicted frame would have significant errors if compared to the actual next frame in the original series; the pixels do not all move uniformly within a given block, and there will be areas of the "new" frame containing information not present in the original frame. But these errors may be compensated for to a large degree. Since it may safely be assumed that both the source encoder/compression hardware and the receiver can produce identical "predicted" frames through the above process, the encoder will also compare this frame to the actual next frame in the series. This may be done by simply subtracting one frame from the other – the resulting non-zero values represent errors between the two. Such an "error frame," though, clearly requires less data transmitted than a complete "original" frame. Static portions of the image, or those moving areas where the translation of blocks does accurately describe the results, will show zero error – and so the error frame may most often be expected to be mostly zeroes. Thus, transmitting only the motion vectors and the error information should permit the receiver to generate a predicted frame that is a very good approximation to the "real" image which would otherwise have been transmitted at that time.

In the actual MPEG compression standards, three types of frames are recognized, as follows:

- The *I-frame*, or "intra" frame. This is a fully self-contained, compressed version of one frame in the original series. Simply put, the I-frame is the basis for the generation of all predicted frames by the receiver, until the receipt of the next I-frame. I-frames are not the result of any motion prediction; they are, instead, the "starting point" for the next series of such predictions.
- The *P-frame*, or "predicted" frame. P-frames are generated by the receiver, using the motion vector and error information supplied in the transmitted data, per the above description. In the transmission, multiple P-frames may be produced between I-frames, by sending additional motion vector and error information for each.
- The *B-frame*, or "bidirectionally predicted" frame, also known as the "between" frame. B-frames are produced by applying both forward motion prediction, based on the latest I-frame, and backward prediction, based on the P-frame generated from that I-frame and the additional motion and error information. B-frames may be viewed as interpolations between I- and P-frames, or between pairs of P-frames.

These are shown in Figure 12-5.

240 THE IMPACT OF DIGITAL TELEVISION AND HDTV

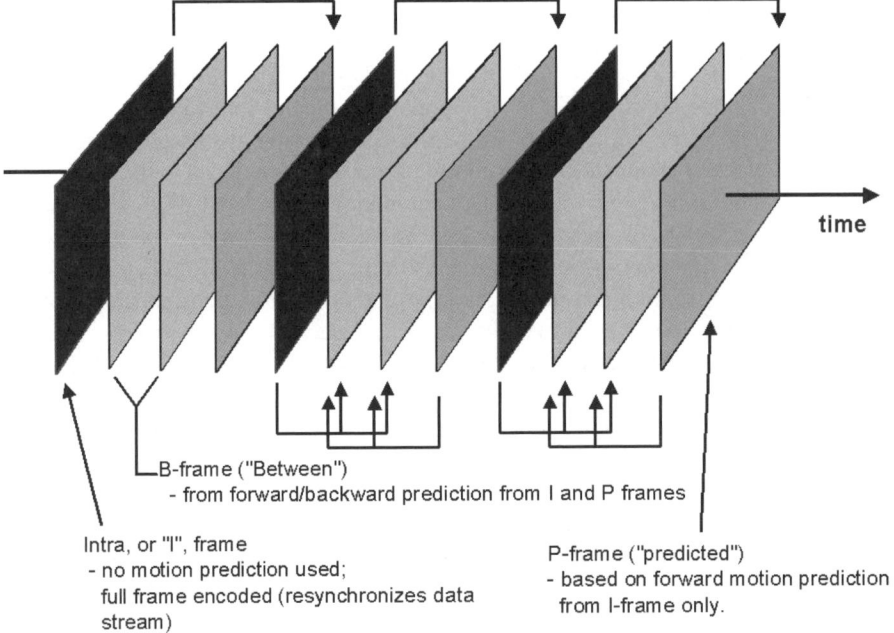

Figure 12-5 The stream of I, P, and B frames in a digital television transmission.

The number of P- and B-frames between I-frames is not fixed; these may be varied depending on the needs of the data channel, the source material, and the desired image quality. Clearly, a transmission of a static or nearly static image can be sent with relatively few I-frames without suffering a visible loss in image quality. Conversely, video containing a high degree of complicated motion may require fewer P- and B-frames between I-frames, and may have to demand higher compression (and the resulting loss in image quality) in the I-frame to compensate. Permitting the compression system to be adaptive in this manner allows it to continually adjust itself for the best overall image quality possible in a given situation.

This is not to say that the MPEG-2 compression method, as implemented in various HDTV, DBS, and other digital-television systems, always delivers a displayed image that is indistinguishable from the original. Trying to fit a very large amount of information into "too small a pipe" is never done without some impact on the quality of the transmission, and in practice compression artifacts and other errors can become quite visible. Momentary signal loss, or periods when the system cannot adjust its behavior rapidly enough, or those situations in which there simply is not sufficient channel capacity for the task at hand, will all result in visible errors. Due to the nature of this compression method, the most commonly seen errors include the visibility of the "block" structure, especially around areas of high detail and rapid motion, such as rapidly moving edges, or momentary corruption of the image (again in a visibly "blocky" manner) until enough new data has been received to "rebuild" it.

12.7 Digital Television Encoding and Transmission

One of the major benefits of using digital encoding for video transmission is that it can serve to completely decouple the image source, processing, transmission, and decoding and display processes. In the original analog TV systems, all of these are maintained in precise lock-step synchronization; the signal timing and the scanning format must remain absolutely identical throughout the chain. This is not the case in digital television. Given the mandatory processing, compression, decompression, etc., at various points in the system, it can always be assumed that sufficient frame storage and digital processing capability exists throughout so as to make these separable. The transmitted image does *not* have to use the same frame rate as the original source material. The displayed image format does *not* have to be the same as that of the transmission. And so forth. This advantage of a digital system is generally not recognized in the computer industry – since it has been a fact of life there since the beginning, it is not seen as remarkable. And in the early days of the HDTV standards development, the possibilities inherent in digital transmission were not always recognized by some in the television industry (and so there were often demands for timing and format compatibility with the existing systems, were none was truly required).

This decoupling of the major portions of the system also permits novel approaches in the transmission itself. In digital television systems, the image information is treated as just a "stream of bits" – and so long as this information is delivered to the receiver in time to keep the generation of displayable images there going, everything works fine. As long as capacity permits, additional information could be transmitted along with the video data, regardless of whether or not it is related to the video transmission. This additional data can be the audio programming associated with the television broadcast, supplementary data (such as subtitling, text-based information services, program guides, etc.) or even a completely separate video transmission. What enables this is the use of *packetization*, along with the transmission of *headers and descriptors*, in the digital transmission.

In the ATSC digital television system approved in the US, for example, all data is sent in fixed-length packets of 188 bytes each, including a 4-byte header. The header comprises one sync byte, which is the first byte in the packet, and three bytes of information that describe the content of the packet. These identify the type of information carried (video, audio, ancillary data, etc.) and the program stream to which it belongs (which permits multiple programs to be carried in a single transmission channel. The European DVB system employs a similar system of packetization, again permitting multiple data types and programs. Note that such systems completely avoid the problem of mutual interference between the "video", "audio" and other supplemental signals that plagued the earlier analog standards. There is only one signal transmitted, not multiple signals trying to share the same channel, and the data for each is completely separable by the receiver.

A significant difference remains between the ATSC and DVB systems in terms of the actual transmission method used. The ATSC specification requires the use of a vestigial-sideband AM transmission, with eight possible signal levels encoding three bits of information per transmitted symbol (referred to as an "8-VSB" system). This provides a capacity of approximately 19.4 Mbit/s of data at a 10.76 MHz symbol rate. (The transmitted data rate is not 3× the symbol rate, due to additional data overhead imposed by the addition of parity bytes for error correction and subsequent encoding. The raw bit rate of the encoded stream is 32.28 Mbit/s). A higher data-rate version of this system, using 16 levels to encode 4 bits per symbol (16-VSB) is also defined for cable transmission and other modes not subject to the constraints of conventional terrestrial broadcast.

In contrast, the DVB specifications define several different transmission methods, depending on the specific application. These include the following:

- DVB-T, which is the terrestrial broadcasting standard. This provides a roughly 24 Mbit/s capacity in standard 7 or 8 MHz television channels, and uses a Coded Orthogonal Frequency-Division Multiplexing (COFDM) transmission method. COFDM is a multiple-carrier approach in which the data to be transmitted is divided among many (thousands, in the DVB standard) of precisely spaced carrier signals, each of which is modulated at a relatively low rate.
- DVB-C, which is the DVB standard for cable transmission. As was the case for the 8-VSB/16-VSB versions of the ATSC system, this standard takes advantage of the expected higher SNR in the cable environment to provide a higher data capacity. DVB-C supports up to 38.1 Mbit/s transmission, using a 64-state (6 bits/symbol) Quadrature Amplitude Modulation (64-QAM) signal.
- DVB-S, for satellite transmission. DVB-S is a configurable standard, allowing variation in several parameters to meet the needs of different satellite transponders (differing bandwidths and power levels). The maximum capacity is again about 38.1 Mbit/s, using quadrature phase-shift keying (QPSK) modulation.

Outside of the differences in encoding and modulation, the various DVB systems are compatible and transmissions may readily be converted from one to the other, within capacity limitations.

The VSB vs. COFDM conflict is among the primary differences between the US ATSC and European DVB systems. Other differences exist, notably in the audio encoding systems used and the supported frame rates and image formats/aspect ratios, but these would be relatively easy to harmonize compared to the incompatibility of the transmission methods. Proponents of the COFDM technique can point to several advantages, including improved immunity to "multipath" reception problems, reduced inter-symbol interference, and resistance to potentially interfering signals from existing NTSC/PAL/SECAM broadcasting. However, the VSB system has already been deployed in the US, and this represents a not-insignificant investment. As of this writing, there is still some hope for resolution of this conflict such that a single worldwide standard may ultimately be possible. In the meantime, however, uncertainty over the future path for HDTV is causing some slowdown in its adoption.

It is important to note that both of these major digital television standards are *transmission* standards only. Neither establishes any requirements on the display device; there are no set specifications for display timing or refresh rate, for example, as was absolutely mandatory with analog broadcasting. Such things are completely up to the receiver or display manufacturer, and the forces of the market.

12.8 Digital Content Protection

One major disadvantage of the digital approach, at least from the perspective of the originators of programming, is that having material in digital form generally makes it relatively easy to generate unlimited, "perfect" copies of this material. When the material in question is copyrighted programming, this raises some very serious concerns. The owners of these copyrights, therefore, have imposed requirements on the digital video industry to ensure that their property remains protected from such unauthorized use or copying. The most common means of achieving this is through encryption of the digital video transmission, and then ensuring that it can only be decrypted by authorized devices (which is presumed *not* to include unauthorized recorders!).

There are currently two encryption schemes endorsed by the major content providers and the consumer electronics industry. These are the High-bandwidth Digital Content Protection (HDCP) specification, developed by Intel for use under the Digital Visual Interface standard (see Chapter 10), and Digital Transmission Content Protection (DTCP), also developed by Intel and others in the so-called "5C" ("Five Companies") group (the other members being Panasonic/Matsushita, Toshiba, Sony, and Hitachi). As the name of HDCP implies, the main difference between the two in terms of functionality is their intended application. HDCP was developed for a very-high-data-rate display interface (DVI). DTCP, on the other hand, requires more processing to decrypt, and so is better suited to the lower rates of standard-definition digital television applications, such as DVD players and digital VCRs.

While the exact details of either of these encryption standards is beyond the scope of this book, we can describe their general operation. Both actually provide two levels of protection for the copyrighted content. First, receiving devices are authenticated as being authorized to receive the data, through the exchange of "key" values which are kept secret by the manufacturers. (The systems allow for the revocation of individual key values, should the security of any become compromised.) Once the authorization of all receivers in the system is determined, the transmission of encrypted data is allowed to proceed. Encryption of the data by the transmitter, and decryption by the receiver also relies on the use of secret "key" information, although through somewhat different processes in the two systems.

In the HDCP scheme, the authentication process ends with the initialization of a pseudo-random number generator (PRNG) in both the transmitter and receiver. The PRNG produces a 24-bit value, the HDCP cipher, which is bitwise exclusive-ORed with the 24 bits of data to be carried over DVI's TMDS link. (In the case of a dual-link DVI interface, two separate ciphers are produced, and the XOR process duplicated for the two 24-bit links.) At the receiver, whose PRNG was similarly initialized, the identical cipher value is produced and bitwise-XORed with the encrypted stream to recover the original data. The cipher value is changed during the vertical blanking interval, such that each frame is transmitted and received using a different code. Note that the authentication process may be re-initiated (and so the HDCP cipher sequence reset to new pseudo-random values) at any time. This can be done at random intervals, to insure the continued authorization of the receiver, or at any time the transmitting device might have reason to believe the link to be compromised.

The HDCP system is strongly tied to the DVI (Digital Visual Interface) specification, and through it to the TMDS electrical interface standard (see Chapter 10). While HDCP is currently authorized for use only within DVI-compliant systems, the encryption and decryption hardware has become a de-facto standard feature of TMDS transmitter and receiver designs.

As noted above, the DTCP system is more compute-intensive, and so generally cannot be used with extremely high-data-rate transmissions. DTCP also involves the exchange of key information as part of the device authentication process, after which the encrypted content may be sent by the source and decrypted by the receiver based on these keys. However, a more complex encryption/decryption process is used, with Hitachi's "M6" cipher as the baseline algorithm. (Other ciphers may optionally be employed, but all DTCP-compliant devices must be capable of this baseline to ensure interoperability.) As with HDCP, the key values are regularly changed over the course of the video transmission, under the control of the source device, and there is a capability for revoking the keys of unauthorized or compromised devices.

In addition to the basic security features provided by both systems, the DTCP specification also provides for copy control, and can be set to permit unlimited copying, no copying whatsoever, or a "copy-once" authorization. This is done via a set of control bits in the header of each data packet set. The "copy –once" setting permits the legitimate copying of programming in certain situations (for example, the recording of a broadcast program for later playback, i.e., "timeshifting"). However, DTCP-compliant devices are required, when producing the first-generation copy, to reset the appropriate control bits in that material so that the fact that it has already been copied once is noted. Further copying of that material would then be prevented by the DTCP system.

Enforcement of these systems, in the sense of requiring devices to support them and comply with their restrictions, is achieved through licensing agreements between the owners of the fundamental technology and the manufacturers of the equipment. For instance, a manufacturer of a DVD player would be required to support DTCP decryption on that player's digital inputs, and HDCP on a supposed DVI output, through the terms of their license agreement for the basic DVD and DVI technologies. DTCP was originally developed for use with the IEEE-1394 serial interface, and is officially recognized by the organizations promoting that interface as the content-protection system of choice for it. However, DTCP is also usable with other digital interface systems, and there are no restrictions preventing its use outside of a 1394 environment. HDCP, on the other hand, is currently restricted to use only with the DVI display interface standard, and its use is mandatory in certain DVI applications under terms of the relevant licensing agreements.

12.9 Physical Connection Standards for Digital Television

As should be apparent from the preceding discussions, a major advantage of digital techniques over the earlier analog television standards is the decoupling of the major components of the overall system. Similarly, digital television has never been strongly tied to a particular transmission system or medium. There is as yet no true standard physical interconnect which is solely used for digital television (at least in the consumer market), and there may never be. Instead, digital TV, under a "bits are bits" philosophy, is more likely to share transmission systems, media, and physical connections with other markets and services.

In the case of consumer television products, such as video recorders, disc players, camcorders, etc., the IEEE-1394 interface has emerged as the de-facto standard for interconnection within the home system. This interface has also, of course, been widely adopted for home-networking standards, and has seen some adoption in the personal computer market (although not, to date, achieving the acceptance of the USB interface). This makes it possible, if not likely, that future digital television products will be just a part of an interconnected digital home system. More on this in a moment.

In terms of the display interface, the situation at present is somewhat unclear. Historically, the television industry has not followed the "separate display" model common in the personal computer market, and so has had few examples of systems provided a true display-only interface (as opposed to connections carrying baseband or RF-modulated video, which must still be demodulated/decoded by the receiver). This has begun to change somewhat with the introduction of digital TV devices and home recording/playback equipment. The possibility of multiple video sources in the home system has led to a growing number of television receivers that can also serve as "monitors" (relying on other products to perform the demodulation tasks for some inputs), or even truly dedicated monitor products with no video decoding capability of their own. This is especially true for the initial HDTV and digital DBS systems. Knowing that the customer is unlikely to want to give up a perfectly good existing television simply to receive these broadcasts, both have followed the model of providing a "set-top box" which can be connected to such receivers. This box has the required demodulating, decoding, and decrypting hardware, and outputs standard analog video to the receiver over the same connections as used for the earlier analog systems (such as the "S-Video", "F", or SCART connector standards; see Chapter 8). In the future, it will become desirable to have a digital interface to the display, as has become the case in the PC industry. It is currently expected that a form of the DVI interface will become the standard here, bringing with it the HDCP content-protection system.

12.10 Digital Cinema

While HDTV and digital video in general was once seen as a potential threat to the cinema, these same techniques are now driving a major revolution in the "film" industry. Digital production, editing, distribution, and ultimately presentation of motion pictures is expected to become the norm over the next 10–15 years, and significant steps have already been taken in each area.

A fully digital model offers significant advantages over the traditional film-based system, but has had to pass several hurdles in order to become a serious contender for this business. Among the chief benefits of this new system is the potential for much more efficient and lower-cost distribution of motion pictures (the term "films," while convenient, is hardly appropriate). The cost of the production and distribution of literally thousands of physical copies of a film is hardly insignificant; consider the costs of making and shipping just one 30 kg, multiple-reel copy to your local theatre. In a all-digital system, this would be replaced with secure, direct, simultaneous transmission, either via satellite or cable, of the material to all outlets. Each would receive an essentially perfect copy, and one that will not degrade with multiple showings. And, at the end of the run, there is no problem with the return or disposal of a physical film. Nor is there a problem with ensuring that this material – a valuable asset to the studio that produced it! – can be safely archived for future distributions or other uses.

Digital techniques also provide the motion picture producer with several potential advantages over traditional film, including simpler (and lossless) editing and duplication, and a clearly easier path for the incorporation of computer-generated effects. It even becomes possible to seamlessly merge "real-world" and computer-generated backgrounds, objects, and even characters. This has been demonstrated in numerous films that relied, either wholly or in part, on computer graphics for their content. Further developments along these lines include a move to actually shooting using high-definition digital video cameras, rather than shooting to film and then transferring the results to the digital environment. This approach is being pioneered by several, most notably George Lucas and Lucasfilm, Ltd.; Lucas' fifth "Star Wars" film (the second episode of that series' "prequel" trilogy), to be released in 2002, is the first in which all principal photography was done via digital video cameras.

(New 24 fps, progressive-scan cameras, using the 1920 × 1080 format, were developed by Sony and Panavision for this effort.) Earlier, Lucas had also made digital-cinema history by showing the previous film (*Star Wars Episode One: The Phantom Menace*) in four specially equipped digital cinemas in New York and Los Angeles in mid-1999. This marked the first commercial showing of a motion picture using digital projectors instead of film.

The projection equipment itself remains an obstacle in the path of the all-digital model for the cinema. Replacement of existing film projectors will require a significant investment on the part of theatre owners, but this is expected to proceed through a combination of immediate upgrading by those who desire "leading-edge" status, followed by the eventual replacement of aged film equipment by digital in other venues. Currently, digital video projection equipment suitable for commercial use has been demonstrated by Hughes-JVC and Texas Instruments (Figure 12-6). The Hughes-JVC projector (using their D-ILA, or "digital image light amplifier, devices) is in fact now being offered commercially. The TI projector is to date only a demonstration prototype, using their Digital Light Processing/Digital Micromirror Device (DLP/DMD) technology. TI, being a component producer, will no doubt commercialize this technology through one or more partnerships with outside projector manufacturers.

Figure 12-6 Texas Instruments' prototype DLP (Digital Light Processing) cinematic projector. (*Picture courtesy of Texas Instruments, Inc.; used by permission.*)

There have also, of course, been concerns as to how the delivered image quality of an all-digital system will compare with that of traditional film. As compared with most electronic display and projection technologies, film projection has generally enjoyed advantages in resolution, color, and contrast, and these have posed significant challenges to the establishment of an all-electronic system. However, great strides have been made in the color and contrast of electronic projection systems, and HDTV-like resolution is proving to be at least adequate in many venues. The resolution advantage of standard film is not as great as it may at first glance appear, in any case. While 35 mm film is capable of capturing images at very high resolution by electronic-display standards, much of this potential is lost in the actual delivered product. Degradation of the as-delivered resolution results from a number of factors, including the inevitable losses in the copying processes and in film registration and other errors, primarily at the projection step. When this is coupled with the fact that the digital projection system offers an extremely stable image, free of any the visible artifacts of film damage or wear, the overall viewing experience in the all-digital system can be at least comparable and often superior to traditional cinema.

12.11 The Future of Digital Video

Hopefully, it is clear at this point that the advent of digital television broadcasting has brought considerably more than just a higher-resolution image on the home TV screen. By its nature, digital broadcasting provides numerous capabilities that may even overshadow the concept of "high-definition" television per se. Higher image quality is certainly a desirable thing, but in many applications may be unnecessary. The other advantages of a digital transmission system and digital encoding represent the foundations of a much more pervasive revolution in home entertainment.

In fact, true "high-definition" televisions may remain a relatively small share of the market for some time to come. As was noted in Chapter 8, the original broadcast television standards were developed with a goal of achieving resolution suited to the expected screen size and viewing distances of the time. The analysis that went into these remains valid, and great increases in the line count, etc., beyond the "standard definition" level do not truly result in a significant improvement unless larger display screens are used. And while the average television certainly does use a much larger screen today than in the 1950s, it is still not large enough to really take advantage of a 1000-line format. The standard-definition modes of the new digital standards, such as 640×480 or 720×576, will provide a very noticeable improvement over standard analog broadcasts, as they can potentially provide higher delivered resolution with none of the problems of those earlier systems. Many broadcasters, in fact, see more potential in the SD formats than in HD programming. The digital systems permit multiple standard-definition programs to be broadcast in a single channel, rather than a single HD program, and this can mean greater revenues for the broadcaster at a lower investment.

The ability of digital transmission systems to carry content other than simply video is also likely to result in significant changes in several consumer markets. By treating video programming as "just bits", the same as any other content, and through the use of the packetized formats discussed here, the "television" transmission system can become the delivery vehicle for a much wider range of services. The reverse is true, as well – "television" programming can now be easily provided by any other carrier with a sufficiently high-capacity channel into the consumer's home. The lines between the television broadcaster, the cable-TV provider, the internet service provider, and even the traditional "telephone company" are becoming hazy indeed. This trend is also expected to have an impact on the appliances used by the consumer to access these sources of information. If "television", "telephone" and "computer" services are all delivered over a common carrier, what is the nature of the device connected to that line? Significant convergence amongst the various "information appliances" used in the home can reasonably be expected, although not necessarily in the same forms as were predicted in the 1990s for "PC/TV" convergence.

One final change that must yet occur to enable the full "digital video revolution" in the home is the establishment of high-capacity *bidirectional* channels within the home and between the home and the greater outside community. To date, the standard telephone service has been the only bidirectional information channel connecting the home to the outside world. This has been used as the "upstream" path by several other services, such as programming-on-demand cable and satellite TV systems and by high-speed internet access providers. This has been acceptable only while the usage model has remained one of far more data entering the home than leaving it; typically, for instance, the outbound information has consisted only of requests for information (web pages or television programs) that would then require the much-higher-capacity inbound channel for delivery. This could easily change to some degree, as among the changes happening now is the provision of more power to the home user to act as a content *source*, rather than just a consumer. Digital photography and video recording products make for "homemade" content that, while often viewed in the home, will also need to be shared with family and friends.

In the final chapter, we examine future trends for display devices and interfaces in general, and look at some further needs and issues which have yet to be addressed.

13

New Displays, New Applications, and New Interfaces

13.1 Introduction

Throughout this book, we have examined the development of both display technology and display interfaces from their early beginning to the current state of the industry. We should not, of course, expect the current state of affairs to represent the ultimate end of display development, and so it is reasonable at this point to consider what the future may bring.

Predicting the future is a notoriously hazardous occupation, at least in terms of being able to make predictions that will not be later shown to be slightly off the mark (at best) or even laughably wrong. If nothing else, then, this final chapter may have some entertainment value for those reading it in the days to come. But if we can agree to some basic trends and desires within the industry, and temper these with the limitations we know will still apply, we should be able to make at least some general statements regarding what we can reasonably expect. In some cases, we will even be able to go beyond this – efforts are underway right now toward some very specific goals, and we should be able to judge which of these are likely to be successful and result in new products and applications.

Several of our starting points – our initial assumptions regarding the needs and wants of various markets and their customers, the possibilities and limitations of new technologies, etc. – may seem obvious and even simplistic, but they're still worth repeating here if only to see them as part of the complete environment:

- First, electronic displays will obviously remain a very important part of our lives, and can reasonably be expected to grow into areas and applications not yet seen. This may be seen as a part of the overall trend to "smarter" products: as the cost of adding "intelligence," in the form of digital processors, storage, and connectivity, to products continues to decrease, even the most basic and mundane of tools and appliances are becoming "smart." And the more information a given device is capable of handling, for and on behalf of its user, the greater is the need to communicate some of that information to the user. As part of that process, those devices in which we already expect information display devices will grow more sophisticated; the product which has a simple, monochrome, text-only display this year may offer color and/or graphical imagery next, or even full-motion video capability. (An excellent example of this may be seen in the cellular telephone market – beginning with very simple displays capable of little more than showing a telephone number, these products have now developed into full-fledged information terminals, often with very sophisticated display capability.)

- With an increase in the amount of information being made available to, and handled by, each of us, comes the need to display more of this information at once. In terms of the display devices we need, this translates to requirements for larger displays (ideally, up to the physical limits imposed by the situation, rather than by the technology), and, within a given size of display, the ability to show more information – denser pixel formats, or "higher resolution." Color is also a major factor in being able to quickly and easily handle large amounts of information, so we can reasonably expect that the demand for color displays (over monochrome) can only increase in all applications.
- The trend toward more and more portable devices will continue in a wide range of markets, and with this will come further demands on display devices from the standpoints of power consumption, size, weight, and viewability under a wide range of possible conditions. And, even when not directly impacting the display technology itself, portability has serious implications for the display interface. A device cannot really be seen as very "portable" if, for instance, it must be tethered to a fixed location by a bulky cable.
- The line between "information" and "entertainment" products continues to grow less distinct, and it appears likely that with this convergence, display devices will be less and less specialized for one or the other application. This does not, however, have the ultimate effect of making things easier on the display – on the contrary, it may mean that even more products will be expected to display full-color, full-motion imagery as well as very high-quality "static" text and graphics. This also should not be interpreted as meaning that the "convergence" of certain product types should be expected to happen as was predicted in the past. We still should not necessarily plan on televisions which are also personal computers, with a relatively few distinct products performing multiple roles. – it doesn't have to be "convergence" in this manner. Instead, it may be more logical to expect the array of products incorporating sophisticated display capabilities to grow even wider.
- Finally, we must always keep in mind a point made very early in this book – that no matter what else is said of it, the display interface is ultimately a human interface, and the goal is always an image to be seen by people. The human viewer is very much a part of the display system, and that simple fact brings with it an enormous number of requirements and limitations. And, in many cases, the ideal display device will remain one that can produce a convincing visual perception of "reality"; i.e., display an image of such quality that a human viewer cannot distinguish the image from the "real thing". This has implications in terms of the display resolution, color capabilities, and in an area not examined in depth to this point – the illusion of depth, of a "three dimensional" im-

age. But there will remain serious constraints that may make the goal of true "virtual reality" unattainable in any practical system.

13.2 Color, Resolution, and Bandwidth

Through the 1980s and 1990s, one of the most notable changes in both the computer display and television industries was the increase in "resolution" – placing more and more pixels before the viewer, as either higher-pixel-count image formats on the PC, or in the improvements brought by digital and HDTV in television. Coupled with this has been an increase in the average sizes of the displays in these two markets. Where a 25 inch television was once considered to be "large", it is today actually below average, and the range of available products has been extended to 60 inch diagonals and beyond with the introduction of projection television systems. HDTV has increased the "resolution" provided by the TV as well, from the original 525 or 625 line systems to image formats with over 1000 active lines per frame. A similar change has occurred on the desktop; gone are the early 13 inch and 14 inch monitors, with a typical 640×480 pixels. Today's personal computer or workstation display averages 17 inches in diagonal size, and is displaying either the 1024×768 or 1280×1024 format. The high end of this market is at 21 inch diagonal and higher, with image formats up to at least 2048×1536 pixels no longer uncommon.

Should we expect this trend to continue? Increasing the resolution of the display would seem to be something that is always desirable, a case of bigger numbers always being better. However, we can expect to run into limits, both in terms of what is technically feasible in a practical display (and display interface!), and even in what is desirable. There is little sense in providing more pixels than the viewer can see under the expected viewing conditions.

As was discussed in Chapter 2, human visual acuity (in terms of luminance only) peaks at approximately 10–30 cycles per visual degree, and may generally assumed to be limited to not more than 50–60 cycles per visual degree. A simple rule of thumb, then, is that humans are limited to seeing details no finer than approximately 1 arc-minute of the visual field at any time. At typical viewing distances for desktop monitors and similar devices (usually on the order of a half-meter or so), this translates to a maximum usable resolution in the range of perhaps 100–200 dots (pixels) per centimeter, or 300–600 dots per inch. Resolution improvements beyond these limits will result in only subtle increases in the perceived image quality, if that. And obviously, the resolution limits for displays typically seen from greater distances will be correspondingly lower.

At the present time, "desktop" and other "close-up" displays such as PC monitors typically provide between 30 and 50 dots per cm (about 80–120 dpi) of resolution. Television and similar "large-screen" displays, usually viewed by one or more persons from a much greater distance, are significantly worse in terms of absolute resolution – perhaps 15–20 dots/cm, or 40–50 dpi. For the "desktop" sorts of applications, we could clearly desire significantly greater resolution than is currently available. The need for greater resolution in television is less clear, if one assumes current screen sizes and viewing distances (as was pointed out in Chapter 8, the existing TV standards were originally developed with the limitations of visual acuity in mind). However, if we expect TV screen sizes to increase – or if we expect "TV" to extend into the realm of cinematic presentation – then greater resolution will be required there as well.

But ignoring, for the moment, the practicality of display devices that could provide these resolution levels, we have to look at the impact of such increases on the interface requirements. As was noted in Chapter 8, the basic interface capacity requirements are set by the image format, the number of bits transmitted for each pixel (i.e., the "depth" used to represent luminance and color), and the frame rate. We cannot reasonably expect either the required "bit depth" or frame rates to decrease from present levels, and so increasing the resolution along each axis by a factor of 2–3 represents an increase in the required data rate of 4–9 times! As an example, providing a "300 dpi" image on something similar to a current 21 inch diagonal PC monitor would require the following:

$$120 \text{ pixels/cm} \times 40 \text{ cm (horizontal image size)} = 4800 \text{ pixels}$$

$$120 \text{ pixels/cm} \times 30 \text{ cm (vertical image size)} = 3600 \text{ pixels}$$

A 4800 × 3600 pixel image, at even 60 frames/s, and using the typical 24 bits/pixel representation, represents a data rate of approximately 25 Gbits/s! If this were to be transmitted in uncompressed form, using the existing TMDS digital interface standard (see Chapter 10), at least 16 data lines (pairs) would have to be provided by the interface! While this might still be within the realm of practicality, it clearly shows that the interface may be a significant limiting factor in extending display resolution in the short term.

The above analysis assumed that the current frame rates and "pixel depths" would remain adequate in future displays. We can also question if this is a reasonable assumption, especially given a "bigger is always better" mentality on the part of many consumers. But a brief analysis would say that there is certainly less reason to expect significant increases in these areas, as compared with the want or need to simply add more pixels. As noted in earlier chapters, the dynamic range of human vision (in terms of luminance contrast) is at best a few hundred to one. Therefore, eight bits per sample of luminance (or of each primary color) is generally considered adequate for "photorealistic" imaging, especially if a non-linear encoding can be employed. If the luminance encoding must be done in a simple linear fashion, it can be argued that up to perhaps 10–12 bits for luminance (or per color) could be justified – but this would at most represent a 50% increase in data capacity requirements over the present norm. Similarly, there is little need for extreme increases in the transmitted frame rate, particularly if we expect (and we do) the traditional CRT display to give way to less flicker-prone technologies. With the question of display flicker no longer a driving concern, the frame rate issue is determined solely by the requirement to show convincing motion – and we already find satisfactory results in this regard using frame rates in the upper tens of frames per second range. (This analysis ignores the possible need to increase the "frame" rate for other reasons, such as the use of field-sequential color or stereoscopic display, but these are addressed later in this chapter.)

From this, we could conclude that it might be reasonable to expect that, at some future time, display interfaces will be needed which provide approximately an order of magnitude greater capacity than those currently in widespread use. This is certainly within the realm of practicality, even with present interface technologies; as mentioned above, this could be achieved by simply adding additional data paths to an existing digital standard. (This might be a relatively expensive interface, but it is certainly achievable.) Further increases in the per-channel data rate that can reasonably be expected of these interfaces would, of course, reduce the number of channels that must be added. However, there are at least three additional questions that must be addressed before we can draw any conclusions as to the shape of future interface standards. First, can we reasonably expect displays providing this level of performance (resolution, etc.) to exist? Next, if such high-performance products are expected to appear in the market, is it reasonable to expect that the basic nature of the display interface will remain unchanged – i.e., that the interfaces supporting these displays will be similar to those we have now, only with greater capacity? Finally, are there other forces – new display products, new usage models, etc. – which might drive additional changes from the present display interface models?

13.3 Technological Limitations for Displays and Interfaces

Looking first at the expected limitations of display technology, we find that not only are higher-resolution displays possible – they have already been produced. As of this writing (late 2001), several examples providing 200–300 dpi resolution have been produced at least as prototypes, and in some cases have seen limited volume production. Cost reductions, and therefore increased volumes, will no doubt follow. However, increases in resolution significantly beyond this level are seen as unlikely, due to the diminished return in terms of further image quality improvements. We can therefore reasonably expect displays supporting the roughly 4k × 3k pixel format used as an example above, and possibly even somewhat higher in large screen sizes.

The ultimate limits on any display's resolution arise from the limitations in the manufacturing processes used to create that display, and to some extent from the limitations inherent in the materials used in the display and in the nature of the technology itself. For example, the limits on resolution in an LCD result from the limitations of the photolithographic processes used to create the features – transistors, electrodes, color filters, etc. – which make up the structure of the display, but also from the needed LC cell gap size and the size of the LC molecules and the mode in which they are made to operate. Other limitations may be imposed elsewhere in the system. In the case of projection displays, the number or size of the pixels on the display device itself may not be the most important consideration. Instead, the ability of the optical system, including the projection lenses or even the screen itself, to resolve these pixels may become the limiting factor controlling system resolution. For example, practical projection optics may be able to resolve display features (pixels) down to perhaps 5–10 μm on a side before being limited by diffraction effects. As a result, while smaller features could certainly be produced in devices such as a liquid-crystal-on-silicon (LCoS) microdisplay, there is no point in attempting to use this to produce higher-resolution displays. The 1–10 μm range seems to be a common – although coincidental – lower limit on the practical feature size in a number of display technologies.

We wind up, again, at roughly the same conclusion as before. Through the 1980s and 1990s, image formats – and display devices – slowly increased to the point of one to two thousand pixels per side being common. There appears to be no hard limit through the next order of magnitude – to between five and ten thousand pixels per side – but little practical possibility or need for increases beyond this point for the vast majority of applications. But as interface capacity requirements go up as the square of the resolution, this is probably for the best – for we would surely find ourselves more quickly limited in our ability to drive a given display than to produce it in the first place! A 10k × 10k display at 60 frames/s would require a 144 Gbit/s interface, per the above.

It is arguable, however, that this analysis is too simplistic, on at least two points. First, in calculating the required data capacity of the interface, we have not taken into account the possibility of compressing the information. As the experience of digital HDTV has shown, a very high level of compression is possible without given up perceived image quality, and we can reasonably expect – as the cost of processing power continues to drop – digital compression techniques to be extended to other display applications. Our analysis of the required and practical display resolutions was also overly simplified. For one thing, the resolution limit imposed by human visual acuity, as described above, actually applies only to a very small area of the visual field; there is really no need to fill the complete field with, say, a 300 dpi display, as most of this information would be wasted. Or at least it would be if we expected the viewer's gaze to be fixed on a particular area – so the "brute force" argument that comes up at this point is that the entire display has to present 300 dpi imagery, since you cannot tell where the viewer will be looking next. But perhaps a compromise solution, that would significantly reduce the load on the interface, is still possible.

Consider a display such as the one imagined in Figure 13-1. Here, a very large surface is capable of 300 dpi, full-color, full-motion imagery, but we realize that there is really no need to continuously update the complete display at this level of quality. Small, static areas might be showing 300 dpi images, but can be updated relatively slowly (simply because they *are* static). But there is little need to present fast-moving objects, etc., at this resolution, since the eye cannot distinguish this level of detail in such images. In some areas, then, the display might trade off resolution for frame rate, and still be driven with a relatively low data rate. The basic concept involved here is one of *conditional updating* of the display – rather than sending the contents of the entire display over and over, as is done in the traditional display

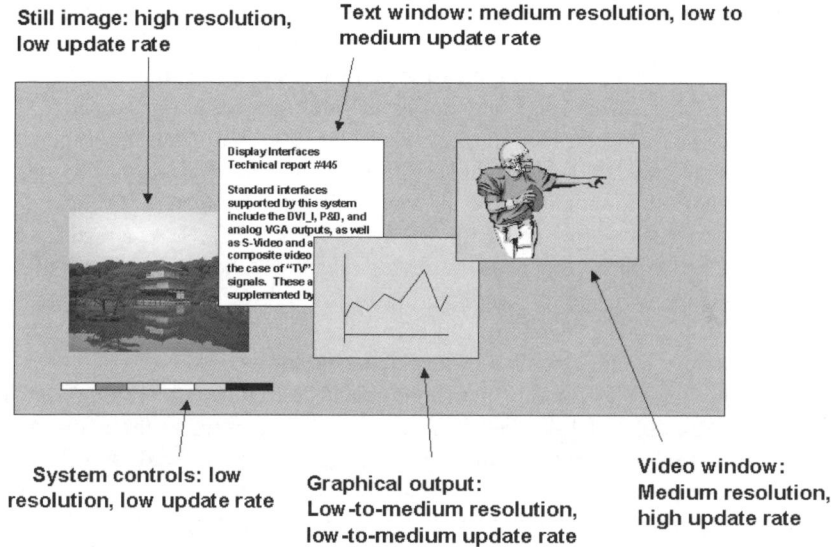

Figure 13-1 A hypothetical large-area, widescreen, very high resolution display device of the future. This example shows that not all of the total displayed image needs to be transmitted by the source at the same rate, assuming that the display itself provides the "intelligence" and storage required to manage such mixed content.

interface, only that data that requires changing is sent at any one time. This makes for much more efficient use of the available capacity of the interface, and is examined in more detail in a later section of this chapter.

13.4 Wireless Interfaces

The expected growth in portable information devices also implies a growth in the need to communicate image information to and from such products, *without* physical connections – in other words, a *wireless interface*. Traditionally, "wireless" implied RF transmissions, although such connections can certainly (at least over short distances) also be achieved via infrared (IR) or similar optical links. But wireless connections also make for a unique set of problems, especially when it comes to the transmission of image data.

Wireless interfaces face challenges in several areas, including the power needed to support them, the available capacity of the wireless channel, and what might generically be called "containment" issues – owing to the fact that such communications occur over a shared medium (the visible, IR, or RF spectrums), there are concerns regarding both transmission security and the possibility of mutual interference between two or more systems. The power and capacity issues are to some degree two aspects of the same problem. Increasing the capacity of a given wireless channel, while maintaining the same range (the permissible distance between transmitter and receiver) will in general require a more powerful transmitter. However, power is clearly a major concern in any portable product, which typically must provide its own self-contained power source in the form of batteries (or be limited to the power available from sources which are themselves "wireless", such as solar cells). This problem is to some degree lessened in an application which does not require the same channel capacity for both transmission and reception; e.g., a portable appliance that can display video, but not transmit it.

The power required to maintain a wireless interface at the same degree of reliability (error rate, etc.) as a comparable wired connection is higher primarily due to the fact that most of the transmitted power in any wireless transmission never reaches the receiver. Even in the case of such interfaces using a highly directional signal (such as is the case with IR links, and even more so with laser transmission), little of the energy consumed by the transmitter is available to the receiving device. And directional transmissions significantly reduce one of the advantages of wireless connections – the ability to communicate regardless of the location or orientation of the device at either end of the channel. A very highly directional wireless link may require such precision in location and "aiming" so as to render the connection less convenient than a direct physical connection. (Successful "point and shoot" interfaces, such as the IR transceivers used in many personal digital assistants (PDAs), use a relatively broad beam so as to make the interface usable – at the expense of range that could be achieved with a narrower, but therefore more alignment-critical, path.) In most cases, the wireless link is required to be practically omnidirectional; communications must be maintained regardless of the relative orientation of transmitter and receiver. An obvious example of this is the cellular telephone. In such cases, it is clear that the vast majority of the transmitted power has been "wasted" if judged by the fraction actually reaching the receiver. Simply put, the nature of wireless interfaces is such that channel capacity and the ability to maintain communications has been traded for convenience and mobility.

256 NEW DISPLAYS, NEW APPLICATIONS, AND NEW INTERFACES

As such, wireless display interfaces have to date been restricted to short-distance applications, with relatively low data rates. The obvious exception to this has been broadcast television, and even more so with the introduction of high-definition, digital TV transmissions. With image formats comparable to those used by the desktop PC monitor, we are tempted to say that HDTV shows that wireless interfaces could be used for practically any display. However, television also points out another significant difference between the typical applications for wired vs. wireless transmissions. Wireless is best suited to a "one source, many receivers" model – in other words, *broadcasting*. In such applications, it is generally far easier to use a wireless path than to run a separate physical connection to each receiver. But we must recognize that this is true only if each receiver does wish to access exactly the same information as all the others. Providing unique content for each receiver, in the wireless model, has traditionally required a separate channel (in the sense of a unique portion of the available broadcast spectrum), an approach that becomes rapidly unworkable as the number of unique transmissions increases.

But this ignores another development in the wireless arena, one that also had its roots in a common "wired" practice. Many of the wireless interfaces now coming to market, or under development, use a networked approach similar to that used for some time in wired computer-network systems. In general, such approaches provide for a means of addressing transmissions such that they will be accepted only by a single intended receiver, a group of receivers, or as a broadcast to all devices on the network. There are also often protocols and command structures provided so as to allow for bidirectional communications, permitting the individual devices on the network to send data back to the host or even to communicate with each other directly (as in direct "peer to peer" communications). Various network topologies – loops, "stars," tree-like designs with nodes and branches, etc. – are possible, depending on the needs of the system. The transmission of data via a networked system generally involves some form of "packetization," as discussed in Chapter 12 – the division of the data stream into separate, standalone segments, each of which will provide information on the type of data being carried, the intended receiver or receivers (in the packet address), and other supplemental information and commands, usually distinguished from the data itself through placement in a predefined location within the transmission, known as the *packet header*. All receiving devices must be capable of observing the headers of the various packets being transmitted, in order to determine which are intended for that receiver and how to handle those that are.

However, it is very important to remain aware of one basic fact – use of a networked approach does *not* increase the available data capacity of a given wireless channel. If anything, the effective capacity is reduced, as the protocol, headers, error-handling, etc., of the networking system consume capacity that might otherwise be used for the data itself. A typical wireless network may impose an overhead of 30–50% of the available "raw" capacity. What networking *does* do is to permit a much more efficient use of the available capacity, in a multiple-receiver (i.e., multi-display), multiple-data-stream situation. This advantage is realized, however, only if sufficient "intelligence" is present in the receiver – in the display device – to permit for efficient "packaging" of the image information. At the very least, this generally implies support for some degree of image compression. Beyond this, the conditional-update concept, and the transmission of "higher-level" graphical information (e.g., the transmission of graphics commands and "primitives" rather than simply transmitting all the pixels) may be used for further efficiency gains. But we now face another tradeoff; the primary motivation for the use of a wireless interface was to enable portable devices. Portable,

Table 13-1 An overview of selected short-range wireless interface standards.

Standard	Max. data rate (Mbps)	Throughput (Mbps)	Band
Bluetooth™ 1.0 (IEEE 802.15.1)	1	0.72 (max.)	2.4 GHz
Bluetooth™ 2.0 (under development)	2–12	1.4–8.4	2.4 GHz
IEEE 802.11a	up to 54	32	5 GHz
IEEE 802.11b	11	5–7	2.4 GHz

however, implies limitations – often severe – on the available power, and this argues against placing more of the processing burden (whether for decompression or graphics rendering) on the portable display device. Future wireless display standards (or for that matter, standards for any wireless, portable devices) must make a careful tradeoff here, between the efficient use of the wireless channel and the demands this may place on the portable unit.

A summary of several of the wireless network or point-to-point communications systems either now available or under development is shown in Table 13-1. Note that the available capacity of these ranges from well under 1 Mbit/s, to a high of perhaps 50 Mbit/s (at least at present). As has been demonstrated by the existing digital HDTV standards, a channel supporting tens-of-megabits data rates is sufficient for the transmission of high-resolution, full-motion color video, but only if some rather sophisticated compression techniques are employed. And we must keep in mind that HDTV uses the "broadcast" model; if the available capacity of any of these systems has to be shared by multiple video streams, the quality of transmission (in terms of the pixel format, color depth/quality, and/or frame rate) will likely suffer in at least some of them. Still, we should expect that wireless links will be capable of supporting reasonably high-quality image transmissions, at reasonable frame rates (certainly enough for fairly high-resolution "still" image displays, as in document review and most "web"-type applications) even in a networked situation.

13.5 The Virtual Display – Interfaces for HMDs

Another factor in the field of portable devices is the growing availability of high-resolution "virtual" displays – meaning those in which the display screen is not viewed directly, or at least in which the image appears to be in a location, and at an apparent size, different than is actually the case. By far the most common example of this type of display at present are the various projection types, but a new display type is entering the market aimed specifically at the needs of portable/mobile users. This is the "near to eye" display, generally based on an LCoS or similar microdisplay component, and exemplified by the "eyeglass" or "wearable" head-mounted display (HMD).

HMDs may be used to provide the performance of a relatively large-screen desktop monitor in a portable application, by using magnifying optics to place a high-resolution image at the appropriate *apparent* distance and location within the user's field of view. Full-color, full-motion video, with an appearance equivalent to a large-screen (20–40 inch diagonal) display viewed from typical distances, is relatively easy to achieve. To date, the cost of high-resolution display devices has limited large-volume HMD products to barely standard-TV quality (although some very small-volume, high resolution products have been made). However, the introduction of low-cost LCoS devices have made practical displays supporting at least SVGA (800 × 600) reasonably priced, and there is every reason to expect that such products will be extended to the larger formats.

Such displays have two primary applications. The first is the one described above; providing high-resolution, "large-screen" capability to portable products, which could not support a direct-view display of the required size. However, HMDs – when designed to provide images to both eyes simultaneously – also are the simplest way to provide the user with an "immersive" display experience – a simulation of reality, with the displayed images filling the visual field. Providing separate images to each eye can also make for a simulation of a three-dimensional visual environment, and this approach is certainly one of the easiest and most straightforward for achieving this.

In the case of an HMD being used in a portable application, the source of the images might be either internal (as in the case of a portable computer, or a self-contained video playback system) or external (via, perhaps, a wireless network as described above). But in any application, either as part of a portable system or even with a conventional PC or workstation, there has to date been no interface standard defined specifically for the needs of these displays. In many cases, the HMD is supplied as a part of the complete system, and uses a proprietary interface to the "host" product. In others, the HMD is connected to a standard display interface – such as the "VGA" output of a PC, or an "S-Video" connection in entertainment products – via a converter (usually external) which translates such signals into the format required by the display itself.

There are two major differences between the interface needs of head-mounted displays and their more conventional counterparts. First, and most obvious, this type of display product will typically require an external source of power. It might be possible to operate an HMD from batteries, but if the display is to use a wired interface anyway it is typically preferable to obtain power from the host system (even if that unit itself is a portable device). This is not a severe burden on the interface, since through the use of LCoS microdisplays the power requirements of these products are relatively small – often less than 1 W per eye.

The other requirement of these displays is not so much a demand on the physical interface as on the electronics that drive it. The majority of microdisplays suitable for use in HMDs employ field-sequential color, as spatial color (with separate color subpixels and filters) requires too much space for use in such devices. But, of course, there are currently no display interface or image transmission standards that supply the image in this form. So the interface electronics must perform a conversion from a spatial or "parallel-color" form, such as RGB, into field-sequential color for the HMD. This requires the use of a frame buffer in the interface, although by being somewhat clever with the use of the memory it does not have to be large enough to contain a full frame of RGB (as one of the fields can begin being delivered to the display as it arrives). Note that the capacity required of the interface does not increase with the use of field-sequential color – there is still the same amount of information to be transmitted per frame – but a higher clock rate is generally needed, with less information sent per clock. For instance, where a spatial-color interface might transmit 24 bits per pixel at, say, a 30 MHz pixel rate, the field-sequential equivalent must use a 90 MHz pixel rate and send only the 8 bits per pixel for the color of the current field.

The Video Electronics Standards Association (VESA) is, as of this writing, well along with the development of a standard defining the physical and electrical interface for a head-mounted or "eyeglass" type displays. This specification, which should be completed some time in 2002, is expected to define a physically small connector, suited to the expected portable applications such as personal DVD players, laptop and pocket computers, and similar devices. The interface will provide field-sequential color via a serial digital interface, most likely one of the differential systems described in Chapter 10, along with power, display identification, and audio support.

13.6 The Intelligent Display – DPVL and Beyond

In each of the applications examined so far, it has been noted that an additional measure of "intelligence," of information processing capability, has been required of the display or interface over the traditional display device. The support of higher resolutions requires support for compression and decompression, if unreasonably high interface capacity requirements are to be avoided. Adding more sophisticated displays to portable, wireless devices brings similar needs, to enable support for these with the limited capacities of a wireless channel. Head-mounted displays typically require at least sufficient "intelligence" in the interface for the conversion of the image and transmission format into something suited to this display type.

This points to a trend that has already started to influence the shape of display interface standards. With a much wider range of display technologies, applications, and formats, the traditional interface approach – in which the interface specifications were very strongly tied to a single display type and image format – no longer works well. The original television broadcast standards, arguably among the first truly widespread display interfaces, were clearly designed around the assumption of a CRT-based receiver and a wireless transmission path. The analog PC display interface standards began in a similar manner, and then had to be adapted to the changing needs of that industry, in supporting a wider range of display types and formats than originally anticipated.

The advent of digital interfaces has been seen by some as necessary to support certain display technologies, but as was discussed in Chapter 6, there is really little difference between the "digital" and "analog" interface types as long as the traditional form of connection is implemented in either. The true advantage of the digital class of interfaces is not in its better compatibility with any particular display type, but rather that it has enabled the use of digital storage and processing techniques. The use of these has the effect of decoupling the display from the interface, or, from a different perspective, permitting the interface to carry image information with little or no attention to the type, format, or size of the display on which the image will ultimately appear.

The Digital Packet Video Link (DPVL) proposed standard (discussed in Chapter 10) represents the first significant step away from the traditional display interface. In "packetizing" the image information, DPVL brings networking concepts onto digital interfaces originally used simply to duplicate the original analog RGB model. This has the effect of moving the physical interface between the "host system" and the "display" farther back along the traditional chain (Figure 13-2). Or, again to look at the situation from the other side, the traditional "display interface" now is buried within the display product, along with a significant amount of digital processing and storage hardware.

In the future, then, it is likely that the "intelligent display" model will dominate many markets and applications. Rather than seeing dedicated video channels between products, image information will be treated as any other, and transmitted along general-purpose, very-high-capacity digital data interfaces. Video information has the characteristic of being time-

260   NEW DISPLAYS, NEW APPLICATIONS, AND NEW INTERFACES

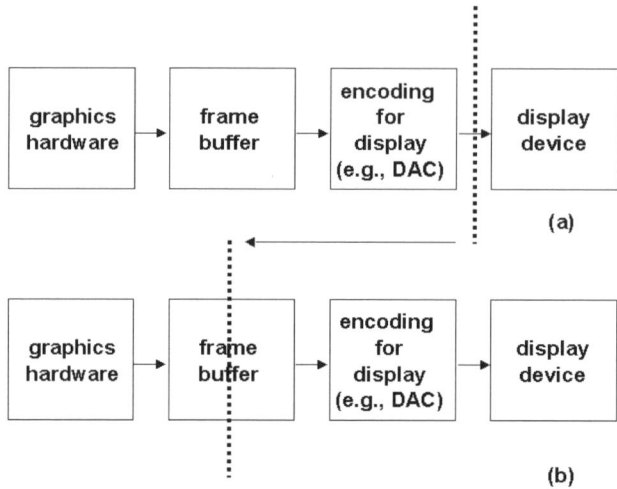

Figure 13-2   The effect of systems such as DPVL, or others which move the display interface to more of a "network" model, is to move the location of the display interface more "upstream" in our overall system model. In (a), the dotted line shows the traditional location, within the system, of the physical display connection – between an output specifically intended for a particular display type or device, such as an analog video connection, and the display itself. Making the display into a networked peripheral, and permitting such features as conditional update, etc., requires more "intelligence" and image data storage within the physical display product – and hence moves the display interface logically closer to the source of this data (b).

critical in nature – the data for each field or frame must be received in time for that frame to be assembled and displayed as part of a steady stream – but this is not unique to video. Audio samples share this characteristic, and data channels intended to carry either or both must therefore support isochronous data transmission. But this is becoming more and more common, particularly in digital standards for home networking and entertainment.

We must also consider the question of just how much intelligence is likely to be transferred to the "display" product in the future, vs. it being retained in a host product such as a PC. So far, we have still considered the interface as carrying information corresponding to complete images, albeit compressed, packetized, or otherwise in need of "intelligent" processing prior to display. But should we also expect the display to take over the task of image composition and rendering as well? You may recall that, in the very early days of computer graphics, this was often the case – the graphics hardware was packaged within the CRT display, and connected to the computer itself via a proprietary interface. However, this was really more a case of packaging convenience for some manufacturers; there has to date never been a widely accepted standard "display" interface, between physically separate products, over which graphics "primitives" or commands for the rendering of images were passed in a standard manner. Obviously, it would be possible to develop such a standard, but two factors argue against it. First, the history of graphics hardware development has shown much more rapid progress than has been the case in displays, in terms of basic performance. As a result, display hardware has tended to have a much longer useful life prior to obsolescence than has graphics hardware, and it has made little sense to package the two together. There has also generally been a need to merge in images from other sources – such as television video – with the locally generated computer graphics prior to display. And, until recently, the high-speed, bidirectional digital channel needed to separate the graphics hardware from the rest of the computer simply was not available in a practical form – meaning an interface that could be implemented at a reasonable cost, and which would operate over an acceptably long distance.

And, at this point, there is little motivation for placing the rendering hardware within the display product in most systems. Instead, we can reasonably expect a range of image sources – as well of sources of information in other forms, such as audio and plain text and numeric information – all to transmit to an output device which will process, format, and ultimately display the information to the user. In this model, there is no distinct "display interface" between products; instead, images, both moving and still, are treated as just another type of data being conveyed over a general-purpose channel.

13.7 Into The Third Dimension

One particularly "futuristic" display has yet to be discussed here. A staple of the science-fiction story, the truly three-dimensional display has long been a part of many people's expectations for future imaging systems. To date, however, displays which have attempted to provide depth information have seen only very limited success, and in relatively few specialized applications. Most of these have been stereoscopic displays, rather than providing a truly three-dimensional image within a given volume. And for the most part, these have been accommodated within existing display interface standards, often by simply providing separate image generation hardware and interfaces for each eye's display. At best, accommodating stereoscopic display within a single interface is done by providing some form of flag or signal indicating which image – left or right – is being transmitted at the moment, and then using an interleaved-field form of transmission (Figure 13-3).

Several different approaches have been tried, with varying degrees of success, to create a true volumetric three-dimensional display. Most of these fall into two broad categories: the projection of light beams or complete images onto a moving screen (which sweeps through the intended display volume, synthesizing an apparent 3-D view from these successive 2-D images), or, more recently, the synthesis of holographic images electronically. Examples of the former type have been shown by several groups, including prototype displays demonstrated by Sony and Texas Instruments. The latter class, the fully electronic holographic display – again a favorite of science-fiction authors – has also been demonstrated in very crude form, but shows some promise. But a truly useful holographic display, or for that matter any three-dimensional display suitable for widespread use, faces some very significant obstacles.

The mechanical approaches to obtaining a three-dimensional image are burdened with the same problems as have faced all such designs – the limited speeds available from mechanical assemblies, especially those in which significant mass must be moved, and the associated problems of position accuracy and repeatability (not to mention the obvious reliability concerns). These products have generally taken the form of a projection surface, often irregularly shaped, being swept through or rotated within the volume in which the "3-D" image is to appear. In the examples mentioned above, for instance, Texas Instrument's display creates images by projecting beams of light (possibly lasers) onto a translucent surface rotating within a dome; Sony's prototype 3-D display uses a similar approach, but with a more-or-

262   NEW DISPLAYS, NEW APPLICATIONS, AND NEW INTERFACES

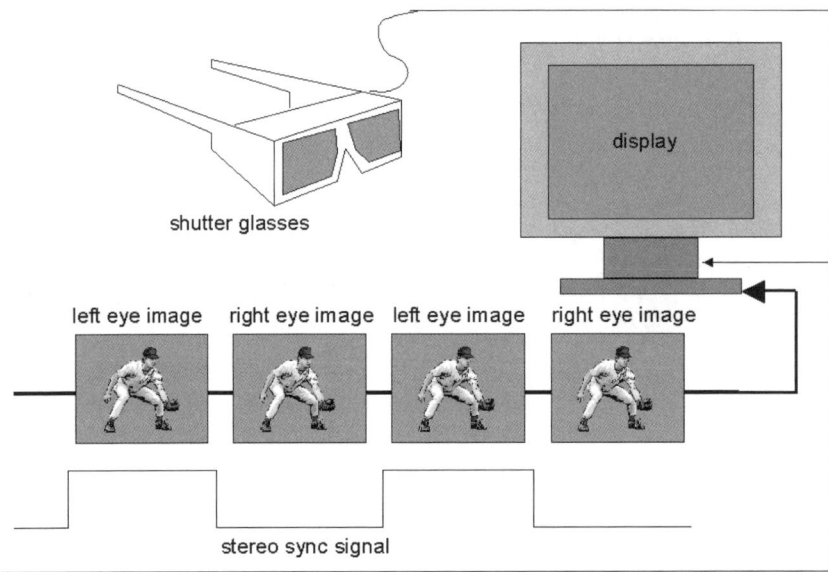

Figure 13-3 Stereoscopic display. One common means of display stereoscopic imagery is to use an ordinary two-dimensional display, but transmitting left- and right-eye images as alternating frames. Special glasses are used that shutter the "wrong" eye, based on the state of a stereo synchronization signal that cycles at half the frame rate.

less conventional CRT display as the image source. Others have employed similar approaches. The images produced to date, from any such display, do appear "three-dimensional", but all suffer from poor resolution (as compared with current 2-D displays), color, and/or stability.

Holographic imaging through purely electronic means would seem to be the ideal solution. Most people are already somewhat familiar with holograms; they have appeared in wide range of popular media, although it must be admitted that their use has been primarily restricted to novelties and a very few security applications (as in the holograms on many credit cards). They produce a very convincing three-dimensional effect, from what appears to be a very simple source. Surely adapting this technology to electronic displays will result in a practical 3-D imaging system!

However, a quick overview of the principles behind holography will show that this is far easier said than done. A hologram is basically a record of an optical interference pattern. In the classic method (shown in Figure 13-4), a coherent, monochromatic light source (such as a laser) is used to illuminate the object to be imaged, as well as providing a phase reference (via a beam splitter). If the light from the object is directed to a piece of film, along with the reference beam, the film records the resulting interference pattern produced by the combining of the two beams. After developing the film, if the resulting pattern is again illuminated by the same light source, the rays reflected from the original object are recreated, providing a three-dimensional image of that object.

Holograms have many interesting properties – which we will not be able to examine here in depth, no pun intended – but the key point, from the perspective of creating an electronic

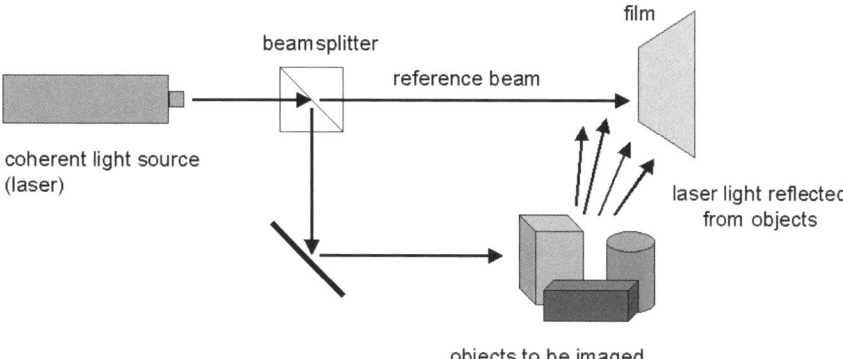

Figure 13-4 Holography. In the classic hologram process, a coherent light source, such as a laser, creates two beams of light – one used as a reference, and the other reflected from the objects to be imaged. Film captures the interference pattern created when the two beams meet.

display, is the complexity of the stored image. Fine details in the original translate to high-frequency components in the resulting interference pattern, across the entire area of the hologram. The ability of the film to adequately capture these is a limiting factor in determining the spatial resolution of the resulting holographic image. It is commonly known that film holograms of this type may be divided into several pieces, with each piece retaining the ability to show the complete original image, as if by magic. What is not commonly understood is how this occurs. A somewhat simplistic explanation is that the nature of the hologram stores image information across the entire area of the original film, but in a transformed manner that is, in a sense, the opposite of a normal photograph. Coarse detail, such as the general shape of the object, is stored redundantly across the entire hologram, and is retained if the hologram is later divided. But fine details are stored as information spread across the original, and will be progressively degraded as the area used to recover the image is reduced. Reproducing extremely fine detail requires the storage of a highly accurate image of the interference pattern, over a large area. In terms of creating or transmitting a hologram electronically, this implies a very large amount of information to be handled for each 3-D image. (It also implies an imaging device with resolution capabilities similar to those of photographic film – resolution at least in the upper hundreds, if not low thousands, of pixels per centimeter – but that is yet another problem.) An image area just ten centimeters square might easily require 100 million pixels or more to be a marginally practical holographic imager; assuming our usual eight bits per sample of luminance, the information required for a single frame is truly staggering.

This points out the major difficulty, purely from an interface perspective, with any three-dimensional display system. Even stereoscopic display, which simply involves separate two-dimensional images (one per eye) doubles the capacity requirements of the interface, if the other aspects of the image (pixel format, color depth, frame rate, etc.) are to be maintained vs. a more conventional 2-D display. A true volumetric display potentially increases the information content by several orders of magnitude; consider, for instance, a 1024 × 1024 × 1024 "voxel" (volume pixel) image space, as compared with a more conventional 1024 × 1024 2-D image. To be sure, the data-handling requirements for such displays can be reduced considerably by applying a little intelligence to the problem; for instance, there is no need to transmit information corresponding to samples that are "hidden" by virtue of being located inside solid, opaque objects! It is also certainly possible to extend the compression techniques employed with two-dimensional imagery into the third dimension. But the storage and processing capabilities required to achieve these steps literally see a geometric increase above what was needed for a comparable two-dimensional image system. The numbers become very large just for a static 3-D image, and would grow still further if a series of "3-D" frames ("boxes"?) were to be transmitted for motion imagery. We must therefore conclude that the notion of three-dimensional display systems, while certainly attractive, will remain impractical for all but some very high-end, specialized applications for the foreseeable future.

13.8 Conclusions

In this final chapter, we have examined some of the possible future directions for display products and the expected impact on display interface requirements for each. To summarize these:

- While there may be some need for higher resolution in many applications, we can also expect the current display formats to remain popular for quite some time. The higher resolution displays, when and where they do come to market, will require significantly different types of display interfaces than simply the repeated transmission of the raw image data that has characterized past approaches.
- Portable products can reasonably be expected to become much more sophisticated, and so will require more sophisticated displays; portable displays will become larger (to the limits imposed by the need for portability itself), providing higher resolution, and more and more often the ability to support full-color images and often motion video. However, the nature of the portable product demands a wireless information between that product and the source of the displayed information.
- These two factors argue for a continued transfer of "intelligence" – image processing and storage capability – to the device or product containing the display. This will be used to support more sophisticated compression techniques, increased reliance on local image processing, etc., primarily in order to make more efficient use of the capacity of the interface between the product and its host system. The "host system" may not even exist as a single entity, but rather be, in reality, a distributed set of image sources, all accessed through a general-purpose data network. In such a situation, no real "display interface" exists except within the display product itself, between the image-processing "front end" and the actual display. Coming standards such as the Digital Packet Video Link (DPVL) specification represent the first step along that path.
- Some new display types will succeed, and will place new demands on the physical and electrical interfaces to be used on many products – with a clear example being the development of high-resolution, "head-mounted" or "eyeglass" displays, for portable products and other applications. Others, such as a true three-dimensional display system, are not likely to impact the mainstream for the foreseeable future, due to the nature of the obstacles on their particular road to commercial practicality.

The one overall conclusion that one can always draw from looking over the history of electronic imaging, and studying the trends as they appear at present, is that our conclusions and predictions made today will in many cases turn out to be completely wrong. As recently as 40 years ago, practically all electronic displays were CRT-based; the personal computer, its "laptop" offspring, the cellular telephone, or even the digital watch and pocket calculator – each of which were driven by, and in turn drove, display technology development – were not even imagined. Even color television was still in its infancy. The history of the display industry has shown it to be remarkably innovative and surprising. While we would like to think we have a good understanding of where it will be in even five or ten years, new developments and the new products and markets that drive them have a tendency to ignore our predictions and expectations, and go where they please. At the very least, it seems safe to say that the delivery of information via electronic displays will be with us for quite some time to come, and with it the need for ever better and more reliable means of conveying this information – the display interface will remain, truly, the interface between the "information superhighway" and its users.

Glossary

Active-Matrix
A drive technique used to high brightness and/or contrast ratio in high information content display panels, by placing an active element at each pixel in the array. The most common type of active-matrix display is based on a technology known as *thin-film transistor*, or TFT. The two terms, active matrix and TFT, are often used interchangeably.

Analog
"Analog" refers to systems which encode variations in one quantity as *analogous* variations in another; for example, in analog video systems, changes in light intensity ("brightness") are represented by changes in the voltage level of the video signal. Note that "analog" does not necessarily mean the same thing as "linear" or "continuous".

Aspect Ratio
The ratio of the physical width to the height of an image or display. Most current displays and standard image formats have a 4:3 aspect ratio; high-definition television uses a 16:9 ratio. Aspect ratio is sometimes stated with respect to unity; i.e., 1.33:1, and 1.78:1 for the above, respectively.

ATSC (Advanced Television Standards Committee)
The ATSC may be considered the successor to the original NTSC; it is an advisory committee established by the US Federal Communication Commission to help define and recommend an advanced digital television standard. The US digital television broadcast standard is often referred to as the "ATSC standard".

ATV (Advanced Television)
ATV is the generic name used by the Federal Communications Commission and other bodies to refer to television systems, such as the "ATSC" digital television system, which provide performance or features beyond those of the original broadcast television standard.

Bandwidth
In the strict sense, "bandwidth" refers to the range of frequencies which are available for the transmission of information, or which will be passed by a given device (as in the bandwidth of an amplifier). The term "bandwidth" is often misused to refer to the rate at which information is conveyed over any given communications channel or interface, in bits/second. See "bit rate", "channel capacity".

Bit
A binary digit; the fundamental unit of information in a binary (two-state) system.

Bit Depth
See Color Depth.

Bit Rate
The rate at which information is conveyed over a communications channel or interface, in terms of bits per second. Often mislabeled "bandwidth". See "bit" and "bandwidth".

Brightness
The *perceived* intensity of light emitted from a source or reflecting from a surface. See "luminance".

Candela/Candela per square meter
The candela is measure of the intensity of light, equal to one lumen of luminous flux per steradian. The standard unit for *luminance* is the candela per square meter (cd/m^2), sometimes referred to by the obsolete term, "nit". Another obsolete unit for luminance is the foot-Lambert (ft.-L); one foot-Lambert is equal to 3.426 cd/m^2.

Channel Capacity
The maximum rate at which information can be transmitted over a given communications channel or interface, generally given in bits per second. Channel capacity is often referred to as "bandwidth" in casual conversation; technically, channel capacity is determined by several factors including the bandwidth of the channel in question.

Characteristic Impedance
A value, in ohms, which gives the relationship expected between voltage and current on a transmission line in the absence of reflections. While generally expressed as a pure resistance, the characteristic impedance of a line does not describe its loss characteristics, but rather establishes the above relationship and the value of impedance which will properly terminate the line at either the source or load ends, or at any other junction. From another perspective, this is the impedance seen "looking into" a transmission line, and is the impedance of the line itself if the line is either infinitely long or terminated by a lumped network of the same impedance.

COFDM (Coded Orthogonal Frequency Division Multiplexing)
A modulation technique in which transmitted data is distributed among a large number of relatively closely spaced individual carrier frequencies; adjacent carriers are 90° out of phase ("orthogonal") with respect to each other, to minimize mutual interference. In the DVB ter-

restrial digital television broadcast standard, COFDM with 2048 or 8192 separate carriers is used; this is one of the major incompatibilities between the DVB system and the competing US "ATSC" system.

Color
The visual perception which distinguishes different wavelengths or frequencies of visible light. Color is often specified through reference to three parameters: *hue* (e.g., red, yellow, blue, etc.), *saturation* (e.g., red vs. pink), and *value* (i.e., the lightness or darkness of a color, as in the distinction between various shades of gray between white and black).

Color Depth
The number of bits used to represent color and/or luminance in storing or displaying an image. Also called "bit depth" or "pixel depth", color depth determines the number of possible colors that can be displayed by the system. A "24-bit" system, for example, has a color depth of 2 to the 24th power (about 16.7 million) colors. Note that additional bits of storage may be used for other information not related to the color of the pixel, such as blinking or transparency values, so that the total number of bits provided per pixel by a frame buffer may not be the same as the true color depth.

Color Gamut
The range of colors that can be produced by a given display. In terms of representation within a 2-D color space (such as the CIE xy diagram), this is often indicated by plotting the location of the three primary colors used in the display; the color gamut is then all colors within the triangle thus created.

Color Space
Any of a number of three-dimensional spaces in which colors may be described as a set of values corresponding to the three dimensions or axes. Examples include RGB color systems, the CIE XYZ space, or a hue-saturation-value (HSV) color model. Color spaces may also be distinguished according to whether or not they are *perceptually uniform*, meaning that equal geometric distances covered within the space correspond to similar changes in perceived color by the normal human observer.

Color Temperature
A means of identifying the color of nominally "white" light sources, by identifying the temperature of a theoretical "black-body" radiator which would emit light of the same color. Color temperature is normally given in degrees Kelvin (K).

CCIR
Comité Consultatif International en Radiodiffusion (in English, the International Consultative Committee on Radio Broadcasting); an international standards group which created many standards in the areas of radio and television broadcasting and related fields. The CCIR was later subsumed into the ITU, as ITU-R.

Contrast Ratio
The ratio of the luminance of a bright ("full white") pixel or area to a dark ("full black") pixel or area on a given display or image. Often simply referred to as *contrast*.

Convergence

In CRT displays, "convergence" refers to the alignment of the three electron beams (red, green, and blue) such that they combine to produce a single point or area of apparently "white" light as seen by the user. A *misconverged* CRT display will show fringes of color around the edges of "white" objects.

CMY/CMYK

The standard set of subtractive primaries (cyan, magenta, and yellow), also used to refer to systems employing these primaries and thereby a subtractive-color method of color encoding. As such systems generally cannot produce an acceptable "black" color from the three primaries alone, a fourth channel is usually added for black, and the result referred to as CMYK.

Digital

"Digital" refer to systems which generate, store, and/or process data in terms of numeric values, most commonly using a binary (1 or 0) representation. "Digital" should not be confused with "sampled" or "discrete".

DDWG (Digital Display Working Group)

The group responsible for the Digital Visual Interface (DVI) specification. The seven original "promoter" companies include Compaq Computer Corp., Fujitsu Ltd., Hewlett-Packard Company, Intel Corp., International Business Machines Corp., NEC Corp., and Silicon Image, Inc.. The DDWG also includes an "Implementer's Forum," which is a much larger group of companies involved in using and promoting the DVI standard.

DFP

"Digital Flat Panel," a simple digital-only display interface based on a 20-pin MDR ("micro-delta-ribbon") connector, and first defined by Compaq Computer Corp.. The DFP specification was later adopted as a standard by the Video Electronics Standards Association (VESA).

DVI (Digital Visual Interface)

The specification produced by the Digital Display Working Group. Similar to the VESA "Plug & Display" standard in that it supports both digital and analog interfaces and employs a very similar physical connector, the DVI specification differs in offering two TMDS channels and no supplemental interfaces such as USB or IEEE-1394.

Driver

An interface circuit that conveys information or signals at higher level (usually in terms of voltage or current) to a receiver or load. An example in many flat-panel display types are those circuits which translate the digital video information to signals required to drive the row or column lines in the panel.

DVD (Digital Versatile Disc or Digital Video Disc)

A digital recording medium, similar in appearance to the audio CD or CD-ROM, but with much greater data capacity. In video applications, DVDs use MPEG-2 compression and provide approximately 135 min of video per side. The video data rate is 1–10 Mbits/s.

EIA
The Electronic Industries Association, a group today comprising several subsidiary organizations, and which has traditionally been one of the major private-sector sources of standards for all forms of electronic equipment and applications, including television and radio broadcasting practices.

Emissive Display
Used to refer to a display technology that emits light, as opposed to one that controls or modulates light provided by a source external to the display elements themselves. The CRT and the OLED display types are both examples of emissive displays.

EVC
The "Enhanced Video Connector" standard defined by the Video Electronics Standards Association (VESA). The EVC standard defined an advanced analog-only interface system, intended as a replacement for the earlier "VGA" industry standard. The EVC definition was never widely adopted, and it is best seen today as the predecessor of the P&D and DVI specifications.

Eye Pattern
A test pattern commonly used to judge the quality of a "digital" signal or transmission; waveforms of similar but opposite-sense transitions are overlaid on an oscilloscope or similar device, and the amount of open area in the resulting "eye" is assumed to correspond to the degree to which the states may be distinguished in time and amplitude.

Field
In display systems, "field" often refers to a portion of the complete image or *frame*. For example, in an interlaced display, each frame is made up of two fields, one containing the odd-numbered scan lines and the other the even lines. Another relevant example is *field-sequential color*, in which each complete full-color *frame* is separated into red, green, and blue *fields*.

Field-Sequential Color
See "sequential color".

Flat Panel Display (FPD)
An electronic display, typified by a flat screen formed by an array (usually orthogonal) of basic light-controlling or emitting devices, such as electroluminescent devices, light-emitting diodes or liquid crystal cells.

Flicker
The perception of rapid, periodic changes in the brightness of large areas of a display; the perception of flicker is determined by the display luminance, size, viewing distance, refresh rate, and the sensitivity of the individual viewer, along with the specific characteristics of the display device itself.

272 GLOSSARY

Foot-Lambert
A unit of luminance equal to $1/\pi$ candela per square foot. Use of the foot-Lambert is generally discouraged today, in favor of using cd/m^2. See "Candela/candela per square meter."

Frame
In display systems, "frame" generally refers to the smallest set of information which constitutes one complete, full-color image. It may best be understood by considering motion-picture film; here, each individual picture or image is called a *frame*. "Frame rate" is an important consideration in imaging systems, but may not always be the same thing as the display's *refresh rate*.

Gamma
A quantity used to indicate the non-linearity of a display device, in terms of its output response (luminance) vs. input signal level. In the simplest model, luminance is related to the input signal level by a simple exponential relationship where gamma (γ) is the exponent:

$$Y = KI^\gamma$$

where Y is the output luminance, I is the level of the input signal, and K is a scaling factor. For CRT displays, gamma is usually in the range of 2.0–2.7. A gamma of 1.0 indicates a linear response.

HDTV (High Definition Television)
A generic term used to refer to any television system having a higher "resolution" (larger image format) than the current 525-line or 625-line broadcast standards. A number of different image formats have been referred to as "HDTV", including 1920×1035 and 2048×1152 pixels. The US digital television standard provides two formats above "standard definition", or "SDTV": these are 1280×720 pixels, and 1920×1080 pixels. HDTV systems generally provide a 16:9 image aspect ratio.

Hue
In color science, "hue" refers to the perception associated with the dominant wavelength of light in a given source or reflected from an object; in simpler, common language, "hue" is what is being described when something is referred to as "reddish" or "bluish." Along with *saturation* and *value*, hue is one of three parameters which may be used to describe color in an intuitive "color space."

IEEE-1394 (aka FireWire™)
A serial bus standard that allows for the connection of up to 63 devices and transmission speeds ranging as high as 400 Mbits/s in its original version.

Illuminance
A measure of the light energy reaching a surface, in terms of the amount of energy per unit area. The standard unit of illuminance is the *lux*, which is one lumen per square meter.

Intensity
The "strength" of a light source, in terms of the amount of luminous flux being emitted over a given solid angle; i.e., two sources could be emitting the same total luminous flux, but have different *intensities* depending on the spatial distribution of that flux and the location of the point of measurement. *Intensity*, as opposed to luminance, is not corrected for the characteristics of human vision. The standard unit of intensity is the *candela* (see definition above).

ISO (International Standards Organization)
An international group of national and other standards bodies. For example, ANSI (American National Standards Institute) is a member of ISO.

ITU
International Telecommunication Union, an organization comprising representatives from numerous national regulatory agencies and related groups, which sets international standards for broadcasting and telecommunications.

Jitter
The instantaneous temporal instability of a given signal, with respect to either a reference point in time for that signal itself, or to a separate reference signal, or the average value over a given period for such instability.

Kelly Chart
A version of any of the various two-dimensional color coordinate diagrams (such as the 1931 CIE *xy* diagram) in which the area of the chart has been divided into regions with assigned color names; after K. L. Kelly, the color scientist who first presented such a chart.

Light-Emitting Diode (LED)
A semiconductor device that emits light when current is passed through it. An LED operates by having a sufficiently wide band-gap such that the energy emitted when a charge carrier passes through the device falls into the visible spectrum.

Liquid Crystal (LC)
A class of materials, typically organic compounds, which exist in the liquid state at normal temperatures but which also exhibit some ordering of the molecules. In the case of the LC materials used in electronic displays, the molecules are generally in the shape of long rods which are electrically polar, and which exhibit some degree of optical anisotropy (i.e., the effect these molecules have on incident light depends on the orientation of the molecule with respect to the light).

Liquid Crystal Display (LCD)
A large class of display devices employing liquid-crystal materials (see above) to control the transmission or reflection of light. Liquid-crystal displays use a wide variety of LC modes and affects to achieve this.

Lumen
The standard unit of measurement of the rate of emission of light energy. A typical candle produces about 13 lumens; a 100-W bulb generates 1200 (see candela).

Luminance
The intensity of light emitted from a surface, per unit of area, and corrected for the standardized spectral response of human vision (which distinguishes it from the formal use of the term *intensity*). The standard unit of measurement is the candela per square meter (cd/m^2, see definition above).

MacAdam Ellipses
Roughly oval or elliptical areas, as appearing on a two-dimensional color coordinate diagram, which describe the locus of just-noticeable-color-differences (JNCDs), to a given fraction of viewers in the general population, from the center color point. These are often shown larger than the actual JNCD locus for clarity; 10× is a common scaling. Named after D. L. MacAdam, and taken from his work in the 1940s which described typical color-difference sensitivities in the general population of human obsevers.

MicroCross ™
The pseudo-coaxial connector design used in several PC-industry video interfaces; originated by Molex Corp. ("MicroCross" is a trademark of Molex Corp.)

Microdisplay
Most generically, any display device with a diagonal size of under 2.5–5 cm (1–2 inch), in common usage of the term. Microdisplays are commonly used in either direct-view (through magnifying optics) applications, often referred to as "near-eye" applications from the physical location of the display, or in projection applications through the use of high-intensity light sources and projection optics.

MPEG (Motion Picture Expert Group)
Refers to an ISO working group that develops standards for digital video compression.

Nematic
A phase or mode of a liquid crystal material in which the long axes of the molecules are aligned with one another but without further organization. The "twisted-nematic" (TN) mode, in which the molecules' tendency to so align results in their a helical or "spiral staircase" arrangement (due to other influences built into the device) is the basis for most common liquid-crystal displays.

NTSC (National Television Standards Committee)
The original NTSC was an advisory committee established by the US Federal Communications Commission, which directed the development of broadcast television standards in the 1940s and 1950s. Today, "NTSC" is often used to refer to the US color TV standard itself (although it more properly refers only to the color encoding method used by that standard). The US color TV standard uses a format of 525 lines per frame, transmitted as approx. 60 interlaced fields (of 262.5 lines each) per second. The color encoding technique used is too complex to describe here. This standard is used in North America, parts of South America, and Japan.

Overshoot/Undershoot
The condition in which a waveform exceeds its desired "steady-state" value, following a transition to that value. Overshoot (or undershoot, which is the same thing following a negative-going transition) may be measure in terms of either the absolute amplitude of the excursion beyond the desired value, or as a percentage of that value or of the transition itself.

PAL (Phase Alternating Line)
A color TV standard, originating in Europe, commonly using a format of 625 lines per frame, transmitted as 50 interlaced fields (of 312.5 lines each) per second. The PAL color system is closely related to, but not totally compatible with, the US "NTSC" system. This video standard is used in Europe, Australia, China, and some South American and African countries. A common "digital" version of PAL uses a sampling format of 768 pixels × 576 lines.

Passive Matrix
A common type of flat-panel display in which the pixel array consists of a simple grid of horizontal (row) and vertical (column) electrodes, with the pixels themselves defined by the intersection of these, but without active control or drive elements at these locations. In LC displays, the passive matrix types typically do not have as broad a viewing angle as active-matrix (TFT) displays, and have slower response times.

Pixel
Contraction of "picture element." Physically, this term is often used to refer to the smallest individually addressable unit of an image that can be rendered or displayed. In strict usage with respect to imaging, a "pixel" is a single point sample of an image, and as such has neither size nor shape.

Pixel Density
The number of picture elements per unit of distance (e.g., pixels per inch or pixels per centimeter). See also "resolution".

P&D
The "Plug & Display" interface standard, defined by the Video Electronics Standards Association (VESA), including both digital and analog interface specifications. The P&D standard is best viewed now as a predecessor to the Digital Visual Interface (DVI) specification, which it closely resembles.

QAM (Quadrature Amplitude Modulation)
A variation of conventional amplitude modulation in which two versions of the same carrier, 90° out of phase with respect to each other ("in quadrature") are amplitude-modulated by different signals. This results in a combined signal from which the two original baseband signals may still be recovered separately by the receiver. QAM is used, for example, in modulating the two color signals onto the chroma subcarrier in the "NTSC" color broadcast system. (The original names for these signals, "I" and "Q," refer to this modulation technique, as they identify the *in-phase* and *quadrature* signals, with respect to the original chroma subcarrier signal.)

276 GLOSSARY

Resolution
The ability of a display device, image sensor, or viewer to discriminate detail in an image; resolution is most often stated in terms of cycles or lines per visual degree, or similar units indicating the number of intensity or color changes per unit distance in the visual field, which can be discriminated under given conditions. In common usage, "resolution" has also come to be used in reference to what is properly called the "image format" or "addressibility" (i.e., the horizontal and vertical pixel count), as in "a *resolution* of 1024 × 768".

RGB (Red, Green, Blue)
The common set of additive primaries, as used in electronic displays such as CRT monitors, or used to refer to systems employing this set and the resulting encoding of color into RGB values. This is in contrast to the *reflective* or *subtractive* primary set (cyan, magenta, and yellow, commonly referred to as "CMY" or "CMYK" with the addition of black) as used in printing.

Ringing
The damped oscillation which may occur following an abrupt transition between states in a signal or waveform. Ringing is typically characterized by the degree of overshoot it causes, and the settling time (the amount of time, measured from either a defined point on the transition itself or the first overshoot/undershoot peak, which is required for the signal to settle such that it remains within defined limits from that point on – e.g., "settling to within 5% of full scale").

Saturation
In color science, "saturation" describes the "purity" of a given light source in terms of the dominant wavelength; the closer the light comes to being purely of a single frequency, the greater the "saturation" of that color is said to be. For example, a light source which emits nothing but light of a wavelength of, say, 505 nm might be said to be a "100% saturated green." The color commonly called "pink" might also be considered a "low-saturation red." Along with *hue* and *value*, saturation is one of three parameters which may be used to desribe color in an intuitive "color space."

Sequential Color
Any system in which the primary colors or color fields are presented to the viewer as separated in time, rather than in space; in a *field-sequential color* display, for example, the user sees separate red, green, and blue images presented in rapid succession, to create the illusion of a single full-color image.

SECAM (Sequentiel Couleur avec Memoire)
The color television standard used in France, Eastern Europe and some African and Middle Eastern countries. It most often is transmitted using an image format similar to the PAL systems (625 lines/frame at a 50 Hz field rate), but using a completely incompatible color encoding technique in which the color difference components (R-Y, B-Y) are transmitted on different lines.

Skew
The stable (or nearly so) temporal difference between supposedly aligned portions of two or more signals.

SMPTE
The Society of Motion Picture and Television Engineers, a professional society which has also been extremely active in the establishment of technical standards for the television and motion picture industries.

SDTV (Standard Definition Television)
A term used in the context of the new digital "HDTV" broadcast standards to refer to image formats roughly comparable to the previous analog broadcast systems. In the US digital TV systems, two "SDTV" formats are in normal use: 640 × 480 pixels, and 720 × 480 pixels, the latter of these being displayable as either a 4:3 or a 16:9 image. European systems using the 625/50 format in analog broadcast most commonly use 720 × 576 or 768 × 576 for the "standard definition" digital format.

SPWG (Standard Panels Working Group)
A consortium formed by five companies – Compaq Computer Corp., Dell Computer Corp., Hewlett-Packard Co., IBM, and Toshiba – which produces standards for the mechanical dimensions, mounting, and electrical interfaces for LCD panels (to date, primarily for panels intended for the notebook computer market).

Transflective
A class of displays, most commonly of the liquid-crystal type, which can operate using either reflected or transmitted (i.e., through the device) light. This is most commonly achieved through a design similar to the common reflective LCD, but in which the reflective layer will also pass some amount of light from a source located behind the panel.

Transmissive Display
Any display devices which operates by controlling light passing through the display device proper; in this class of display, the light source and the viewer are on opposite sides of the display device, which then acts as a "switch" or "filter" (or more commonly, an array of switches or filters) to produce the viewed image.

VESA (Video Electronics Standards Association)
An industry standards organization developing display interface, timing, and related standards, primarily for the PC and workstation markets.

Viewing Angle
The largest angle or angular range over which one is able to acceptably view an image, generally defined in terms of minimum acceptable contrast ratio, color variation, or other measure of image quality. Viewing angle may be stated in terms of the total included angle over which the measure of minimum acceptability is met, or similarly as an angular measurement referenced to a line normal to the display surface.

278 GLOSSARY

USB (Universal Serial Bus)
A "plug-and-play" serial interface, typically used between a computer and add-on devices (such as keyboards, displays, printers, etc). As originally defined, USB provided a maximum rate of 12 Mbit/s, limiting its use to slow-speed peripherals and limited digital audio. The more recent USB 2.0 specification permits much faster operation, up to 480 Mbit/s.

Value
In color science, "value" refers to the intensity or "lightness" of a light source; for example, a change from black to gray to white, with no change in the "hue" of these shades, is an example of increasing *value*. Along with *hue* and *saturation*, value is one of three parameters which may be used to describe color in an intuitive "color space."

VSB (Vestigial Side Band)
A modulation method closely related to the conventional amplitude modulation (AM) and single-sideband-suppressed-carrier (SSBSC) types; in a vestigial-sideband AM transmission, one of the two sidebands is reduced in amplitude (as in SSB) but not suppressed altogether; also unlike SSBSC, the full carrier is retained. VSB modulation is used in nearly all analog television broadcast systems, and a variation of this method (8-VSB, with eight discrete amplitude levels) is used in the US digital television broadcast system.

VGA
"Video Graphics Adapter," originally used to refer to the complete specification for a graphics hardware subsystem introduced by IBM for its personal computer products in 1987. Today, "VGA" is used to refer either to the 640 × 480 image format (also first introduced as part of that IBM definition) or, more commonly, the 15-pin high-density D-subminiature connector used for video output within that system and still a popular de-facto standard within the industry today.

Bibliography, References, and Recommended Further Reading

Printed Resources

Fundamentals, Human Vision, and Color Science

Color in Electronic Displays," L. D. Silverstein, SID Seminar Lecture Notes, Symposium of the Society for Information Display, 1995.

Color Science: Concepts and Methods, Qualitative Data and Formulae, by G. Wyszecki and W. S. Stiles, Wiley, Chichester, 1982.

Colour Engineering: Achieving Device Independent Colour, P. Green and L. W. MacDonald (Eds), Wiley, Chichester, 2002.

"Designing Flicker-Free Video Display Terminals", J. E. Farrell, E. J. Casson, C. R. Haynie and B. L. Benson, *Displays*, July, 1988.

"Fundamentals of Color," D. Brainard, SID Short Course Notes, Symposium of the Society for Information Display, 1998.

Handbook of General Psychology, B. Wolman, ed., "Psychophysics of Vision" by F. A. Mote and U. T. Keesey, Prentice-Hall, Englewood Cliffs, NJ, 1973.

Optoelectronics Applications Manual, by the Applications Engineering Staff of the Hewlett-Packard Optoelectronics Division, Hewlett-Packard & McGraw-Hill, New York, 1977.

Precise Color Communication: Color Control from Feeling to Instrumentation, booklet published by Minolta Co., Ltd., Osaka, Japan, 1994.

"Predicting Flicker Thresholds For Video Display Terminals", J. E. Farrell, B. L. Benson, and C. R. Haynie, *Proceedings of the Society For Information Display*, Vol. 28, No. 4, pp. 449–453, 1987.

Raster Graphics Handbook, Conrac Division, Conrac Corporation, 1980.

"Understanding Visual Perception and its Impact on Computer Graphics," by B. Guenter, E. Davis, J. Ferwerda, G. Meyer and L. Thibos, Course Notes from the 18th International Conference on Computer Graphics and Interactive Techniques (SIGGRAPH), 1991.

Using Computer Color Effectively: An Illustrated Reference, by L. G. Thorell and W. J. Smith, Prentice-Hall, Englewood Cliffs, NJ, 1990.

"Visual Display Terminals (VDTs) Used For Office Tasks – Ergonomic Requirements – Part 3: Visual Displays", International Standards Organization Document No. 9241, 1991.

Display Technology

"An Introduction to CRT Displays," by C. Infante, Digital Equipment Corporation, Westford, MA, 1990.

Display Systems: Design and Applications, by L. W. MacDonald and A. C. Lowe (Eds), Wiley, Chichester, 1997.

Electronic Display Measurement: Concepts, Techniques, and Instrumentation, P. A. Keller, Wiley, Chichester, 1997.

Flat-Panel Displays and CRTs, L. E. Tannas, Jr. (Ed), Van Norstrand Rheinhold, New York, 1985.

Handbook of Display Technology, by J. A. Castellano, Academic Press, San Diego, CA, 1992.

Liquid Crystal Displays: Addressing Schemes and Electro-optical Effects, by E. Lueder, Wiley, Chichester, 2001.

Projection Displays, by E. H. Stupp and M. S. Brennesholtz, Wiley, Chichester, 1999.

Reflective Liquid Crystal Displays, by S. Wu and D. Yang, Wiley, Chichester, 2001.

The Cathode Ray Tube: Technology, History, and Applications, by P. A. Keller, Palisades Press, New York, 1991.

Television Broadcast Standards and Digital/High-Definition Television

"Channel Compatible DigiCipher HDTV System", Massachusetts Institute of Technology/ American Television Alliance. Proposal submitted to the ATSC/FCC, 1992.

"Digital Spectrum Compatible HDTV," Zenith Corporation and American Telephone & Telegraph, Proposal submitted to the ATSC/FCC, 1992.

Digital Television Fundamentals, by M. Robin and M. Poulin, McGraw-Hill, New York, 1998.

"Electrical Performance Standards – Monochrome Television Studio Facilities," EIA Standard RS-170, Electronics Industries Association, November, 1957.

"Electrical Performance Standards for High Resolution Monochrome Closed Circuit Television Camera," EIA Standard RS-343A, Electronic Industries Association, September, 1969.

Electronics Engineers Handbook, D. Fink (Ed), "Television and Facsimile Systems" by W. Hughes et al., McGraw-Hill, New York, 1975.

"HDTV Status and Prospects," by B. J. Lechner, SID Seminar Lecture Notes, Symposia of the Society for Information Display, 1994, 1995, and 1996.

"High Definition Television Technology," by C. Poynton, L. Thorpe, C. Pantuso, C. Caillouet and G. Reitmeyer, Course Notes from the 18th International Conference on Computer Graphics and Interactive Techniques (SIGGRAPH), 1991.

"High Definition TV Analog Component Video Interface," EIA/CEA-770.3-C, Electronics Industries Association/Consumer Electronics Association, August, 2001

Raster Graphics Handbook, Conrac Division, Conrac Corporation, 1980.

Reference Data for Radio Engineers, 6th edition, Howard W. Sams & Co., 1977.

"Report of the Task Force on Digital Image Architecture," Society of Motion Picture and Television Engineers, White Plains, New York, 1992.

"Standard Definition TV Analog Component Video Interface," EIA/CEA-770.2-C, Electronics Industries Association/Consumer Electronics Association, August, 2001.

"Studio Encoding Parameters of Digital Television for Standard 4:3 and Wide-screen 16:9 Aspect Ratios," CCIR-601 (ITU-R BT.601-4), CCIR/International Telecommunications Union, 1990.

Tube: The Invention of Television, by D. E. Fisher and M. Jon Fisher, Counterpoint Press, Washington, DC, 1996.

Computer Display Interface Standards

The following standards are published by the Video Electronics Standards Association, Milpitas, CA:

"Digital Visual Interface" specification, Vers. 1.0, Digital Display Working Group, 2 April 1999.

"Industry Standard Panels: 13.3", 14.1", and 15.0" Mounting and Top Level Interface Requirements," Vers. 2, Standard Panels Working Group, September 2001.

"Open LVDS Display Interface (OpenLDI) Specification," vers. 0.95, National Semiconductor Corporation, May, 1999.

VESA and Industry Standards and Guidelines for Computer Display Monitor Timing, Vers. 1 Rev. 0.8, September 17, 1998.

VESA Digital Flat Panel (DFP) Interface Standard, Vers. 1, February 14, 1999.

VESA Display Information Format (VDIF) Standard, Vers. 1.0, August 23, 1993

VESA Display Power Management Signalling (DPMS) Standard, Vers. 1.0 Rev. 1.0, August 20, 1993.

VESA Enhanced Display Data Channel (EDDC) Standard, Vers. 1, September 2, 1999.

VESA Enhanced Extended Display Identification Data (EEDID) Standard, Release A, Rev. 1, February 9, 2000.

VESA Flat Panel Display Interface Standard (FPDI-1B) Vers. 1.0, Rev. 2.0., August 18, 1996.

VESA Flat Panel Display Interface Standard (FPDI-2) Vers. 1, February 14, 1998.

VESA Generalized Timing Formula (GTF) Standard, Vers. 1.1, September 2, 1999.

VESA M1 Display Interface System Standard, Vers. 1.0, August 16, 2001.

VESA Monitor Control Command Set Standard, Vers. 1.0, September 11, 1998.

VESA Plug & Display Standard, Vers. 1, June 11, 1997.

VESA Plug & Display – Analog (P&D-A) Physical Connector Standard (formerly the Enhanced Video Connector Standard), Vers. 1, Rev. 3, July 13, 1998.

VESA Video Signal Standard, Vers. 1, Rev. 1, March 29, 2000.

Other Interfaces and Standards

Belden 1998 Master Catalog, Belden, Inc., 1998.

"Digital Packet Video Link for a Next-Generation Video Interface, and its System Architecture," J. Mamiya et al., *Digest of Technical Papers*, International Symposium of the Society for Information Display, 2000.

"Gigabit Video Interface: A Fully Serialized Transmission System for Digital Moving Pictures," by H. Kikuchi, T. Fukuzaki, R. Tamaki and T. Takeshita, Sony Corporation, 1998.

IEC 61966-2-1, Ed. 1.0, "Multimedia systems and equipment - Colour measurement and management - Part 2-1: Colour management - Default RGB colour space – sRGB", International Electrotechnical Commission, 1999.

IEEE Std. 1394-1995, Institute of Electrical and Electronic Engineers, 1995.

"Specification ICC.1:2001-12, File Format for Color Profiles", Vers. 4.0.0, International Color Consortium, 2001.

Universal Serial Bus Specification, Rev. 2.0, published by the USB promoters group (Compaq Computer Corp., Hewlett-Packard Co., Intel Corp., Lucent Technologies, Microsoft Corp., NEC, and Philips), April 27, 2000.

On-Line Resources

"Online" or "net" resources are becoming more and more important to the overall exchange of technical information in today's world, and while certainly not (yet) at the point where such resources can be seen as completely replacing print media, information available via the World-Wide Web can be a valuable resource for those studying any technical field. However, the nature of web-based resources is such that one must be cautious regarding the accuracy and completeness of certain sources, and there is in any event no guarantee that a source that is available online today will remain so in the future. The following sites, however, represent a selection of those which, in the author's opinion, both provide signficant and accurate information in this field, and which are likely to continue to be reliable sources in the future. In many cases, information obtained from these was invaluable in the writing of this text, and in making certain that the information presented here was as current as possible.

Standards Organizations and Similar Groups

Digital Display Working Group (DDWG) – a consortium of seven companies, which developed and published the Digital Visual Interface (DVI) specification. The DVI 1.0 specification may be downloaded from this site.

http://www.ddwg.org

Digital Video Broadcasting Project – a consortium of industry and government organizations working on developing digital television standards for worldwide use.

http://www.dvb.org

Electronic Industries Association (EIA) – Site includes links to information on obtaining EIA standards documents, and the EIA standards process, along with links to other EIA member organizations such as the CEA.

http://www.eia.org

Institute of Electrical and Electronics Engineers (IEEE) – Official site of the IEEE, with links to the IEEE Standards Association (IEEE-SA), the current distribution source for IEEE standards.

http://www.ieee.org

International Color Consortium – information and specifications regarding ICC profiles, etc.

http://www.color.org

sRGB official website – Provides technical information on the sRGB specification and its applications, along with links to other color standards and related organizations. Hosted by Hewlett-Packard Co.

http://www.srgb.com

The Society for Information Display – The leading professional organization for those working in the field of electronic displays and related technologies.

http://www.sid.org

Standard Panels Working Group – a group of computer and display manufacturers working on standards for LCD panels. Site hosted by DisplaySearch. SPWG 1.0 and 2.0 specifications may be downloaded from the site.

> http://www.displaysearch.com/SPWG

1394 Trade Association – The industry group responsible for promoting the IEEE-1394 serial digital interface. Note: the 1394 standard itself is not available from this site, as it is an IEEE standard. However, there are several "white papers" and other sources of technical information at this site.

> http://www.1394ta.org

Universal Serial Bus (USB) official website – A site established by the USB promoters to distribute news and information regarding the Universal Serial Bus. The USB specification may be downloaded from this site.

> http://www.usb.org

The Video Electronics Standards Association (VESA) – a group of over 120 companies in the computer, display, and graphics hardware industries, working on display interface and related standards. Note: some standards are available from this site, although on-line access to the complete set of VESA standards and other documents is limited to VESA members.

> http://www.vesa.org

Other Recommended On-Line Resources

Charles Poynton's "Frequently Asked Questions…" papers on gamma and color. Very complete, concise, and readable explanations of many of the most commonly asked questions regarding various aspects of image generation and presentation on electronic displays, by a leading consultant in the field.

> http://www.inforamp.net/~poynton/GammaFAQ.html
>
> http://www.inforamp.net/~poynton/ColorFAQ.html

On-line technical memos by Dr. Alvy Ray Smith, a longtime leader in the field of computer graphics. Of particular interest here is the memo entitled "A Pixel is *NOT* a Little Square…," which provides an excellent explanation of the nature of pixels and image sampling in general, in both two and three dimensions.

> http://alvyray.com/Memos/default.htm

OpticsNotes.com – An online resource providing information in the field of optics and photonics.

> http://www.opticsnotes.com

Index

Academy standard   126
achromotopsia   19
active-matrix displays   63
additive primaries   36
Advanced Television System Committee (ATSC)   132, 226, 241–242
Advisory Committee on Advanced Television Service (ACATS)   225–226
American Standard Code for Information Interchange (ASCII)   163
Apple Computer   124, 196, 219
Apple Display Connector   196
analog interfaces
   defined   105–106
   for fixed-format displays   111
   and noise   107–109
   performance measurement   113–121
   television standards   141–157
anatomy, human eye (diagram)   15
astigmatism   21–22

bandwidth   12, 89, 106
   (see also *channel capacity*)
beam mislanding, see *color purity*
beam-penetration color CRT   58
black-body curve   43
blanking period   131
"blind spot"   18
BNC connector   159
Braun, Karl   55

cable impedance   92

cable loss   96–98
"cathode rays"   55
cathode-ray tube (CRT)   55–61
   projection   79
CGA (Color Graphics Adapter)   168
channel capacity   12, 84–85, 107, 233
character-generator systems   164–165
chroma burst   150
chromatic aberrations   23
chromostereopsis   24
closed-circuit video   144
coaxial cables   94
CODFM (Coded Orthogonal Frequency-Division Multiplexing)   242
color
   as a perception   34
   blindness   18
   CRT   57–60
   difference signals   149
   encoding   49
   gamut   45
   matching functions   40
   ovals, see *MacAdam ellipses*
   purity   60
   spaces   37
   temperature   42–43
Columbia Broadcasting System (CBS)
   color system   146
comb filter   152
Comité Consultatif International en Radiodiffusion (CCIR)   143
   CCIR channel standards   156–157

286 INDEX

CCIR-601 digital video standard 231–232
Comité Consultatif International Teléphonique et Telégraphique (CCITT) 130, 132
Committee Internationale de l'Eclairage (CIE) 39
   1931 xy color diagram 41–42
   1976 u'v' color diagram 46–47
   L*u*v* color space 46
   standard 2° and 10° observers 40, 42
   tristimulus values 40
Compaq Computer Corp. 193, 218
component vs. composite video 157–158
compression 86–88, 233–240
conditional update 199
cones (retinal receptor cells) 16
   distribution of 17
   sensitivity of 17
connectors
   general concerns 100–102
   (also see by specific type)
convergence 60
cornea (of human eye) 14
critical fusion frequency 26
crosstalk 90

DC balancing 89
DC restoration 144
deflection yoke 56
DeForest, Lee 55
"delta E star" (ΔE^*) 47
dichromats (red-green color blindness) 19
differential signals 91
digital cinema 245–247
digital content protection 242–244
Digital Display Working Group (DDWG) 194
Digital Flat Panel (DFP) interface 181, 193
digital interfaces
   for CRT displays 112–113
   defined 105–106
   and noise 107–109
   performance measurement 121–122
Digital Micromirror Device™ 76, 77, 246
Digital Packet Video Link (DPVL) 201, 216, 259–260
Digital Transmission Content Protection (DTCP) 243–244
Digital Video Broadcasting (DVB) project 227, 229, 241–242
Digital Visual Interface (DVI) 181, 194–196, 243

direct broadcast by satellite (DBS) 89, 224
discrete cosine transform (DCT) 235–237
display applications 80
DDC (Display Data Channel) 207–208
DDC-CI 214
display identification 203–210
display system 1
DPMS (Display Power Management Signaling) 213–214

EDID (Extended Display Identification Data) 207–210
Edison, Thomas A. 55
EGA (Enhanced Graphics Adapter) 168
electroluminescent (EL) displays 71
Electronic Industries Association (EIA) 130, 145, 185
   EIA-644 standard 185
   RS-170 and RS-343 standards 144–146, 171, 173
emissive displays 54
encryption 88
ergonomics 31–32
error correction 88
"eye diagram" or "eye pattern" 103, 121–122
Enhanced Video Connector (EVC) 177
Eureka-95 project 225, 226

"F" connector 159
Farrell, Joyce E. 26, 31
field 9
field-emission displays (FEDs) 73
field-sequential color 51, 146
"FireWire™" see IEEE-1394
"Five Companies" (5C) group 243
"flagging" 174–175
flat-panel displays (FPDs)
   general description of 61
   (see also by specific technology)
flicker 26
fovea 16, 18
frame 8
frame buffer 2, 9
frame rate conversion 63, 199
frequency synthesis 129

gamma correction 2
Gigabit Video Interface (GVIF) 191
Goldmark, Peter 147

INDEX   287

HDCP (High-Definition Content Protection)   194, 243
HD-MAC   226
HDTV/high-definition television   223–229
   formats and rates   227–229
   history of   224–226
"Hercules" graphics   167
head-mounted displays (HMDs)   257–259
Hewlett-Packard Company (HP)   26, 179, 211, 218
Hitachi   201, 243
"Hi-Vision"   224–225
holograms/holography   262–263
HSV color model   37–39
hue   37, 148
Hughes-JVC digital image light amplifier (D-ILA)   246
humor, aqueous   15, 16
humor, vitreous   15, 16
hyperopia, 21–22

IEEE-1394/"FireWire™" interface   191, 198, 219–221, 244
impedance, characteristic   93
in-plane switching (IPS) LCD   66–67
incandescence   42
information rate   84
Intel Corp.   194, 213, 218
intercarrier buzz   143
interlaced scanning   87, 131
International Business Machines (IBM)   166, 168, 176, 177, 201
   IBM PC   166–167
International Color Consortium (ICC)
   ICC profiles   211
International Standards Organization (ISO)   27
   ISO Standard 9241-3   26, 30–31
International Telecommunications Union (ITU)   130
IRE units   145
iris (of human eye)   14

jitter   117

Kell factor   142
Kelly, K.L.   49
Kelly chart   49

Land, Edwin   35
lightness   37
linearity (in signals)   118–119

liquid-crystal displays (LCDs)   64–69
   color   69
   idealized response curve   66
   transmissive vs. reflective   67–69
Low-Voltage Differential Signaling (LVDS)   185
Lucas, George/Lucasfilm, Ltd.   245–246
luminance   8
luminance-chrominance encoding   50

MacAdam ellipses   48
matrix addressing   62
Matsushita   243
MCCS (Monitor Control Command Set) standard   214–216
metamerism   35–36
MDA (Monochrome Graphics Adapter)   167
microdisplays   54, 75–78
Microsoft Corp.   211, 218
MPEG (Motion Picture Experts Group)   234
   MPEG-2 system   234–240
MUSE (multiple sub-Nyquist encoding)   224–225
myopia   21

"N" connector   160
National Television System Committee (NTSC)   147
   color-encoding system   147–153
   decoder   152
   encoder   151
   possible overmodulation with NTSC color system   153
   timing changes for color   150
negative modulation   143
NHK   224

"OpenLDI" specification   185
organic light-emitting devices (OLEDs)   72
overshoot/undershoot   115

packet video   200, 241
PanelLink™, see TMDS™ interface
passive-matrix displays   63
perceptual uniformity in color spaces   46, 47
"Peritel" connector   161
Phase Alternating Line (PAL) color encoding system   154–155
phase-locked loop (PLL)   129
picture element, see *pixel*
pixel   6, 53, 142, 166

pixel clock   12
pixel rates for common formats   86
plasma displays   69–71
"Plug & Display" standard   179–181
primary colors   36
projection displays   78–80
Pulfrich effect   25
pupil (of human eye)   15

quadrature amplitude modulation (QAM)   149, 242

RAMDAC   170–171
raster scanning   6
RCA (Radio Corporation of America)   140, 147
   color television system   147
   connector   158
   tricolor CRT   58, 147
refresh rate   9
   minimum for "flicker-free" appearance   26–28
resolution (also see spatial format)   7, 142, 251
retina (of human eye)   14
retrace   57
RGB encoding   50
rise/fall time   115
rods (retinal receptor cells)   16
   distribution of   17
   sensitivity of   17
Roosevelt, Franklin D.   140
run-length encoding   234

S-Video   160,
   connector standard   160–161
sampling rate, selection of   230
sampling structures   230–233
   under CCIR-601   231
Sarnoff, David   147
saturation   37
SCART connector   161
sclera (of human eye)   15
SDTV (standard-definition TV)   228–229
sequential color   51, 146
settling time   115
"setup" (pedestal) level   145
shadow mask, CRT   59
shielding   95–96
Shannon, Claude   12, 85
Sharp Electronics   200
signal-to-noise ratio (SNR)   85
Silicon Graphics/SGI   176, 177

Silicon Image, Inc.   188
   "PanelLink™"/TMDS digital interface   89, 188–191, 194, 196
skew   103, 116
SMA/SMC connectors   160
Society of Motion Picture and Television Engineers (SMPTE)   130, 132
   SMPTE-253M   146
Sony Corp.   58, 191, 217, 243
   "i.Link™"   217
   Gigabit Video Interface (GVIF)   191
   "Trinitron™" color CRT58   60
spatial color   51
spatial format   7
   selection of   124
Spindt, Charles   74
Spindt cathode   74
"square" pixels or formats   7, 142
sRGB standard   211–212
standard illuminants   44
stereopsis (stereo vision)   24–25
stereoscopic displays   261–262
subtractive primaries   36
synchronization pulses   131
   compositing of   173–175
Sequential Colour Avec Memoire (SECAM) color system   155–156
Standard Panels Working Group (SPWG)   186
   specification summary   187–188
Sun Microsystems   176, 177

television
   digital   197
   high-definition, see *HDTV*
   resolution of   140, 142
termination, cable   98–100
termination, connectors   101
Texas Instruments   76, 246
thin-film transistor (TFT) LCD   65
"13W3" connector   176–177
"3:2 pulldown"   127
time-domain reflectometer (TDR)   121
TMDS (Transition Minimized Differential Signaling)   89, 188–191, 194, 196
Toshiba   201, 243
transmission-line effects   119–121
tristimulus values   40
"twinlead" cable   93–94
twisted-nematic (TN) mode   64–65
twisted-pair cable   93–94
Universal Serial Bus (USB)   193, 198, 217–219

value   37
variable-length coding   234
vector scanning   4
vertical linear alignment (VLA) LCD   66–67
vestigial-sideband modulation (VSB)   142–143, 241–242
VGA standard   128, 167, 168
  ID system   169, 204
  timing and format   169
Video Electronics Standards Association (VESA)   114, 130, 132, 135, 170, 177, 200, 212
  Specific standards:
  Digital Flat Panel   193
  Digital Packet Video Link   200, 216
  Display Data Channel (DDC)   207–208
  Display Information File (VDIF)   205
  Display Power Management Signaling (DPMS)   213–214
  Extended Display Identification Data (EDID)   207–210
  Enhanced Video Connector (EVC)   177–179
  Flat Panel Display Interface (FPDI)   184
  Generalized Timing Formula (GTF)   135–137

Monitor Control Command Set (MCCS)   214–216
Plug & Display (P&D)   179–181, 191–193
timing standards   131–133
Video Signal Standard (VSIS)   114, 172
video signals
  level standards for   170–171
  nomenclature   134
visible spectrum   34
visual acuity   19–22
visual degrees   8
visual pigments   18
visual system, human
  anatomy   15–19
  chromatic aberrations in   23
  dynamic range   22–23
  non-linearity vs. luminance   23
  spectral response   17
  temporal response   25–29
voxel (volume pixel)   10

wireless interfaces   255–257
work function   73
workstation display standards   173
WWV/WWVH (time/frequency standards)   128